Contributions to Statistics

V. Fedorov/W.G. Müller/I.N. Vuchkov (Eds.) Model-Oriented Data Analysis,
XII/248 pages, 1992

J. Antoch (Ed.) Computational Aspects of Model Choice,
VII/285 pages, 1993

W.G. Müller/H.P. Wynn/A.A. Zhigljavsky (Eds.)
Model-Oriented Data Analysis,
XIII/287 pages, 1993

P. Mandl/M. Hušková (Eds.)
Asymptotic Statistics
X/474 pages, 1994

P. Dirschedl/R. Ostermann (Eds.)
Computational Statistics
VII/553 pages, 1994

C.P. Kitsos/W.G. Müller (Eds.)
MODA4 – Advances in Model-Oriented Data Analysis,
XIV/297 pages, 1995

H. Schmidli
Reduced Rank Regression,
X/179 pages, 1995

Wolfgang Härdle
Michael G. Schimek (Eds.)

Statistical Theory and Computational Aspects of Smoothing

Proceedings of the
COMPSTAT '94 Satellite Meeting
held in Semmering, Austria
27–28 August 1994

With 63 Figures

Physica-Verlag

A Springer-Verlag Company

Series Editors
Werner A. Müller
Peter Schuster

Editors
Professor Dr. Wolfgang Härdle
Center for Computational Statistics
Institut für Statistik und Ökonometrie
Wirtschaftswissenschaftliche Fakultät
Humboldt-Universität zu Berlin
Spandauer Straße 1
D-10178 Berlin, Germany

Professor Dr. Dr. Michael G. Schimek
Medical Biometrics Group
University of Graz Medical Schools
Auenbruggerplatz 30/IV
A-8036 Graz, Austria

Co-sponsored by the International Association of Statistical Computing

---

ISBN-13: 978-3-7908-0930-5   e-ISBN-13: 978-3-642-48425-4
DOI: 10.1007/978-3-642-48425-4

Die Deutsche Bibliothek – CIP-Einheitsaufnahme
Statistical theory and computational aspects of smoothing: proceedings of the COMPSTAT '94
satellite meeting held in Semmering, Austria, 27–28 August 1994/Wolfgang Härdle; Michael G.
Schimek (ed.). With contributions by J.S. Marron... With discussions by P. Hall... [Co-spon-
sored by the International Association of Statistical Computing]. – Heidelberg: Physica-Verl.,
1996
    (Contributions to statistics)
    ISBN-13: 978-3-7908-0930-5
NE: Härdle, Wolfgang [Hrsg.]; Marron, James S.; COMPSTAT <11, 1994, Wien>

SPIN: 10534328   88/2202-5 4 3 2 1 0 – Printed on acid-free paper

Preface

The **COMPSTAT '94 Satellite Meeting on Smoothing - Statistical Theory and Computational Aspects** was held on August 27-28, 1994 at the hotel "Erzherzog Johann" in Semmering, Austria. It was the first meeting of this kind and scheduled immediately after the **COMPSTAT '94 Symposium** in Vienna, Austria.

The meeting was hosted by the **Karl Franzens Universität Graz** in Graz, Austria, and organized by its Medical Biometrics Group of the Medical Schools. The European Section of the **International Association of Statistical Computing** (IASC), part of the International Statistical Institute (ISI), co-sponsored it. Members of the Scientific Programme Committee were R. L. Eubank, L. Györfi, W. Härdle and M. G. Schimek (chairman). Session chairs included A. de Falguerolles, R. L. Eubank, T. Gasser, L. Györfi, W. Härdle, R. Kohn, M. G. Schimek and A. van der Linde.

The emphasis of the scientific programme was on invited introductory papers, as well as on invited and/or discussed papers presenting the state-of-the-art of statistical theory and computational aspects with respect to smoothing. There were also a number of contributed papers and presentations of relevant software (S-Plus and XploRe). Parallel session were avoided and the number of participants was limited to fifty to provide an atmosphere for informal discussions of new ideas. The main topic of discussion was recent developments in local regression. The meeting attracted almost fifty scientists from all over the world.

Here we present a selection of papers read at the **COMPSTAT '94 Satellite Meeting on Smoothing**. Only manuscripts not published elsewhere were considered for this volume. All contributions had to undergo a regular review process as usually carried out in scientific journals. Ten of twentysix papers given at the meeting are published here. In addition we have an expository discussed paper by J. S. Marron who was not able to attend. The two contributions on local regression by W. S. Cleveland and C. Loader, B. Seifert and T. Gasser, respectively, were also open to written discussion, printed in this volume together with the rejoinders of the authors. The other papers deal with Bayesian nonparametric regression, invariance problems with smoothing splines, variance estimation, cross-validation, extreme percentile regression, additive models and nonlinear principal components analysis.

We are grateful to the authors, discussants, referees and the editorial assistant who worked together to make the outcome of this successful meeting accessible to a greater audience interested in topics related to smoothing.

Wolfgang Härdle and Michael G. Schimek, Editors
Berlin and Graz, December 1995

Contents

A Personal View of Smoothing and Statistics

J.S. Marron

Department of Statistics, University of North Carolina, Chapel Hill, NC 27599-3260, USA

Summary

Personal views are offered on the foundations of smoothing methods in statistics. Key points are that smoothing is a useful tool in data analysis, and that the field of data based smoothing parameter selection has matured to the point that effective methods are ready for use as defaults in software packages. A broader lesson available from the discussion is: a combination of computational methods and mathematical statistics is a powerful research tool.

Keywords: bandwidth selection, kernel estimation, nonparametric curve estimation, smoothing

1 Introduction

This paper contains some personal views on smoothing and its role in statistics. These views have been shaped by interaction with many people having very diverse opinions. These people are both active methodological statisticians, and also interested non-statisticians. The statisticians include "mainstream smoothers", but many others as well. An important component of my views comes from investigation of the fact that there are many very divergent, yet strongly held, opinions of "smoothing insiders" on issues such as what is the "right" choice of smoothing method.

Section 2 contains what I have learned about choice of smoothing method. Different people have made, and will continue to make, different choices. The main point is that the great majority of choices made are sensible, because there are many noncomparable factors involved in this choice. Diverse personal weightings of these factors result in choices that are different, but equally valid when viewed from an overall perspective.

Section 3 discusses modern bandwidth selection. The main lesson is that recent research has resulted in methods that are quite effective in finding "the right amount of smoothing". They are far better than the current defaults in most software packages, and in the case of kernel density estimation are ready for use in that role.

The discussion of research in bandwidth selection illustrates some important ideas on methodological statistics in general. In Section 4 it is suggested that a combination of mathematical and computational tools provides a particularly powerful approach.

A final point is: "smoothing is a useful method of analyzing data". Reasons for not losing sight of this point are given in Section 5.

2 Choice of smoothing method

A frequent occurrence is that a scientist consults a statistician (or a nonsmoothing statistician goes to a "smoother") and says "I have heard about this smoothing stuff and would like to try it, how do I do it?" A number of different answers could be given depending largely upon whom is asked. The recommended method could be any of:

M1 a kernel - local polynomial method.

M2 a smoothing spline.

M3 a regression - B spline (much different from b).

M4 a Fourier smoother.

M5 a wavelet method.

M6 LO(W)ESS.

All of these methods have their firm adherents, who enjoy espousing the strengths of their personal favorite. It is interesting that the answers can be so different. Much personal discussion with involved parties has yielded the personal conclusion that everybody is right in their different choices. This seemingly contradictory statement is possible because there are several *noncomparable* factors involved. In particular, factors that go into the choice of smoother include:

F1 availability (is it "right there" in your software package?).

F2 interpretability (what does the smooth tell us about the data?).

F3 statistical efficiency (how close is the smooth to "the true curve"?)

F4 quick computability.

F5 integrability into general frameworks (e.g. S/S+).

F6 ease of mathematical analysis.

All of the methods M1 - M6 listed above have differing strengths and weaknesses in these divergent senses. None of these methods dominates any other in all of the senses F1 - F6. Since these factors are so different, almost any method can be "best", simply by an appropriate personal weighting of the various factors involved. Claims of "wretched performance" are also valid, for any of the above methods, when one adopts a suitably narrow viewpoint.

This is not the place to give a complete listing of the relative strengths and weaknesses of M1 - M6, but I shall discuss one case in more detail. The simplest possible naive fixed bandwidth local constant kernel methods have met with substantial criticism. This is justified, because they are relatively weak with respect to F3. I suspect that a contributing factor to this criticism is the popularity of such methods in the theoretical literature, much of which is due to strength in F6. But from a broader viewpoint, this criticism does not provide sufficient reason to rule out their use. In particular, these methods are the best (in my opinion) at F2. Fixed bandwidth local constant kernel methods put the interested analyst in closest possible intuitive contact with the data, because they are simple, understandable, local averages. Note that I am not advocating this estimator as the solution to all problems (in fact it has its drawbacks as does every method, and there are situations where other methods will be much more desirable), but instead am merely pointing out it cannot be dismissed out of hand.

While I appreciate and respect researchers' enthusiasm for their more recently developed methods, I suggest a healthy skepticism be employed with respect to claims of solving all problems. Every method has its shortcomings, and it is important to be aware of these before the method is used.

3 Data based bandwidth selection

The performance of all smoothing methods is crucially dependent on the specification of a smoothing parameter. This choice is a deep issue. While many things are known about "optimal bandwidth selection", they should be kept in proper perspective. There are two important issues to keep in mind. One is that in some situations there is not a single choice which gives the data analyst all the information available in the data. The other is that the classical mathematical definitions of "best" can be quite different from "useful for data analysis". See Marron and Tsybakov (1995) for reasons behind this, and some approaches to addressing this problem..

For these reasons the most useful method of choosing the amount of smoothing is interactive trial and error by an experienced analyst. But this has three important weaknesses.

- It is time consuming. Many statisticians work under time pressure, and are unable to carefully choose a smoothing parameter for every data set analyzed. Also manual choice of smoothing parameter can become very tedious for some more complicated data sets, such as the income data in Park and Marron (1990)

- Interactive choice of smoothing parameter requires some level of expertise. An inexperienced user can miss important features through oversmoothing, and can misinterpret spurious structure as being something important. Even experts are aided by a useful starting point.

- Interactive bandwidth choice is impossible when smoothers are used iteratively. This is essential to most dimensionality reduction methods, including Projection Pursuit, Generalized Additive Modelling, Alternating Conditional Expectations, and Multivariate Additive Regression Splines.

For these reasons, the subject of data-based choice of smoothing parameters is an important one.

Because simplicity allows deepest investigation, most of the research in that area has been done in the context of kernel density estimation. The starting point has been "first generation methods", such as Rule Of Thumb methods and various versions of cross-validation. These methods have their appeal, but also some serious drawbacks. The Rule Of Thumb does not make enough use of the data, and is often very inappropriate. Cross-validation based methods are widely accepted as "too noisy", meaning they can perform satisfactorily for some realizations, but they can be either under or over smoothed, just by luck of the draw, depending on the data realization (although undersmoothing is often a more serious problem).

Hence, there has been substantial effort devoted to surmounting these difficulties, which has been very successful, resulting in "second generation methods". As noted in the review/simulation papers of Park and Turlach (1992), Cao, Cuevas and Gonzales- Manteiga (1994), and Jones, Marron and Sheather (1992, 1994), there are a number of such methods available. The evidence behind these claims is firmly rooted in:

E1 Real data examples, see in particular Sheather (1992).

E2 Simulation studies.

E3 Mathematical analysis.

The whole of this evidence shows that second generation methods of bandwidth selection in kernel density estimation are ready for serious consideration by software developers.

An aside of esoteric historic interest about bandwidth selection is given in the appendix.

4   Methodological research in statistics

The evidence demonstrating the effectiveness of second generation bandwidth selection methods is compelling because it is based on real data examples, on simulation, and on asymptotic analysis. All of these tools have their strengths, and also their weaknesses. Because no method tells the whole story (and in more challenging situations the amount of the full story that is told by any can be woefully small), sensible research should involve an effective combination of as many of these as possible, especially as more challenging problems are encountered.

While this point may seem obvious, not everyone agrees. I have heard extreme viewpoints expressed on both sides of an "asymptotics vs. computation gap". If statistics becomes polarized in this way, the entire discipline will be hobbled. Strength in the field will come from proper recognition of, and respect for, work of a wide variety of types.

5   Importance of smoothing in statistics

From speaking with many statisticians, I have found a significant number who could benefit from the use of smoothing methods, but are not doing so now. I am concerned that debates about choice of smoothing method could obscure the main point that "smoothing is a useful method of analyzing data", as has been made clear for example in many books, including Cleveland (1985, 1993), Eubank (1988), Green and Silverman (1994), Härdle (1991), Hastie and Tibshirani (1990), Müller (1988), Scott (1992), Silverman (1986) and Wahba (1991).

If smoothing insiders argue too vehemently among themselves over minor technical matters, then it is easy for outsiders to lose sight of the major issues. Worse, some will develop false conceptions such as "if the experts argue so bitterly, then it must not be known how to use these methods effectively". In my opinion, the field of robustness has suffered from bickering over minor issues, thereby creating such misconceptions among the broad statistical community. I suspect this is why robust methods have been slow to gain acceptance in the wider statistical community, as exemplified by their rather sparse appearance in software packages.

Again, the key idea here is that smoothing is a useful method of analyzing data, and there are a number of effective ways to do it. While arguments over exactly how to do it become rather emotional at times, it is important that such debates remain in proper context.

6   Conclusion

Here are the main lessons that I hope can be carried away from this article:

1. There are many choices of possible smoothing methods, which are not comparable, because of differing relative strengths and weaknesses. Choice of smoother will always be a personal one, based on personal weightings of these.

2. Real progress has been made in data based bandwidth selection.

3. A combination of mathematical and computational tools provides a powerful approach to statistical problems.

4. Smoothing is an effective method to analyze data.

7 Acknowledgement

This paper was supported by NSF Grant DMS-9203135. N. I. Fisher, R. Kohn and M.P. Wand provided helpful comments.

References

[1] Cao, R., Cuevas, A. and González-Manteiga, W. (1994). A comparative study of several smoothing methods in density estimation. *Comp. Statist. Data Anal.* **17** 153-176.

[2] Cleveland, W. S. (1985) *The elements of graphing data*, Wadsworth, Belmont, California.

[3] Cleveland, W. S. (1993) *Visualizing data*, Hobart Press, Summit, New Jersey.

[4] Eubank, R. L. (1988) *Spline Smoothing and Nonparametric Regression*, Dekker, New York.

[5] Green, P. and Silverman, B. W. (1994) *Nonparametric Regression and generalized linear models, a roughness penalty approach*, Chapman and Hall, London.

[6] Härdle, W. (1991) *Applied Nonparametric Regression*, Cambridge University Press, Boston.

[7] Hall, P. (1980) Objective methods for the estimation of window size in the nonparametric estimation of a density. Unpublished manuscript.

[8] Hall, P. and Marron, J. S. (1991b) Lower bounds for bandwidth selection in density estimation. *Probab. Theory Related Fields* **90**, 149-173.

[9] Hastie, T. J. and Tibshirani, R. J. (1990) *Generalized additive models*, Chapman and Hall, London.

[10] Jones, M. C., Marron, J. S. and Sheather, S. J. (1992) Progress in data based bandwidth selection for kernel density estimation, submitted to *J. Nonpar. Statist.*

[11] Jones, M. C., Marron, J. S. and Sheather, S. J. (1994) A brief survey of modern bandwidth selection methods, submitted to *J. Amer. Statist. Assoc.*

[12] Marron, J.S., and Tsybakov, A.B. (1995), "Visual Error Criteria for Qualitative Smoothing," *Journal of the American Statistical Association*, 90, 499-507.

[13] Marron, J. S. and Wand, M. P. (1992). Exact mean integrated squared error, *Ann. Statist.* **20**, 712-736.

[14] Müller, H.-G. (1988) Nonparametric regression analysis of longitudinal data. Springer Verlag, Berlin.

[15] Park, B.-U. and Marron, J. S. (1990). Comparison of data-driven bandwidth selectors. *J. Amer. Statist. Assoc.* **85** 66-72.

[16] Park, B.-U. and Turlach, B. A. (1992) Practical performance of several data driven bandwidth selectors (with discussion). *Comput. Statist.* **7** 251-285.

[17] Scott, D. W. and Factor, L. E. (1981) Monte Carlo study of three data-based nonparametric probability density estimators. *J. Amer. Statist. Assoc.* **76** 9-15.

[18] Scott, D. W. and Terrell, G. R. (1987) Biased and unbiased cross-validation in density estimation. *J. Amer. Statist. Assoc.* **82** 1131-1146.

[19] Scott, D. W. (1992) Multivariate density estimation: theory, practice and visualization. Wiley, New York.

[20] Sheather, S. J. (1986). A data-based algorithm for choosing the window width when estimating the density at a point. *Comput. Statist. Data Anal.* **1** 229-238.

[21] Sheather, S. J. (1986). An improved data-based algorithm for choosing the window width when estimating the density at a point. *Comput. Statist. Data Anal.* **4** 61-65.

[22] Sheather, S. J. (1992). The performance of six popular bandwidth selection methods on some real data sets (with discussion). *Comput. Statist.* **7**, 225-250.

[23] Silverman, B. W. (1986) Density estimation for statistics and data analysis, Chapman and Hall, New York.

[24] Wahba, G. (1990) *Spline Models for Observational Data*, SIAM, Philadelphia.

8 Appendix: An historical aside

A point of esoteric historical interest is that the field of data based bandwidth selection developed rather slowly for a number of years during the mid 80's because of the following asymptotic reasoning. It is well known that, when smoothers are behaving efficiently (e.g. with respect to some norm), their asymptotic rate of convergence is determined by their "smoothness". Exactly how to quantify this is a fascinating issue (which I shall address elsewhere), but there do exist crude asymptotic surrogates, such as "number of derivatives" and "rate of tail decay of Fourier Transform". An interesting offshoot of this theory is quantification of the idea that the data contain less information about the derivative of a function, than about the function itself. This has serious implications already for "modern bandwidth selection" methods, which depend on trying to estimate derivatives. However, there are deeper conceptual problems. The current asymptotic analysis of modern bandwidth selection methods assumes more "smoothness" than the underlying estimator is capable of using, so some is "left over", and the resulting estimator will be asymptotically inefficient (compared to e.g. a "higher order kernel" method, if one is willing to ignore the bandwidth problem there). This idea held back researchers for quite some time, with some exceptions, including Hall (1980), Scott and Factor (1981) and Sheather (1983, 1985). But the ice was really broken by Scott and Terrell (1987) who showed one could get practical improvements using derivative estimation based methods, over Cross-Validation. That paper opened a flood gate of improvements by others (many of which work much better than Scott and Terrell's BCV), see Jones, Marron and Sheather (1992, 1994) for full details, discussion and references. How can the proven good performance of these methods be reconciled with the above philosophical problems? This is a deep and interesting problem. Here are three tentative ideas on this topic.

The first, as noted in Park and Marron (1990) and Hall and Marron (1991) returns to how one "attaches weights when choosing a smoother", as discussed in Section 2 above. Those who are willing to place more weight on (F2) may not be concerned about squeezing out the last epsilon of efficiency (and the results of Marron and Wand (1992) show this is pretty small all too often) from the estimator. Instead they may be quite satisfied with understanding more clearly what the estimator is telling them about the data, and with the fact that there is a very effective bandwidth choice available.

The second is that our understanding of "how asymptotics kick in" is still quite limited. Marron and Wand (1992) show that the "faster rates of convergence" associated with "more efficient estimators" can take a long time to kick in. Deeper investigation is needed, but there are indications that "what makes bandwidth selection work" kicks in at much smaller sample sizes. An interesting open problem is to develop deeper insights on this issue.

The third is that there is still not complete understanding of the abstract idea of "smoothness". To see that "number of derivatives" is only a clumsy surrogate, think about target functions which have 3.6 but not 3.7 derivatives. If I have a sample of $n = 100$ (or any number), how different are such target functions from those which have 3.7 but not 3.8 derivatives? Clearly what is needed is a more relevant way of measuring "smoothness" in finite samples. Work is under way on this problem, although it is too early to say what the outcome will be. But such ideas may be able to explain why second generation bandwidth selectors work well when current asymptotic folklore suggests otherwise.

Smoothing by Local Regression: Principles and Methods

William S. Cleveland and Clive Loader

AT&T Bell Laboratories, 600 Mountain Avenue, Murray Hill, NJ 07974, USA

Summary

Local regression is an old method for smoothing data, having origins in the graduation of mortality data and the smoothing of time series in the late 19th century and the early 20th century. Still, new work in local regression continues at a rapid pace. We review the history of local regression. We discuss four of its basic components that must be chosen in using local regression in practice — the weight function, the parametric family that is fitted locally, the bandwidth, and the assumptions about the distribution of the response. A major theme of the paper is that these choices represent a modeling of the data; different data sets deserve different choices. We describe *polynomial mixing*, a method for enlarging polynomial parametric families. We introduce an approach to adaptive fitting, *assessment of parametric localization*. We describe the use of this approach to design two adaptive procedures: one automatically chooses the mixing degree of mixing polynomials at each x using cross-validation, and the other chooses the bandwidth at each x using C_p. Finally, we comment on the efficacy of using asymptotics to provide guidance for methods of local regression.

Keywords: Nonparametric regression, Loess, bandwidth, polynomial order, polynomial mixing, ridge regression, statistical graphics, diagnostics, modeling

1 Introduction

1.1 Modeling

Local regression is an approach to fitting curves and surfaces to data by smoothing: the fit at x is the value of a parametric function fitted only to those observations in a neighborhood of x (Woolhouse, 1870; Spencer, 1904a; Henderson, 1916; Macaulay, 1931; Kendall, 1973; Kendall and Stuart, 1976; Stone, 1977; Cleveland, 1979; Katkovnik, 1979; Friedman and Stuetzle, 1982; Hastie and Tibshirani, 1990; Härdle, 1990; Hastie and Loader, 1993; Fan, 1993).

The underlying model for local regression is

$$E(y_i) = f(x_i), \ i = 1, \ldots, n,$$

where the y_i are observations of a response and the d-tuples x_i are observations of d independent variables that form the *design space* of the model. The distribution of the y_i, including the means, $f(x_i)$, are unknown. In practice we must first *model* the data, which means making certain assumptions about f and other aspects of the distribution of the y_i. For example, one common distributional assumption is that the y_i have a constant variance. For f, it is supposed that the function can be well approximated *locally* by a member of a parametric class, frequently taken to be polynomials of a certain degree. We refer to this as *parametric localization.*

Thus, in carrying out local regression we use a parametric family just as in global parametric fitting, but we ask only that the family fit locally and not globally. Parametric localization is the fundamental aspect that distinguishes local regression from other smoothing methods such as smoothing splines (Wahba, 1990), regression splines with knot selection (Friedman, 1991), and wavelets (Donoho and Johnstone, 1994) although the notion is implicit in these methods in a variety of ways.

1.2 Estimation of f

The estimation of f that arises from the above modeling is simple. For each fitting point x, define a neighborhood based on some metric in the d-dimensional design space of the independent variables. Within this neighborhood, assume f is approximated by some member of the chosen parametric family. For example the family might be quadratic polynomials: $g(u) = a_0 + a_1(u - x) + a_2(u - x)^2$. Then, estimate the parameters from observations in the neighborhood; the local fit at x is the fitted function evaluated at x. Almost always, we will want to incorporate a weight function, $w(u)$, that gives greater weight to the x_i in the neighborhood that are close to x and lesser weight to those that are further.

The criterion of estimation depends on the assumptions made about the distribution of the y_i. For example, if we suppose that the y_i are approximately Gaussian with constant variance then it makes sense to base estimation on least-squares. If the parametric family consists of quadratic polynomials, we minimize

$$\sum_{i=1}^{n} w\left(\frac{x_i - x}{h}\right)(y_i - a_0 - a_1(x_i - x) - a_2(x_i - x)^2)^2.$$

In this case, $\hat{f}(x) = \hat{a}_0$. This is a pleasant case. The parameter estimates have a simple closed form expression, and numerical solution of the fitting equations is straightforward. Moreover, provided w and the neighborhood size do not depend on the response, the resulting estimate is a linear function of the y_i; this leads to a simple distribution theory that mimics very closely distributions for parametric fitting, so that t intervals and F-tests can be invoked (Cleveland and Devlin, 1988).

There are many possibilities for the specification of the *bandwidth*, h, for each point x in the design space. One choice is simply to make h a constant. This is *fixed bandwidth* selection; there is one parameter in this case, the halfwidth, h, of the neighborhood. Note that if we are fitting, say, quadratic polynomials locally, then as h gets large, the local regression fit approaches the global quadratic fit. Another is to have h change as a function of x based on the values of x_i. One example is *nearest-neighbor* bandwidth selection; in this case there is one parameter, α. If $\alpha \leq 1$, then we multiply n by α, round to an integer, k, and then take $h(x)$, the neighborhood halfwidth, to be the distance from x to the kth closest x_i. For $\alpha > 1$, $h(x)$ is the distance from x to the furthest x_i multiplied by $\alpha^{1/d}$ where d is the number of numeric independent variables involved in the fit. Thus, as with fixed bandwidth selection, as the bandwidth parameter gets large, the local fit approaches a global parametric fit.

There are many possibilities for the parametric family that is fitted locally. The most common is polynomials of degree p. Later in the paper we will describe polynomial mixing, which enlarges the families of polynomials, replacing the usual integer degree by a continuous parameter, the mixing degree.

Often, we can successfully fit curves and surfaces by selecting a single parametric family for all x; for example we might take p to be 2 so we are fitting quadratics locally. And, we can use a single value of a bandwidth parameter such as α, the nearest-neighbor parameter. Or, we can use an *adaptive* procedure that selects the degree locally as a function of x, or that selects the bandwidth locally, or that selects both locally.

1.3   Modeling the Data

To use local regression in practice, we must choose the weight function, the bandwidth, the parametric family, and the fitting criterion. The first three choices depend on assumptions we make about the behavior of f. The fourth choice depends on the assumptions we make about other aspects of the distribution of the y_i. In other words, as with parametric fitting, we are modeling the data.

But we do not need to rely fully on prior knowledge to guide the choices. We can use the data. We can use graphical diagnostic tools such as coplots, residual-dependence plots, spread-location plots, and residual quantile plots (Cleveland, 1993). To model f we can use more formal model selection criteria such as C_p (Mallows, 1973) or cross-validation (Stone, 1974). For example, Cleveland and Devlin (1988) use C_p to select the bandwidth parameter of nearest-neighbor fitting.

Modeling f usually comes down to a trade-off between variance and bias. In some applications, there is a strong inclination toward small variance, and in other applications, there is a strong inclination toward small bias. The advantage of model selection by a criterion is that it is automated. The disadvantage is that it can easily go wrong and give a poor answer in any particular application. The advantage of graphical diagnostics is great power; we can often see where in the space of the independent variables that bias is occurring and where the variability is greatest. That allows us to decide on the relative importance of each. For example, underestimating a peak in a surface or curve is often quite undesirable and so we are less likely to accept lower variance if the result is peak distortion. But the disadvantage of graphical diagnostics it that they are labor intensive, so while they are excellent for picking a small number of model parameters they are not practical for adaptive fitting, which as a practical matter requires an automated selection criterion. For example, Friedman and Stuetzle (1982) use cross-validation to choose the bandwidth locally for each x. And later in this paper we describe two adaptive procedures, one based on C_p and the other based on cross-validation. Still, when we have a final adaptive fit in hand, it is critical to subject it to graphical diagnostics to study its performance.

The important implication of these statements is that the above choices must be tailored to each data set in practice; that is, the choices represent a modeling of the data. It is widely accepted that in global parametric regression there are a variety of choices that must be made — for example, the parametric family to be fitted and the form of the distribution of the response — and that we must rely on our knowledge of the mechanism generating the data, on model selection diagnostics, and on graphical diagnostic methods to make the choices. The same is true for smoothing.

Cleveland (1993) presents many examples of this modeling process. For example, in one application, oxides of nitrogen from an automobile engine are fitted to the equivalence ratio, E, of the fuel and the compression ratio, C, of the engine. Coplots show that it is reasonable to use quadratics as the local parametric family but with the added assumption that given E the fitted f is linear in C. (It is quite easy to achieve such a *conditionally parametric* fit; we simply ignore C is defining the weights $w((x_i - x)/h)$. In addition, residual diagnostic plots show that the distribution of the errors is strongly leptokurtic, which means that we must abandon least-squares as a fitting criterion and use methods of robust estimation.

Later in the paper we will present other examples to further illustrate and explain the modeling process.

1.4   Why Local Regression?

Local regression has many strengths, some discussed in detail by Hastie and Loader (1993):

1) Adapts well to bias problems at boundaries and in regions of high curvature.

2) Easy to understand and interpret.

3) Methods have been developed that provide fast computation for one or more independent variables.

4) Because of its simplicity, can be tailored to work for many different distributional assumptions.

5) Having a local model (rather than just a point estimate $\hat{f}(x)$) enables derivation of response adaptive methods for bandwidth and polynomial order selection in a straightforward manner.

6) Does not require smoothness and regularity conditions required by other methods such as boundary kernels.

7) The estimate is linear in the response provided the fitting criterion is least squares and model selection does not depend on the response.

Singly, none of these provides a strong reason to favor local regression over other smoothing methods such as smoothing splines, regression splines with knot selection, wavelets, and various modified kernel methods. Rather, it is the combination of these issues that combine to make local regression attractive.

1.5   The Contents of the Paper

The history of local regression is reviewed in Section 2. Then, polynomial mixing is described in Section 3. In the next sections, the four choices required for carrying out local regression are discussed. Section 4 discusses an approach to making judgments about the efficacy of methods of local regression. Section 5 discusses the design of the weight function; the choice here is straightforward since there are just two main desiderata that apply in almost all applications in which the underlying dependence is described by a continuous f, and a single weight function can serve for most of these applications.

Section 6 discusses fitting criteria; there is little to say here because the considerations in making distributional assumptions are the same as those for global parametric regression. Section 7 discusses bandwidth selection and the choice of the local parametric family; the issues here are complex in that there are many potential paths that can be followed in any particular application and we must inevitably rely on the data to help us choose a path. To help convey the salient points made in Section 7, we present three examples in Section 8. In Section 9 we present two adaptive methods; one chooses the bandwidth locally as a function of x and the other chooses the mixing polynomial degree locally as a function of x. In Section 10 we discuss asymptotic theory, in particular, what cannot be determined from asymptotic theory in its current state. Finally, Section 11 summarizes the conclusions drawn in the paper.

2 An Historical Review

2.1 Early Work

Local regression is a natural extension of parametric fitting, so natural that local regression arose independently at different points in time and in different countries in the 19th century. The setting for this early work was univariate, equally spaced x_i; this setting is simple enough that good-performing smoothers could be developed that were computationally feasible by hand calculation. (In our discussion in this subsection we will set $x_i = i$ for $i = 1$ to n.) Also, most of the 19th century work arose in actuarial studies; mortality and sickness rates were smoothed as a function of age.

Hoem (1983) reports that smoothing was used by Danish actuaries as early as 1829, but their methods were not published for about 50 years. Finally, Gram (1883) published work from his 1879 doctoral dissertation on local polynomial fitting with a uniform weight function and with a weight function that tapers to zero. In much of his work he focused on local cubic fitting and used binomial coefficients for weights.

Stigler (1978) reports that in the United States, De Forest (1873, 1874) used local polynomial fitting for smoothing data. De Forest also investigated an optimization problem similar to one studied later by Henderson (1916), which we will describe shortly. Much of De Forest's work focused on local cubic fitting.

In Britain, work on smoothing had begun by 1870 when Woolhouse (1870) published a method based on local quadratic fitting. The method received much discussion but was eventually eclipsed by a method of Spencer (1904a) that became popular because it was computationally efficient and had good performance.

Spencer developed smoothers with several different bandwidths. His 21-point rule, which yields smoothed values for $i = 11, \ldots, n - 10$ has the representation

$$\frac{1}{350}[5][5][7][-1,0,1,2].$$

This notation means the following: first, take a symmetric weighted moving average of length 7 with weights $-1,0,1,2,1,0,-1$; then take three unweighted moving sums of lengths 7, 5, and 5; and then divide by 350. The resulting fit at i is

$$\sum_{k=-10}^{k=10} c_k y_{i+k}$$

where the c_k are symmetric about $k = 0$ and the values of $350c_k$ for $k = -10$ to 0 are

$$-1, -3, -5, -5, -2, 6, 18, 33, 47, 57, 60.$$

We note three crucial properties. First, the smoother exactly reproduces cubic polynomials. Second, the smoothing coefficients are a smooth function of k and decay smoothly to zero at the ends. Third, the smoothing can be carried out by applying a sequence of smoothers each of which is simple; this was done to facilitate hand computation. Achieving all three of these properties is remarkable; Spencer and others spent some considerable effort in deriving such *summation formulas* of various lengths to satisfy the properties.

One might ask why we have put a summation formula such as Spencer's 21-point rule in the category of local fitting. The answer was first provided by an interesting paper of Henderson (1916) on weighted local cubic fitting. Let w_k be the weight function for $k = -m, \ldots, m$. Henderson showed that the local cubic fit at i can be written as

$$\sum_{k=-m}^{m} \phi(k)w_k y_{i+k}$$

where ϕ is a cubic polynomial whose coefficients have the property that the smoother reproduces the data if they are a cubic. (If w_k is symmetric then ϕ is quadratic.) Henderson also showed the converse: if the coefficients of a cubic-reproducing summation formula $\{c_k\}$ have no more than three sign changes, then the formula can be represented as local cubic smoothing with weights $w_k > 0$ and a cubic polynomial $\phi(k)$ such that $\phi(k)w_k = c_k$. For example, for Spencer's 21 term summation formula we can take

$$\phi(j) = (30 - j^2)/175$$

and the weight function, from $k = -10, \ldots, 0$ to be

$$\frac{1}{140}, \frac{1}{34}, \frac{5}{68}, \frac{5}{38}, \frac{1}{6}, \frac{3}{5}, \frac{9}{14}, \frac{11}{14}, \frac{47}{52}, \frac{57}{58}, 1.$$

1000 times these values are

$$7, 29, 74, 132, 167, 600, 643, 785, 904, 923, 1000.$$

Henderson also considered the problem of obtaining the smoothest possible fit subject to reproduction of cubics. Smoothness is measured by the sum of squares of the third differences of the smoother weights, or equivalently, the sum of squares of the third differences of the fit. The closed form solution for the smoother coefficients $c_k, k = -(m-2), \ldots, (m-2)$ is

$$a_m((m-1)^2 - x^2)(m^2 - x^2)((m+1)^2 - x^2)((3m^2 - 16) - 11x^2)$$

where a_m is a term that makes the weights add to 1. From the result in the previous paragraph, this summation formula is equivalent to local cubic fitting with the neighborhood weight function

$$w_k = ((m-1)^2 - k^2)(m^2 - k^2)((m+1)^2 - k^2)$$

for $|k| \leq m - 2$. For large m, this amounts to the triweight weight function $(1 - x^2)^3$. Thus, asymptotic optimality problems of the type considered in Müller (1984) were, for equally spaced x_i, solved exactly by Henderson for finite samples.

The work that became well known was that of Henderson (from the U.S.) and the British smoothing research community — which included accomplished applied mathematicians such as Whittaker (1923), who, along with Henderson (1924), also invented smoothing splines. The influence resulted in the movement of local fitting methods into the time series literature. For example, the book *The Smoothing of Time Series* (Macaulay 1931), shows how local fitting methods can be applied to good purpose to economic series. Macaulay not only reported on the earlier local fitting work, but also developed methods for smoothing seasonal time series.

Macaulay's book in turn had a substantial influence on what would become a major milestone in local fitting methods. It began at the U.S. Bureau of the Census, beginning in 1954. A series of computer programs were developed for smoothing and seasonal adjustment of time series, culminating with the X-11 method (Shishkin, Young, and Musgrave, 1967). This represented one of the earliest uses of computer-intensive statistical methods, beginning life on an early Univac. X-11 did not widely penetrate the standard statistics literature at the time because its methods were empirically based rather than emanating from a fully specified statistical model. However, X-11 became the standard for seasonal and trading-day adjustment of economic time series and is still widely used today. In X-11, a time series is fitted by three additive components: a trading day component described by a parametric function of day-of-week variables and fitted by parametric regression; a trend component fitted by smoothing, and a seasonal component fitted by smoothing. The developers of X-11 used what would become known two decades later as the backfitting algorithm (Friedman and Stuetzle, 1981) to iteratively estimate the components. These iterations are nested inside iterations that provide robust fitting by identifying outliers and modifying the data corresponding to the outliers. Thus X-11 at its inception employed semi-parametric additive models, backfitting, and robust estimation two to three decades before these methods became commonplace in statistics.

While the early literature on smoothing by local fitting focused on one independent variable with equally-spaced values, much intuition that was built up about smoothing methods remain valid for smoothing as a function of scattered multivariate measurements. We will invoke this intuition later in the paper.

2.2 Modern Work

The modern view of smoothing by local regression has origins in the 1950's and 1960's, with kernel methods introduced in the density estimation setting (Rosenblatt, 1956; Parzen, 1962) and the regression setting (Nadaraya, 1964; Watson, 1964). This new view extended smoothing as a function of a single independent variable with equally spaced measurements to smoothing as a function of scattered measurements of one or more independent variables. Kernel methods are a special case of local regression; a kernel method amounts to choosing the parametric family to consist of constant functions.

Recognizing the weaknesses of a local constant approximation, the more general local regression enjoyed a reincarnation beginning in the late 1970's (Stone, 1977; Cleveland, 1979; Katkovnik, 1979). The method can also be found in other branches of scientific literature; for example numerical analysis (Lancaster and Salkauskas, 1981 and 1986). Furthermore, while the early smoothing work was based on an assumption of a near-Gaussian distribution, the modern view extended smoothing to other distributions. Brillinger (1977) formulated a general approach and Cleveland (1979) and Katkovnik (1979) developed robust smoothers. Later, Tibshirani and Hastie (1987) substantially extended the domain of smoothing to many distributional settings such as logistic regression and developed general fitting algorithms. The extension to new settings continues today (Fan and Gijbels, 1994a; Loader, 1995)).

Work on local regression has continued throughout the 1980's and 1990's. A major thrust has been the application of smoothing in multidimensional cases. Here, numerous approaches can be taken: Cleveland and Devlin (1988) apply local linear and quadratic fitting directly to multivariate data. Friedman and Stuetzle (1981) use local linear regression as a basis for constructing projection pursuit estimates. Hastie and Tibshirani (1990) use local regression in additive models. These methods have substantial differences in data requirements and the types of surface that can be successfully modeled; the use of graphical diagnostics to help make decisions becomes crucial in these cases.

Accompanying the modern current of work in smoothing was a new pursuit of asymptotic results. It began in the earliest papers (e.g., Rosenblatt, 1956; Stone, 1977) and then grew greatly in intensity in the 1980s (e.g., Müller, 1987; Härdle, 1990; Fan, 1993; Ruppert and Wand 1994). In Section 10 we comment on the role of asymptotics in local regression.

3 Polynomial Mixing

The most common choice for the local parametric family is polynomials of degree p. However, the change from degree 1, say, to degree 2 in many applications often represents a substantial change in the results. For example, we might find that degree 1 fitting distorts a peak in the interior of the configuration of observations of the independent variables, and degree 2 removes the distortion but results in undue flopping about at the boundary. In such cases we can find ourselves wishing for a compromise. Polynomial mixing can provide such a compromise. The idea has been used in global parametric fitting (Mallows, 1974). The description in this paper of its use for local regression is from Cleveland, Hastie, and Loader (1995).

The mixing degree, p, is a nonnegative number. If p is an integer then the mixed fit is simply the local polynomial fit of degree p. Suppose p is not an integer and $p = m + c$ where m is an integer and $0 < c < 1$. Then the mixed fit is simply a weighted average of the local polynomial fits of degrees m and $m + 1$ with weight $1 - c$ for the former and weight c for the latter. It is easy to see that this amounts to a local ridge regression of a polynomial of degree $m + 1$ with zero ridge parameters for monomial terms except the term of degree $m + 1$. We can choose a single mixing degree for all x or we can build an adaptive method by letting p vary with x. Both approaches will be used in later sections of the paper.

4 To What Do We Look for Guidance in Making the Choices?

In the next sections we turn to making the choices necessary to carry out local regression — the weight function, the bandwidth, the fitting criterion, and the local parametric family. To what do we look for guidance to make the choices? The answer is, as we have emphasized, to treat choices of degree and bandwidth as modeling the data and use formal model selection criteria and graphical diagnostics to provide guidance. What happens through such a process is that we begin to build up a knowledge base about what tends to provide good models and what does not. In the end, it becomes in some sense a statement about the behavior of data rather than the performance of selection methods. We can never have anything quite as definitive as a theorem, but we can from this process get good guidance about what is likely to work as we approach a new set of data.

The development of methods of parametric regression has had a long history of using model selection criteria and diagnostic methods for the common parametric models fitted to regression data (Daniel and Wood, 1971). It is much to the credit of researchers in parametric regression that these methods have become part of the mainstream of statistical practice. Much reporting of what works in practice from this process has greatly strengthened such parametric fitting.

Attention to the needs of data that arise in practice, and an assessment of methods based on what works in practice was also an important part of the early smoothing literature discussed in Section 2. In the coming sections we will draw on this early literature.

5 Selecting the Weight Function

To begin, suppose the data to be analyzed have an f that is continuous, and suppose that if higher derivatives are not continuous then it is not necessary to produce estimates that reproduce the discontinuity.

In such a case we will almost always want to consider weight functions $W(u)$ that are peaked at $u = 0$, and that decay smoothly to 0 as u increases. One alternative is a rectangular weight function, or boxcar. With a boxcar, all observations within a distance h receive weight 1, and those further away receive weight 0. This results in a noisy estimate; as x changes, observations abruptly switch in and out of the smoothing window. A smooth weight function results in a smoother estimate. This was widely appreciated in the early smoothing literature. For example, Macaulay (1931) writes:

> A smooth weight diagram leads to smoothness in the resulting graduation because smoothing by means of any weighted or un-weighted moving average amounts to distributing each observation over a region as long as the weight diagram and of the same shape as the weight diagram.

Second, we will almost always want to use a weight function that is nonzero only on a bounded interval rather than, for example, approaching zero as u gets large. The reason is computational speed; we can simply ignore observations with zero weight.

Given these two constraints, the choice is not too critical; for our examples — which are all cases with a smooth f — we use the tricube weight function, which is used in the loess fitting procedure (Cleveland, 1979; Cleveland and Devlin, 1988),

$$w(u) = \begin{cases} (1 - |u|^3)^3 & |u| < 1 \\ 0 & |u| > 1 \end{cases}.$$

For cases where f cannot be locally approximated by polynomials, it can sometimes be helpful to consider quite different weight functions. One example is the case of discontinuous f, where one-sided kernels can be employed (McDonald and Owen, 1986; Loader, 1993).

6 Selecting The Fitting Criterion

It turns out that virtually any global fitting procedure can be localized. Thus local regression can proceed with the same rich collection of distributional assumptions as have been used in global parametric fitting. And methods for making the choice can proceed as they have for global parametric fitting.

The simplest case, as we have discussed, is Gaussian y_i. An objection to least squares is lack of robustness; the estimates can be quite sensitive to heavy tailed residual distributions. When the data suggest such distributions we can use robust fitting procedures, for example, the iterative downweighting in loess (Cleveland, 1979; Cleveland and Devlin, 1988) implements local M-estimation. Robust fitting is also discussed in some detail in Tsybakov (1986). Another approach, taken by Hjort (1994), is local Bayesian regression.

Other error distributions lead to other fitting criteria (Brillinger, 1977; Hastie and Tibshirani, 1990; Staniswalis, 1988). For example, a double exponential distribution leads to local L_1 regression. Perhaps more interesting is the extension to generalized linear models; for example, binary data. Suppose

$$P(y_i = 1) = 1 - P(y_i = 0) = p(x_i).$$

Then the locally weighted likelihood is

$$\sum_{i=1}^{n} w(\frac{x_i - x}{h})(y_i \theta_i - \log(1 + e^{\theta_i})).$$

with $\theta_i = \log(p(x_i)/(1 - p(x_i)))$. It is sensible in cases like this to model the natural parameters — in this case, the θ_i — by local polynomials and then transform to the quantities of interest — $p(x_i)$.

In density estimation we observe $x_1, \ldots, x_n$ from a density $f(x)$. Loader (1995) uses the local likelihood

$$\sum_{i=1}^{n} w(\frac{x_i - x}{h}) \log f(x_i) - n \int_{\mathcal{X}} w(\frac{u - x}{h}) f(u) du$$

to model the density; it is natural to use polynomials for $\log f(x)$. See also Hjort and Jones (1994) for discussion and further generalizations.

7 Selecting the Bandwidth and Local Parametric Family

In this section we discuss the bandwidth and the local parametric family together because the choices interact strongly. A change from one parametric family to another can often have a dramatic effect on the sensible choice of bandwidth.

The goal in choosing the bandwidth and the local parametric family is to produce an estimate that is as smooth as possible without distorting the underlying pattern of dependence of the response on the independent variables. In other words, we want $\hat{f}$ to have as little bias as possible and as small a variance as possible. Usually, we need to strike a balance between the two, but with the right model selection tools and graphical diagnostics we can in many applications find a parametric family and select the bandwidth to satisfy the needs of the analysis.

Two methods of bandwidth specification are considered in this section — nearest-neighbor and fixed. These are both simple and easily implemented. (In Section 9 we consider adaptive bandwidth selection).

Finally, we will use either ordinary polynomials or mixed polynomials as the parametric family with degrees ranging from 0 to 3, and use a single degree for all x. (In Section 9 we consider adaptive selection of the mixing degree).

For many applications, particularly those with smooth f, they provide sufficient flexibility to get good fits.

7.1 Bandwidth Selection

Nearest neighbor bandwidths are widely used for local regression (e.g., Stone, 1977; Cleveland, 1979; Fan and Gijbels, 1994a). The reason is simple. A fixed-bandwidth estimate often has dramatic swings in variance due to large changes in the density of the data in the design space leading to unacceptably noisy fits; in extreme cases, empty neighborhoods lead to an undefined estimate.

Boundary regions play a major role in bandwidth choice. Suppose the independent variable x is distributed uniformly on $[0, 1]$. When estimating at 0, a bandwidth of h will cover only half as much data as the same bandwidth when estimating at 0.5; moreover, the data are all on one side of the fitting point in the boundary case. Using the same bandwidth at the boundary as the interior point clearly results in high variability. In practice, the situation is even worse, since the data density may be sparse at boundary regions, for example, when the x_i have a Gaussian distribution. And the situation can become much worse in two or more dimensions where most of the data in the design space can lie on the boundary. The boundary-region problem was well known to early researchers in smoothing. For example, Kendall (1973) writes the following about local polynomials:

> This is as we might expect: the nearer the tails, the less reliable is
> the trend point, as measured by the error-reducing power at that
> point. The fitted curve it has been said, tends to wag its tail.

While fixed bandwidth selection can provide good fits in many cases, it does not do so in many others, and nearest-neighbors appear to perform better overall in applications because of this variance issue. Of course, nearest-neighbor bandwidth selection can fail to model data in specific examples. But usually, if nearest-neighbor selection fails it is not fixed bandwidth selection that is needed to remedy the problem but rather adaptive methods. We will take this up in Section 9.

As stated earlier, we should approach any set of data with the notion that bandwidths can range from very small to very large. Even though most asymptotic work is based on an assumption of the bandwidth going to zero, quite large bandwidths are important for practice. For example, for some applications, using nearest neighbor bandwidth selection, we get the best fits for small values of α, say $1/10$ or smaller; this is often the case for a time series when the goal is to track a high frequency component. In other applications, the best fits are provided by large values of α, say, $3/4$ or higher.

One reaction might be that if a large bandwidth can be used then there is probably some global parametric function that fits the data. But this is not necessarily so. Large bandwidths can provide a large number of degrees of freedom and substantial flexibility in our fits. For example, for loess fitting — that is, local fitting with the tricube weight function, nearest-neighbor bandwidth selection with bandwidth parameter α, polynomial fitting of degree p, and d independent variables — the degrees of freedom of the fit is roughly $1.2\tau/\alpha$ where τ is the number the degrees of freedom of a global parametric fit of a polynomial of degree p (Cleveland, Grosse and Shyu, 1990). Thus the global parametric degrees of freedom is multiplied by $1.2/\alpha$ for a loess fit which means, for example, that if $\alpha = 3/4$, the global parametric degrees of freedom are multiplied by 1.6.

7.2 Polynomial Degree

The choice of polynomial degree — mixed polynomial degree or ordinary polynomial degree — is, like bandwidth, a bias-variance trade off. A higher degree will generally produce a less biased, but more variable estimate than a lower degree one.

Some have asserted that local polynomials of odd degree beat those of even degree; degree 1 beats degree 0, degree 3 beats degree 2, and so forth. But in applications, it is sensible to think of local regression models as ranging from those with small neighborhoods, which provide very local fits, to those with large neighborhoods, which provide more nearly global fits, to, finally, those with infinite neighborhoods, which result in globally parametric fits. In other words, globally parametric fitting is the limiting case of local regression. Thus it is obvious that we should not rule out fitting with even degrees; this is no more sensible than ruling out even degrees for global polynomial fitting.

There is, however, one degree that very infrequently proves to be the best choice in practice — degree zero, or locally constant fitting. This was widely appreciated in the early smoothing literature. The term often used was "moving average" or "moving average with positive weights". In fact, the standard was methods that preserved either quadratic or cubic polynomials; for our context this would mean degrees of two or higher. Macaulay (1931) writes the following about moving averages away from the boundary:

For example, a simple moving average, if applied to data whose underlying trend is of a second-degree parabolic type, falls always *within* instead of *on* the parabola. ... A little thought or experimentation will quickly convince the reader that, so long as we restrict ourselves to *positive* weights, no moving average, weighted or unweighted, will exactly fit any mathematical curve except a straight line.

Exactly the same point made by Macaulay holds for local regression generally. For most applications, if a single degree for polynomial fitting is to be chosen for all x, careful modeling of the data seldom leads to locally constant fitting. The general problem is that since locally constant fitting cannot even reproduce a line except in special cases, for example, equally-spaced data away from the boundaries. Reducing the lack of fit to a tolerable level requires quite small bandwidths, producing fits that are very rough. By using a polynomial degree greater than zero we can typically increase the bandwidth by a large amount without introducing intolerable bias; despite the increased number of monomial terms in the fitting, the end result is far smoother because the neighborhood size is far larger.

In fact, local linear fitting often fails to provide a sufficiently good local approximation — when there is a rapid change in the slope, for example a local minimum or maximum — and local quadratic fitting does better. It is for this reason that in the early smoothing literature, the methods that were devised preserved at least quadratic functions because peaks and valleys in f are common in practice.

8   Examples

In this section, we will use three examples to demonstrate modeling the data through the choices of the weight function, the fitting criterion, the bandwidth, and the parametric family. We will use cross-validation, C_p, and graphical diagnostics to guide the modeling.

8.1   Ozone and Wind Speed Data

The first example is measurements of two variables: ozone concentrations and wind speed at ground level for 111 days in New York City from May 1 to September 30 of one year. We will take the cube roots of the ozone concentrations — this symmetrizes the distribution of the residuals — and model cube root ozone as a function of wind speed. We will fit with the following choices: the tricube weight function, nearest-neighbor bandwidth selection, least-squares, and polynomial mixing.

Figure 1 shows the cross-validation sum of absolute deviations against the mixing degrees for four values of α from 0.25 to 0.75 in equal percentage steps. Because of the rough approximation referred to above, equal percentage steps in α tend to make the degrees of freedom of fits for a fixed mixing degree change in equal percentage steps. Figure 2 shows fits. Each column contains fits for one value of α. The top row is the mixing fit with the minimum cross-validation score and the remaining rows show the fits for degrees 3 to 0. Figure 3 shows the residuals for each of the 30 fits in Figure 2. Superposed on each plot is a loess smooth with local linear fitting and $\alpha = 1/3$. Such residual plots provide an exceedingly powerful diagnostic that nicely complements a selection criterion such as cross-validation. The diagnostic plots can show lack of fit locally, and we have the opportunity to judge the lack of fit based on our knowledge of both the mechanism generating the data and our knowledge of the performance of the smoothers used in the fitting.

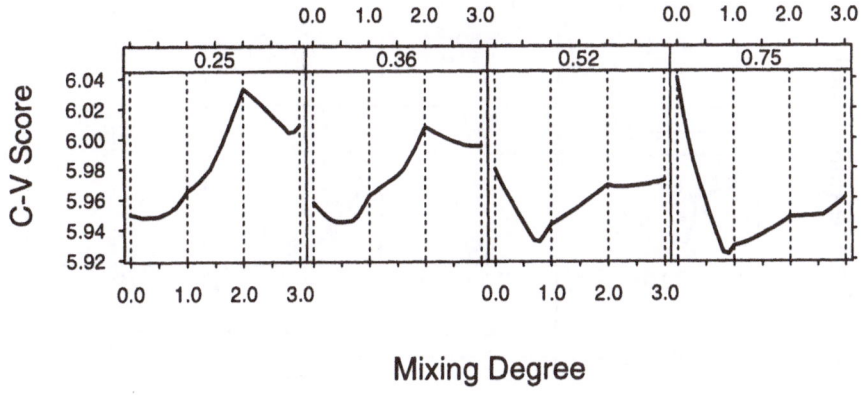

Figure 1: Cross-validation scores for the mixed fits as a function of p.

For the ozone and wind speed data there is a pronounced dependence of the response on the independent variable but overall there are not radical changes in curvature, in particular there are no peaks or valleys. We might expect that locally, a linear family provides a reasonable approximation. The diagnostics and cross-validation show this is the case. For local constant fitting, that is $p = 0$, a small α is needed to capture the dependence of ozone on wind speed without introducing undue distortion. Even for $\alpha = 0.25$, the plot of residuals suggests that there is lack of fit at the smallest values of wind speed, that is, at the left boundary, where there is a large slope. Local constant fitting cannot capture a linear effect at a boundary. The same distortion does not occur at the right boundary because the local slope is zero, and local constant fitting can cope with a zero slope. Despite the distortion, for $\alpha = 0.25$ the cross-validation criterion suggests that overall, local constant fitting is the best fit because for larger values of p the variability of the fit increases rapidly. But the local constant fit is itself quite noisy, far noisier than is reasonable. In other words, we do not have a satisfactory fit for $\alpha = 0.25$.

As we increase α to get a smoother fit, the local constant fit introduces a

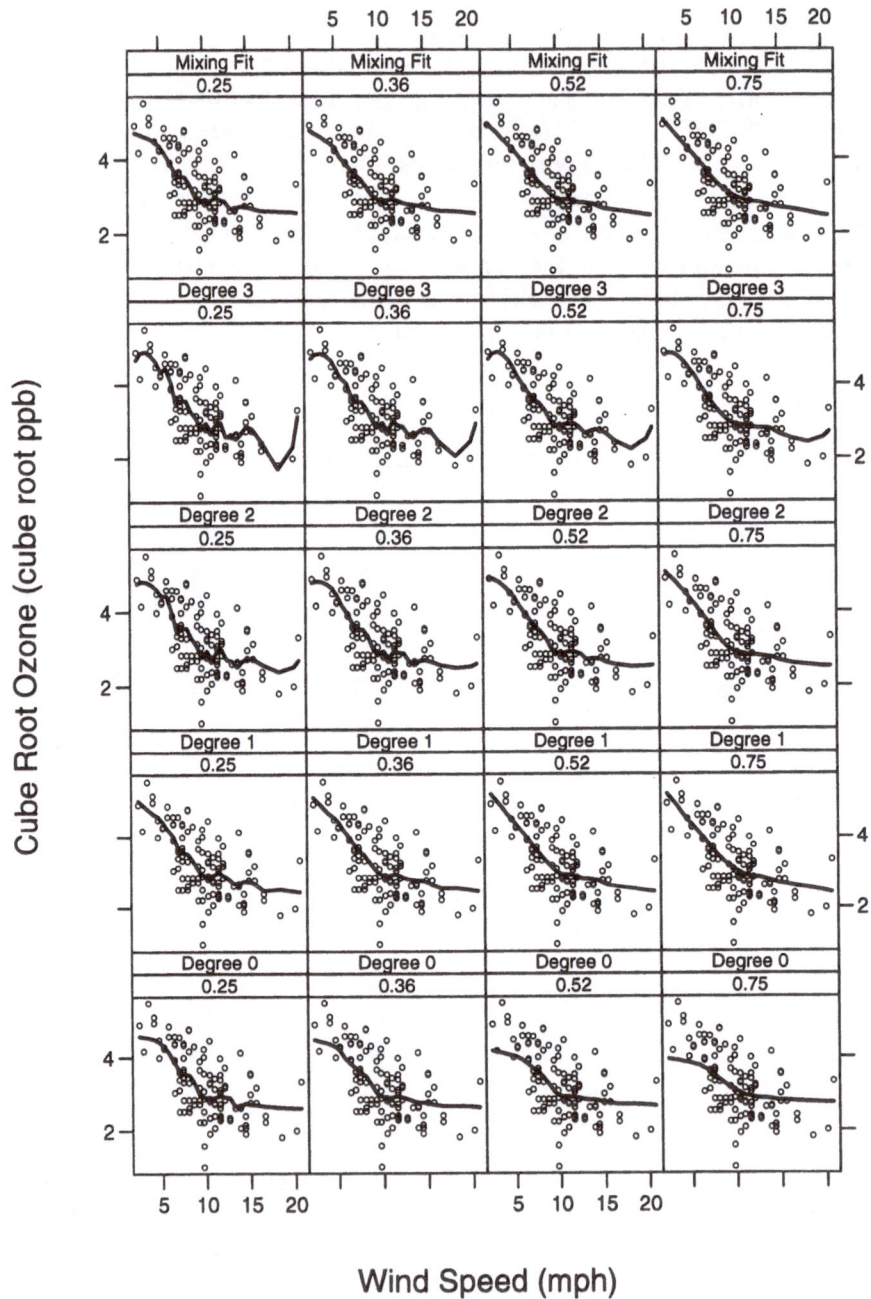

Figure 2: Fits for four nearest-neighbor bandwidths ($\alpha = 0.25$ to 0.75 in equal percentage steps) and five local fitting methods. The top row shows the fits for the mixing method with cross-validation and the remaining rows show local polynomial fits for degrees 3 to 0.

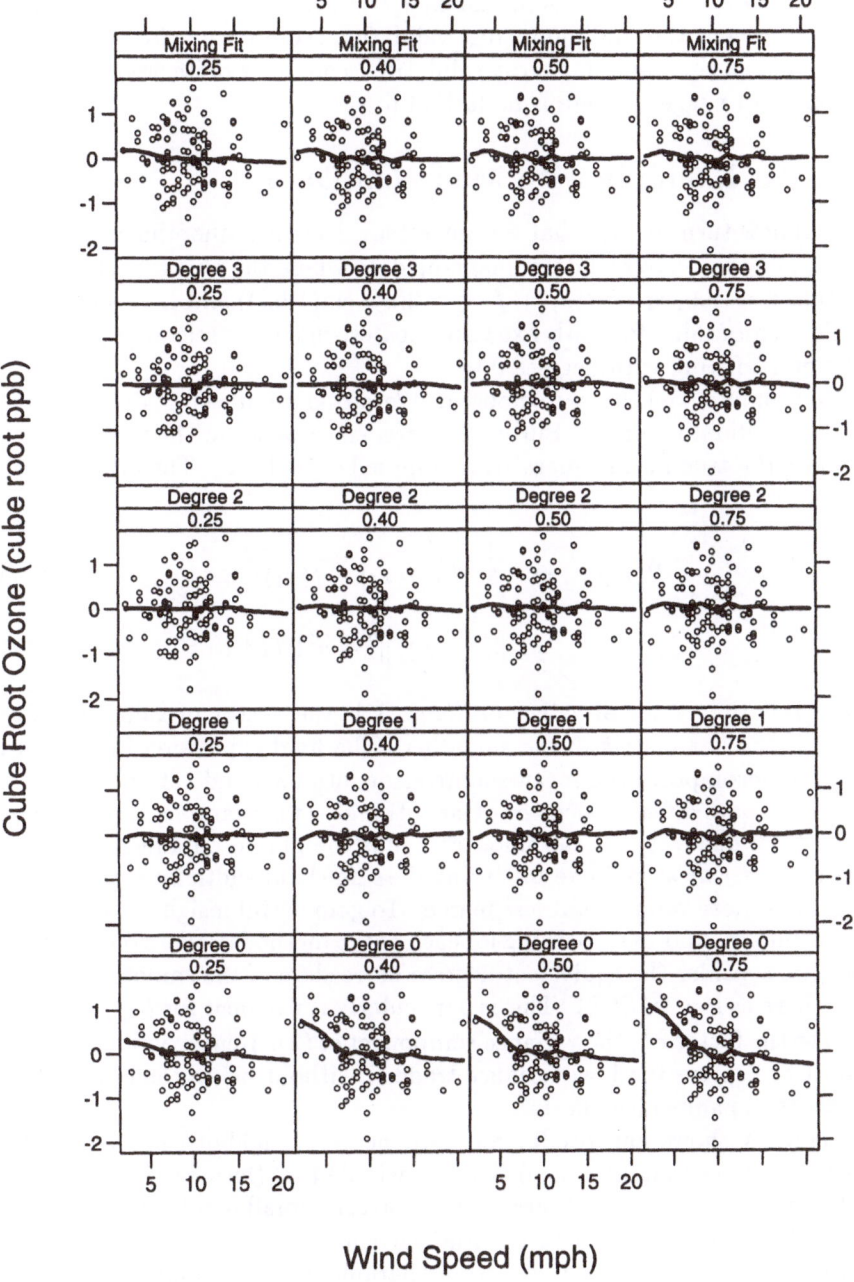

Figure 3: Residual plots for the fits in the previous figure.

major distortion, one that is so bad that cross-validation judges it to be the worst case, and the minimum cross-validation score occurs for p close to 1. For $\alpha = 0.75$, the minimum cross-validation score fit is quite smooth and the residual plot suggests there is no lack of fit.

8.2   An Example with Made-Up Data

We will now turn to data that we generated. To study smoothers, it is useful to use, in addition to real data, made-up data where the true model is known.

Take $n = 100$, $x_i \sim N(0,1)$, $f(x) = 2(x - \sin(1.5x))$ and $\epsilon_i \sim N(0,1)$. We consider smoothing for both fixed and nearest-neighbor selection using local polynomial fitting with degrees 0 to 3.

How should we compare smooths? Were this a real set of data we could use cross-validation or C_p. But in this case, since we know the model, we can compute the true mean-square error summed over the x_i. Thus the criterion is

$$R(\hat{f}, f) = \sum_{i=1}^{n} E((\hat{f}(x_i) - f(x_i))^2) \tag{1}$$

$$= \|(I - L)f\|^2 + \sigma^2 \text{tr}(L^T L), \tag{2}$$

where f is the vector of values of $f(x_i)$, $\hat{f}$ is the vector of values of $\hat{f}(x_i)$, and L is the hat matrix of the smoother. The final expression in the above equation decomposes the mean-square error into bias and variance.

To compare different degrees of smoothing — for example, local constant versus local quadratic — the smoothers must be placed on an equal footing. We cannot get meaningful results using the same bandwidth for each; a higher order fit is more variable but less biased. To gain useful insight, one must use equivalent amounts of smoothing for each of the methods under consideration. Thus, we consider the equivalent degrees of freedom of the smooth, which we define here as $\nu = \text{tr}(L^T L)$. This is particularly convenient for $R(\hat{f}, f)$ since ν is, up to the factor σ^2, the variance component of (2). Hence, a plot of $R(\hat{f}, f)$ against ν displays the bias-variance tradeoff without being confounded with meaningless bandwidth effects.

Figure 4 shows the results for both nearest-neighbor and fixed bandwidths. Degrees ranging from 0 (local constant) to 3 (local cubic) are considered. Points on the right of these plots represent small bandwidths for which the fits have little bias but substantial variance.

Fits for the values of the smoothing parameters that minimize R are displayed in Figure 5. Local constant fitting is clearly unsatisfactory for both fixed and nearest-neighbor bandwidths. The smooths are very noisy because small bandwidths must be used to properly track the boundary effects. For local linear fitting, the improved boundary behavior enables larger bandwidths to be used, and a smoother curve results. However, even in this example with fairly modest curvature, the local quadratic and cubic fits perform better, with the fitted curves being substantially less noisy. Visually, there is not much difference between the fixed and nearest-neighbor fits for higher orders. The best fit, according to Figure 4, is the local quadratic fit with nearest-neighbor bandwidths, but fixed bandwidth selection performs well in this example.

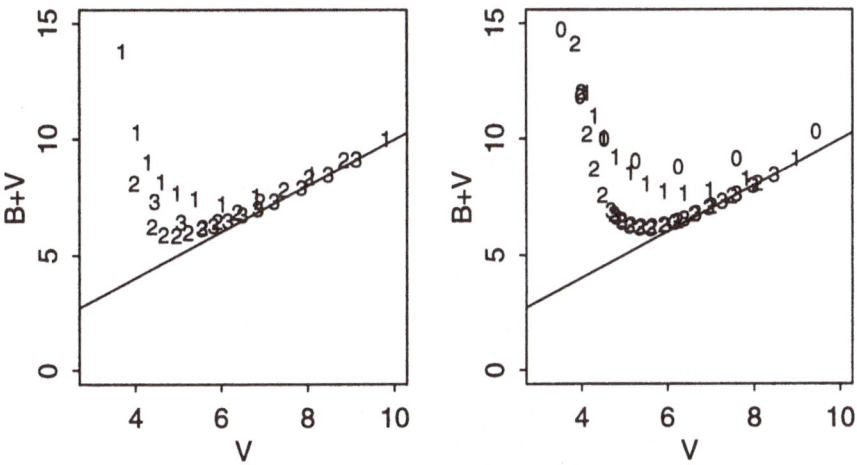

Figure 4: Plots of $R(\hat{f}, f)$ against $\text{tr}(L^T L)$ for nearest neighbor bandwidths (left) and fixed bandwidths (right), for fitting of degrees 0 to 3. (Note 0 is entirely off the scale on the left figure).

It is important to reiterate that exclusive reliance in practice on a global criterion similar to that in the figure is unwise because a global criterion does not provide information about where the contributions to bias and variance are coming from in the design space. In many applications, high bias or high variance in one region may be more serious than in another. For example, a careful look at the right panels of Figure 5 reveals that for degrees one to three the y_i at the minimum and maximum values of the x_i have nearly zero residuals; the fits at these points are very nearly interpolating the data which results in unacceptably high variance. Nearest neighbor bandwidths go some way towards relieving the problem; this accounts for most of the advantage of nearest-neighbors observed in Figure 4 for local quadratic fitting.

As this example suggests, boundaries tend to dominate the fixed vs. nearest-neighbor comparison. Asymptotic comparisons provide no useful guidance in this case. For example, Gasser and Jennen-Steinmetz (1988) use the asymptotic characterization of the nearest-neighbor bandwidth as being proportional to the reciprocal of the density of the x_i in the design space; in our example $h(x) \approx \alpha/2\phi(x)$ where $\phi(x)$ is the standard normal density. At the left end-point $x = -2.736$ and $\alpha = 0.35$ for local linear fit-

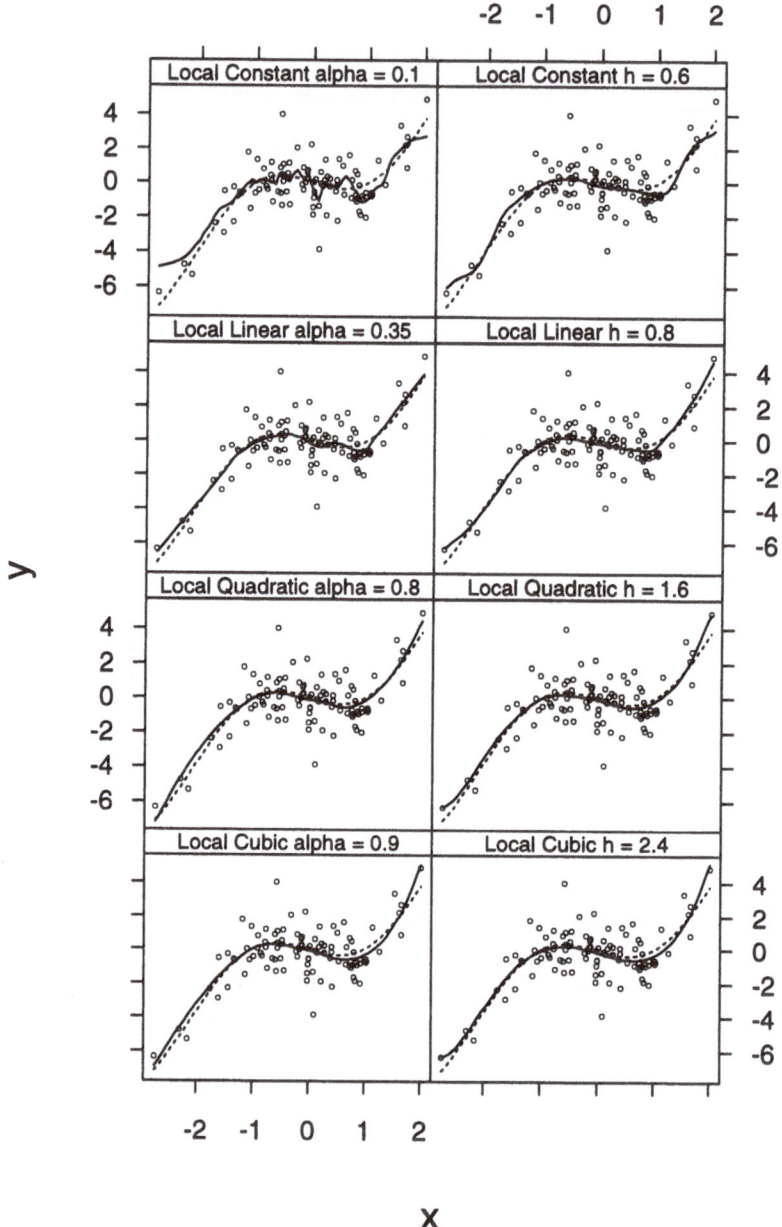

Figure 5: Local fits that minimize mean-square error for nearest-neighbor bandwidths (left) and fixed bandwidths (right). The solid curves are the estimates, and the dashed curves the true function.

ting. The asymptotic characterization yields $h = 18.5$ and all observations receive weights between 0.95 and 1! Clearly, this does not approximate reality, where only 34 observations receive non-zero weights, and many of the weights are small.

8.3   More on the Ozone Data

The ozone and wind speed data studied earlier are just two variables from a multivariate data set that has two other independent variables, temperature and solar radiation Cleveland (1993). The full dataset is 111 daily measurements of wind speed, temperature, radiation, and ozone concentration. Cleveland (1993) found that it is appropriate to fit models that are conditionally parametric in R and W. A C_p plot for this data is shown in Figure 6 for degrees 0 to 3 and a variety of bandwidths. Local constant fitting is vastly inferior to other degrees. Local linear also struggles; its minimum C_p is about 50% larger than for local quadratic. Local quadratic also slightly beats local cubic. The minimum C_p for the local quadratic fitting occurs at an α of 0.4; however, the C_p pattern is quite flat for quadratic fitting and the minimum C_p fit is fairly noisy, so Cleveland elected to use a larger α.

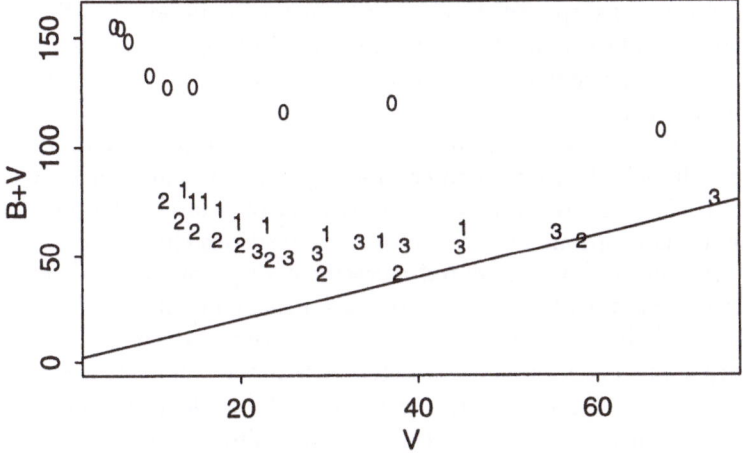

Figure 6: C_p plot for the environmental data, with trivariate predictors.

9   An Approach to Adaptive Fitting Based on Assessing Localization

In the previous sections we have discussed smoothing procedures for which the parametric family fitted locally does not change with x and for which the bandwidth selection parameters of fixed and nearest-neighbor smoothing do not change with x. But for some data sets, particularly those for which the amount of curvature in different broad regions of the design space is quite different, it can make sense to use adaptive methods. Many adaptive bandwidth methods have been suggested in the past (e.g., Friedman and Stuetzle, 1982). Varying the parametric family locally is less common (e.g., Fan and Gijbels, 1994b) but very promising.

We use a general approach to adaptive fitting based on an *assessment of parametric localization*. In Section 1, "parametric localization" was used to describe the basic approach that guides local regression — the approximation of a the true f locally by a parametric function. Using an assessment of localization, we can design methods for adaptive bandwidth selection (Cleveland and Loader, 1995) and adaptive parametric family selection (Cleveland, Hastie, and Loader, 1995).

The important principle is this. *For a particular x and a given neighborhood, the fit at x can be expected to perform well if the locally fitted parametric function adequately approximates the true function over the neighborhood.* Now one might object to this notion on the basis that it is possible for the fit at x to be good even if the fitted parametric function provides a poor approximation at places other than x; but further thought makes it clear that only in degenerate cases can we do well despite a poor approximation. For example, for equally-spaced data away from the boundary locally constant fitting will provide a good fit to a linear f. But this is solely due to the serendipitous result that local linear fitting in such a case happens to be the same as local constant fitting.

Now we will not assess the fit equally throughout the neighborhood. Instead we will weight our assessment based on the weight function used to compute the fit at x. In other words we will have greater tolerance for poor performance at positions far from x than at positions close to x.

With this notion of a weighted assessment of parametric localization as the approach, details of carrying it out are reasonably straightforward. For the fit at x we simply take an automatic selection criterion that we might have used for picking a global parameter such as the α for nearest-neighbors, and we make two changes. First, instead of applying the criterion over the x_i using the smoother, we the apply the criterion using the parametric function fitted at x. Second, we weight the criterion with the neighborhood weights used in the local fit at x. The next two examples illustrate this process.

9.1 Local Polynomial Mixing

We will now use the assessment of localization to develop an adaptive method for choosing the mixing degree for polynomial mixing. The method is discussed in detail by Cleveland, Hastie, and Loader (1995).

In Section 8 we used polynomial mixing and selected a single mixing degree for all x by the cross-validation sum of absolute deviations. We chose a value of p, fitted the mixed polynomial to get the fit $\hat{f}(x_i)$ at each x_i, and then studied the single-observation deletion effect by

$$\sum_{i=1}^{n} |[y_i - \hat{f}(x_i)]/[1 - h_{ii}(x_i)]|,$$

where h_{ii} is the i-th diagonal element of the hat matrix of the mixed polynomial smoother. The value $h_{ii}(x_i)$ comes from the hat matrix for the weighted least squares operator that fits the polynomial at x_i, which is why in the notation for the diagonal element we show the dependence on x_i. We will now localize this procedure to provide an adaptive fitting method.

For a given x we select a candidate p and carry out a weighted cross-validation of the mixed polynomial, $\hat{p}(x)$, fitted at x. Thus the criterion is

$$\frac{\sum_{i=1}^{n} w_i(x)|[y_i - \hat{p}(x_i)]/[1 - h_{ii}(x)]|}{\sum_{i=1}^{n} w_i(x)}.$$

Now $h_{ii}(x)$ is the i-th diagonal element of the hat matrix of the weighted least squares operator that fits the polynomial at x. We compute the localized criterion for a grid of p values from mixing degree 0 to mixing degree 3, and choose the mixing degree that minimizes the criterion.

We will illustrate the method with an example. The data are sickness rates for ages 29 to 79 reported by Spencer (1904b). We will study the percentage change in the rates by smoothing the first differences of the (natural) log rates as a function of age. Fits are shown in Figure 7. Each column is one value of the nearest-neighbor bandwidth parameter. The values increase logarithmically from 0.25 to 2. The top row is the adaptive fit with the mixing degree chosen locally by the above local method. The remaining rows show the fits for integer degrees 3 down to 0, in other words, local cubic fitting to local constant fitting. Figure 8 shows the residuals of these fits; superposed on each plot is a loess smooth with local linear fitting and $\alpha = 1/3$. Figure 9 shows the cross-validation score as a function of age for each of the five bandwidths, and Figure 10 shows the selected values of the mixing degree p, also as a function of age.

The four displays are quite informative. The residual plots in Figure 8 show the bandwidth at which each smoother just begins to introduce distortion (bias) in the fit. For degree 0, even the smallest bandwidth, $\alpha = 0.25$, has a slight distortion for the largest ages. For degree 1, it begins at $\alpha = 0.42$; for degree 2, it begins at $\alpha = 0.71$, and for degree 3 and the adaptive fit it begins at $\alpha = 1.19$. For no bandwidth does degree 0 produce a fit that has no distortion revealed in the residuals plots even though the fit for the smallest bandwidth is noisy. For degrees 1 to 3 the only distortion-free fits are a bit too noisy, although the fits are smooth and the distortion is quite minor for degree 1 with $\alpha = 0.42$, degree 2 with, $\alpha = 0.71$, and degree 3 with $\alpha = 1.19$. (No saving grace comes to degree 0 since it cannot produce a smooth fit without major distortion.) For the locally adaptive mixed fit with $\alpha = 0.71$

we get an excellent fit with almost no distortion and with requisite smoothness. And, of course, the fit has the attractive property that the parametric family was chosen automatically. The fit also has the interesting property that at about 45 years, just near the first age at which the pattern begins its nonlinear behavior, the local mixing method switches from degree 1 fitting to degree 3 fitting.

9.2 Adaptive Bandwidth Selection

We will now use the assessment of localization to develop an adaptive method for choosing the bandwidth for local linear polynomial fitting. The method is discussed in detail by Cleveland and Loader (1995).

For each x and bandwidth h, consider a localized goodness-of-fit criterion

$$\frac{\sum_{i=1}^{n} w_i(x) E(\hat{p}(x_i) - f(x_i))^2}{\sigma^2 \text{tr}(W)}$$

where $w_i(x) = w(h^{-1}(x_i - x))$, $W = \text{diag}(w_i(x))$, and $\hat{p}(x_i)$ are values of the local polynomial, $\hat{p}$, fitted at x. This is estimated by a localized version of C_p

$$C(h) = \frac{1}{\text{tr}(W)}[2\text{tr}(M_2) - \text{tr}(W) + \frac{1}{\hat{\sigma}^2} \sum_{i=1}^{n} w_i(x)(y_i - \hat{p}(x_i))^2]$$

where $M_2 = (X^T W X)^{-1}(X^T W^2 X)$, X is the design matrix, and $\hat{\sigma}^2$ is an estimate of σ^2 from a non-adaptive smooth with a small bandwidth. Roughly, for each fitting point, we choose a bandwidth h with small $C(h)$. Specifically, we choose an interval $[h_0, h_1]$, where h_0 is the distance from x to the kth nearest x_i for small k, and h_1 results in rejection of a goodness-of-fit test at a low significance level. Then, the interval $[h_0, h_1]$ is searched for the right-most local minimum of $C(h)$.

This algorithm is carried out at the vertices of a tree in a manner similar to that of loess (Cleveland and Grosse, 1991) but with a split rule suggested in Loader (1994). Suppose the tree has a cell with vertices $[v_k, v_{k+1}]$, and the adaptive procedure produces bandwidths h_k and h_{k+1} at these vertices. The cell requires further refinement if either bandwidth is small relative to the length of the cell; specifically a new vertex is added if $v_{k+1} - v_k > 0.7 \min(h_k, h_{k+1})$. The function estimate is constructed as a cubic interpolant using the local fits and local slopes at the vertices. Implementation involves only minor modification of existing algorithms, and our present code includes multivariate extensions, based on both rectangular and triangular cells. Computation of $C(h)$ is a straightforward by-product of the local fit; we do not rely on higher order fits or Taylor series expansions to estimate bias.

The top panel of Figure 11 shows a made-up data set with 2048 observations on $[0, 1]$; the response function is one of the four examples considered by

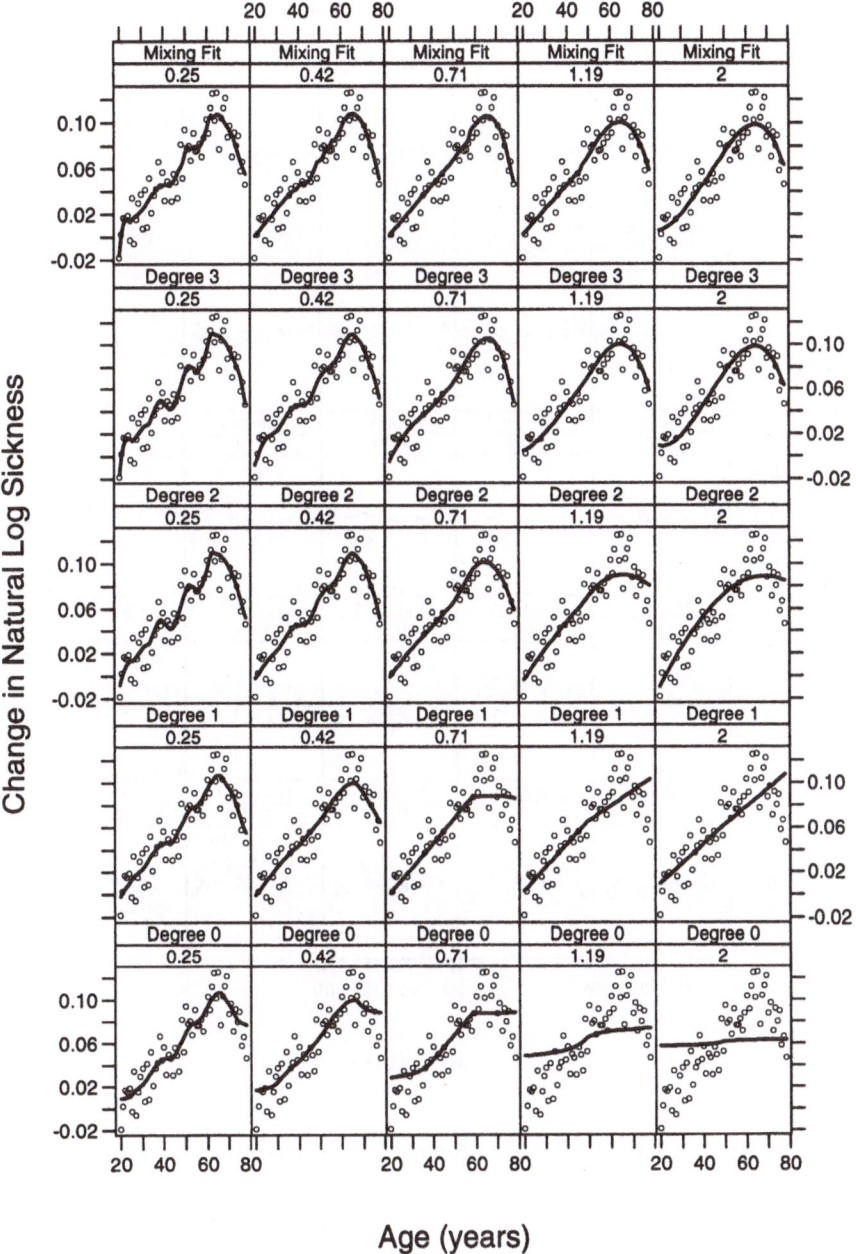

Figure 7: Fits for five nearest-neighbor bandwidths ($\alpha = 0.25$ to 2 in equal percentage steps) and five local fitting methods. The top row shows the fits for the adaptive mixing method, and the remaining rows show local polynomial fits for degrees 3 to 0.

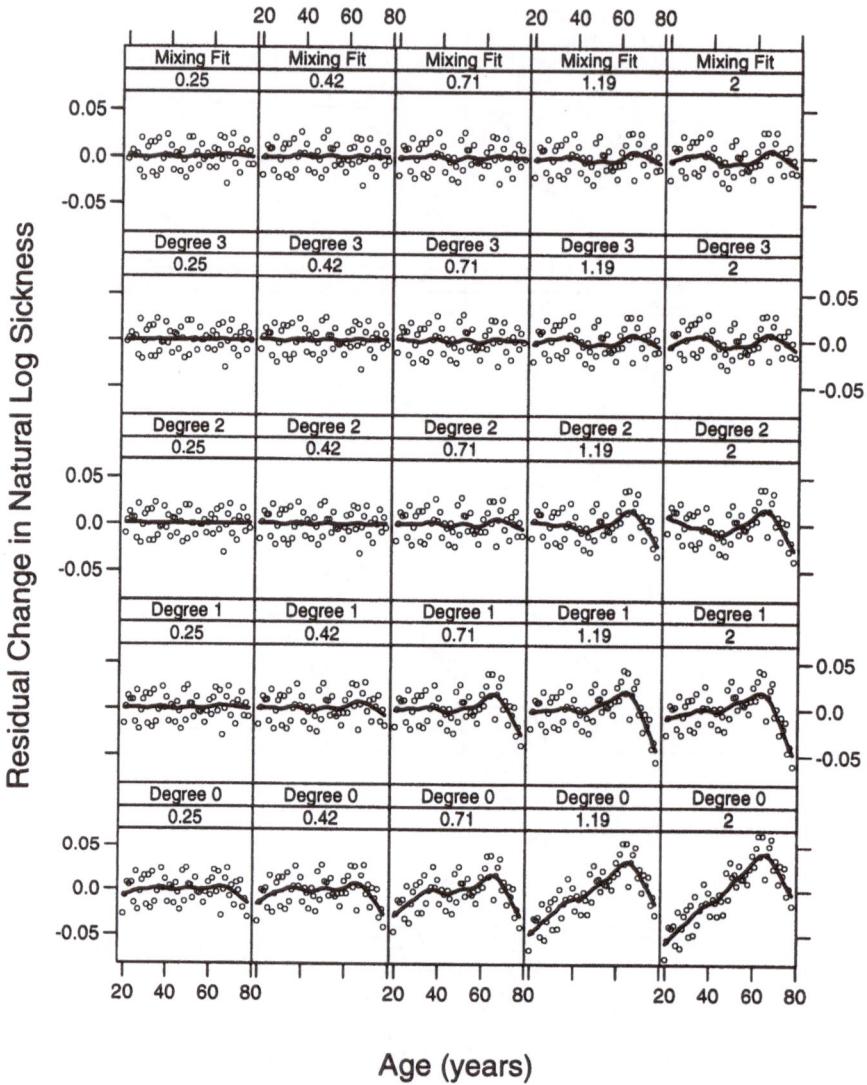

Figure 8: Residual plots for the fits in the previous figure.

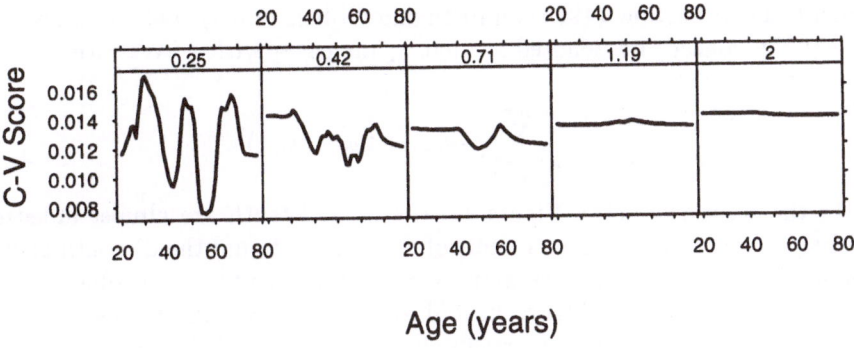

Figure 9: Cross-validation scores for the adaptive mixing fits.

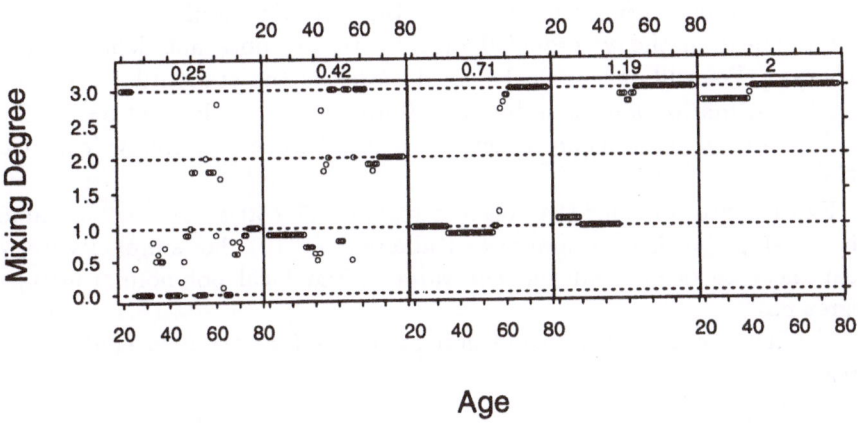

Figure 10: Mixing degrees for the adaptive mixing fits.

Donoho and Johnstone (1994) and Fan and Gijbels (1995). Let the x_i, which are equally spaced, be denoted so that x_i increases with i. We estimate σ^2 by

$$\frac{1}{6(n-2)} \sum_{i=2}^{n-1} (-y_{i-1} + 2y_i - y_{i+1})^2.$$

In the second panel of Figure 11, the fit and truth are almost indistinguishable, apart from small amounts of roughness around the discontinuities. The fit appears better than those previously obtained for this problem, with sharper reproduction of the breaks and less noise. However, the residual plot in the third panel shows large residuals around most of the discontinuities, a clear indication that the fit is not perfect. As mentioned earlier, proper modeling of discontinuous functions requires appropriate choice of the weight function and basis functions (Loader, 1993; Speckman, 1995); this is how one would sensibly proceed for such data in practice. The plot in the fourth panel shows bandwidths at the knots determined by the tree algorithm.

10   Theory

10.1   Some Theory with Minimal Assumptions

Local regression with least squares as the criterion results in an estimate that is linear in the y_i provided model selection does not depend on the response. We will take x to be fixed. This is the standard for practice; that is, estimation and sampling distributions are conditional on x. Such conditioning is the sensible practice, as been argued by many (e.g., Cox, 1958).

Exact expressions for the bias and variance of $\hat{f}$ at x are easily obtained. The development here is similar to Katkovnik (1979). For simplicity we will treat the case of one independent variable and local polynomial fitting of degree p.

If f has a continuous second derivative, Taylor's theorem enables us to write

$$E\hat{f}(x) = \sum_{i=1}^{n} l_j(x)f(x_j) = f(x)\sum_{i=1}^{n} l_j(x) + f'(x)\sum_{i=1}^{n}(x_j - x)l_j(x)$$

$$+ \frac{1}{2}\sum_{i=1}^{n}(x_j - x)^2 l_j(x)f''(\theta_j) \qquad (3)$$

where $(\theta_j - x)(\theta_j - x_j) \leq 0$. For local regression,

$$l(x)^T = (l_1(x), \ldots, l_n(x)) = c(x)^T (X^T W X)^{-1} X^T W$$

where $c(x)$ is a column vector of fitting functions $(\begin{array}{cccc} 1 & x & \ldots & x^p \end{array})^T$, X is the design matrix, with rows $c(x_i)$, and W is a diagonal matrix of the weights

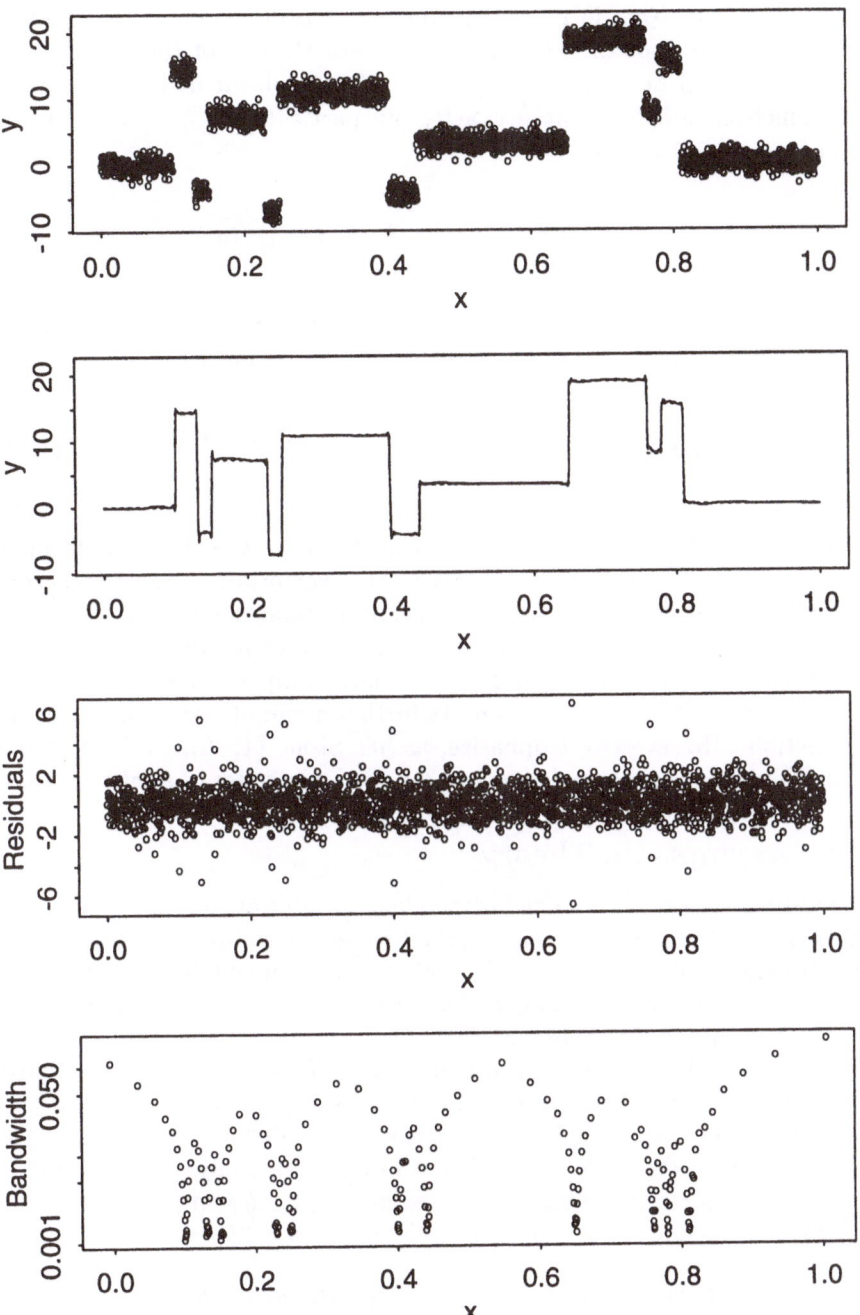

Figure 11: Adaptive bandwidth selection. From the top panel to the bottom the graph shows the data, the adaptive local linear fit, the residuals, and the local bandwidths at the fitted knots.

$w((x_i - x)/h(x))$. Note in particular, $l(x)^T X = c(x)^T$. This is just a mathematical statement of the obvious: If f is exactly one of the fitting functions, then f will be reproduced exactly. If local linear fitting is used, all linear functions are reproduced exactly. In particular, $\sum_{i=1}^{n} l_j(x) = 1$ and $\sum_{i=1}^{n}(x_j - x)l_j(x) = 0$. Hence

$$E\hat{f}(x) = f(x) + \frac{1}{2}\sum_{i=1}^{n}(x_j - x)^2 l_j(x) f''(\theta_j). \tag{4}$$

With fits of degree higher than 1, one can retain more terms in the Taylor series, leading to more precise bias estimates. The variance also has a simple closed form expression

$$\mathrm{var}\hat{f}(x) = \sigma^2\|l(x)\|^2 = \sigma^2 c(x)^T (X^T W X)^{-1} X^T W^2 X (X^T W X)^{-1} c(x). \tag{5}$$

Expressions (4) and (5) are exact; no asymptotics are used. For fits with $p > 0$, terms in the bias expansion involving $f'(x)$ are *gone* for finite samples and *any* values of x_i in the design space, however bizarre. The expressions are easily computed and compared for various estimates in specific examples, and (5) is useful for assessing the variance of the estimate in practice. To make asymptotic analysis of performance and comparison with other estimates tractable, it is necessary to make further assumptions, as we do in the next section. However we emphasize, as has Stone (1980), *such additional assumptions are not required to retain the good properties of local regression.*

10.2 Asymptotic Theory

The results stated below for one independent variable are not new. Tsybakov (1986) and Müller (1987) are among the first to derive these for local regression, although similar expressions for kernel regression and density estimation have been known for much longer. Ruppert and Wand (1994) derive similar results for multivariate predictors.

We must make design assumptions, describing how the design behaves as a function of n. One common assumption is a random design: x_i are independent random variables with density $g(x)$. (This is simply a device, and should not be taken as a signal that in our data analyses we will model our data by random x_i.) Another is fixed design; $i - \frac{1}{2} = n\int_{-\infty}^{x_i} g(u)du$.

When fitting odd order polynomials of degree $p = 2k + 1$

$$E\hat{f}(x) - f(x) = h^{p+1} f^{(p+1)}(x) B_p(w) + o(h^{p+1})$$
$$\mathrm{var}\hat{f}(x) = \frac{\sigma^2}{nhg(x)} V_p(w) + o((nh)^{-1})$$

at points where the design density $g(x)$ is continuous and non-zero. Here, $B_p(w)$ and $V_p(w)$ are constants depending only on p and the weight function

w. Letting $\Lambda_j = \int_{-1}^{1} c(x)c(x)^T w^j(x)dx$

$$
\begin{aligned}
B_p(w) &= \frac{1}{(p+1)!}[\Lambda_1^{-1}\int_{-1}^{1} x^{p+1}c(x)w(x)dx]_1 \\
V_p(w) &= [\Lambda_1^{-1}\Lambda_2\Lambda_1^{-1}]_{1,1}
\end{aligned}
\tag{6}
$$

assuming w has support $[-1,1]$.

For fitting even order polynomials of degree $p = 2k$

$$
\begin{aligned}
E\hat{f}(x) - f(x) &= h^{p+2}(f^{(p+2)}(x) + (p+2)f^{(p+1)}(x)\frac{g'(x)}{g(x)})B_{p+1}(w) \\
&\quad + o(h^{p+2}) \\
\mathrm{var}\,\hat{f}(x) &= \frac{\sigma^2}{nhg(x)}V_p(w) + o((nh)^{-1})
\end{aligned}
$$

at points where $g(x)$ is differentiable. It can be shown that $V_{2k} = V_{2k+1}$; the only difference between successive even and even orders is in the bias term.

A lower boundary point is typically defined as a point x_0 satisfying $g(x) = 0$ for $x < x_0$; $g(x) > 0$ for $x \geq x_0$ and $g(\cdot)$ is right continuous at x_0 (Fan and Gijbels, 1992). For example, $g(x) = I_{[0,1]}(x)$ has a lower boundary at 0, and an upper boundary at 1. At boundary regions, asymptotic results must be modified. Suppose x_0 is a boundary point and $x = x_0 + hv$ for fixed v. Then for even order fitting,

$$
E\hat{f}(x) - f(x) = h^{p+1}f^{(p+1)}(x)B_{p,v}(w) + o(h^{p+1}).
$$

The variance expansions and odd order bias have modified constants $B_{p,v}$, $V_{p,v}$ with forms similar to (6) but with integrals now taken over $[-v,1]$. The biggest difference is in the bias; even order fitting has bias $o(h^{p+1})$ rather than $o(h^{p+2})$. A less obvious difference is that now $V_{2k,v} \neq V_{2k+1,v}$; usually $V_{2k,v} < V_{2k+1,v}$.

How well do these asymptotics perform? Figure 12 studies the simulated data used previously in Figure 5 for fitting local constant through local cubic. The asymptotics suggest the variance (and hence fitted degrees of freedom) should be the same for orders $2k$ and $2k+1$; hence we use $h = 0.8$ for local constant and local linear, and $h = 2.4$ for local quadratic and local cubic. The first problem to note with the asymptotics is the weakness of the setting. Consider the problem of boundary bias. Under the definition of boundaries, the design density has compact support, and is bounded away from 0 on that support. By this definition, the case $x_i \sim N(0,1)$ does not have boundaries! But as we saw earlier, there are substantial boundary effects for this problem. For local constant and linear fitting, the asymptotics are mostly fine. The standard deviation approximation is slightly conservative in most cases. Note however one departure from the asymptotic theory: The local linear method is more variable than local constant near the boundaries. But for local quadratic and cubic fitting, the asymptotics perform very poorly. The bias approximations are completely misleading, with peaks and zeros having little relation to reality.

The limitations of asymptotics have been noted before in various settings. For example, Rosenblatt (1971) in a discussion of optimal kernels states:

> The arguments usually given have been of an asymptotic character and it is a mistake to take them too literally from a finite sample point of view.

Stoker (1993) goes even further and challenges the standard approach to asymptotics, suggesting instead that a bandwidth fixed as $n \to \infty$ gives a more realistic assessment of the performance of estimates.

The clear message then is one of caution: realistic problems are not well modeled by the asymptotic framework, and the asymptotic approximations can be quite poor. For example, it would be a mistake to consider two procedures as equivalent purely on the basis of having the same asymptotic expansions, or to use asymptotics alone to justify smoothing procedures.

Unfortunately the use of asymptotics in the kernel smoothing literature over the past 10-15 years has not been good. Sweeping conclusions are drawn solely from asymptotics; in some cases, procedures are labeled as 'optimal' without any real justification. We present two examples in the next two sections — bandwidth choice and polynomial degree.

10.3 Bandwidth

Over the last decade, much effort has been expended devising methods to 'automatically' select a bandwidth and assessing how close the selected bandwidth is to a theoretically optimal bandwidth, with little regard for the possibility that the optimal bandwidth may be very inadequate. Recently, a major focus has been 'plug in' methods (Hall et al., 1991), in which one directly estimates the bias and variance (or asymptotic approximations to these) and choosing a bandwidth attempting to minimize either pointwise MSE or global criterion such as MISE.

The key problem here is bias estimation. For a heavily biased estimate, bias estimation is a reasonable problem. For the local linear estimate considered in Figure 5, bias estimation essentially amounts to estimating the curvature; this can be accomplished using a local quadratic or cubic fit (higher order kernels have usually been employed in the literature). A plug-in algorithm could work quite well here, if our criterion were how close the selected bandwidth is to an optimal bandwidth.

However, our criterion should not be the quality of the bandwidth selector, but the quality of the resulting estimate. Hence, the available information on curvature should go *directly into the estimate $\hat{f}$* and not just into a bandwidth selector. The resulting smoother should not be heavily biased, and therefore its bias should be difficult to estimate. The problem with plug-in methods

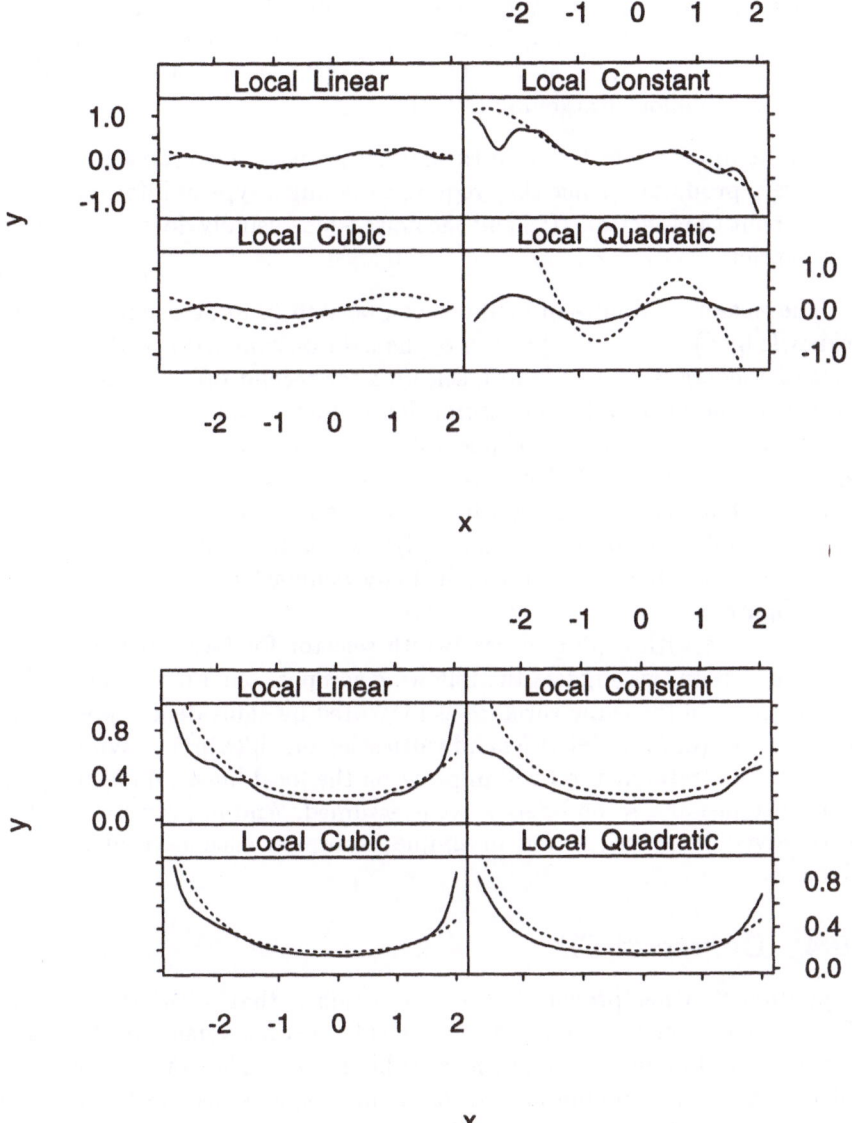

Figure 12: Biases (top panel), standard deviations (bottom panel) and their asymptotic approximations.

now becomes clear: For local quadratic and cubic fits, bias estimation essentially amounts to estimating fourth order derivatives, about which the data contains little or no information. In any case, the asymptotic expressions are very poor. To quote Katkovnik (1979)

> Attempts to select h from theoretical precision estimates are not very productive since they require assigning a type of information about the function $f(x)$ and the noise which usually does not exist apriori.

The weakness of plug-in methods can be seen (and occasionally has been acknowledged) from a comparison of the assumptions with work on optimal rates of convergence. Bandwidth selectors for second order kernel estimates are usually derived under an assumption of four or more derivatives; under this assumption the best possible rate on convergence of $\hat{f}(x)$ to $f(x)$ is $O_p(n^{-4/9})$ (Stone, 1980). But a second order kernel method achieves a rate no better than $O_p(n^{-2/5})$, regardless of how good the bandwidth selector is. While the difference between 2/5 and 4/9 is small, the advantage of going to higher order methods can be convincingly demonstrated through examples (e.g., Figure 5).

To anyone with a plug-in bandwidth selector for local constant or local linear regression, we suggest the following comparison: Fit a local quadratic regression, with the same variance as measured by equivalent degrees of freedom $\mathrm{tr}(L^T L)$ (or for pointwise bandwidth selectors, $\|l(x)\|^2$). Asymptotically, the local quadratic method will improve on the local linear, if anything more than existence of a second derivative is assumed. And improvement will also be observed in practice even in simple examples, such as that studied in Figure 4.

10.4 Degree of Fit

A question that has provoked much discussion is that of the relative merits of successive even and odd orders — local constant versus local linear fitting, and local quadratic versus local cubic fitting. The comparison is quite different at interior points and in 'boundary' regions, as can be seen either through examples or the asymptotics.

Consider the case of local constant versus local linear fitting. As noted earlier, for local constant fitting, bias is of size $O(h)$ at the boundaries, and $O(h^2)$ at interior points. For local linear fitting, the bias is $O(h^2)$ everywhere. A similar comparison holds at higher orders; fitting of order $2p + 1$ has $O(h^{2p+2})$ bias everywhere, and fitting of order $2p$ has bias $O(h^{2p+1})$ at boundaries and $O(h^{2p+2})$ at interior points.

However, this observation alone *is not an argument in favor of odd order fitting*, and *does not imply odd orders have no boundary effects*. Variance has so far been ignored. Local polynomial fitting is much more variable at boundary regions, and this variance increase is more pronounced for odd orders, as shown for one example in Figure 12. What we can conclude is that boundary regions dominate the comparison. For local constant fitting, the boundary bias caused by slope is widely recognized as being unacceptable, and local linear is usually preferable. Our examples support this conclusion. The comparison of local quadratic and local cubic is much less clear cut.

11   Summary of Conclusions

We have had many detailed discussions in this paper with major conclusions embedded in them. Here we simply present the conclusions.

In carrying out local regression we use a parametric family just as in global parametric fitting, but we ask only that the family fit locally and not globally. This is parametric localization.

Four important aspects of local regression are the weight function, bandwidth selection, the local parametric family, and the fitting criterion. The choices of these aspects represent a modeling of the data that can be guided by automatic selection criteria and graphical diagnostics.

Polynomial mixing provides a continuous progression of polynomial fits beginning with local constant fitting and proceeding continuously though higher orders.

The early smoothing literature from the first four decades of the twentieth century provides a number of important insights about the choices (e.g., Spencer, 1904a; Henderson, 1916; Macaulay, 1931). For example it was nearly a given in this literature that for most applications the weight function needed to be smooth, that local constant fitting was inadequate, and that smoothers needed to reproduce exactly (and not just asymptotically) at least a quadratic. Despite this important widely professed intuition, much theoretical smoothing research during the 1980s and early 1990s focused on locally constant fitting and ignored higher order polynomial fitting until the papers of Fan (1993) and Hastie and Loader (1993) appeared. Fortunately, the success of local polynomial methods in practice (Stone, 1977; Cleveland, 1979; Friedman and Stuetzle, 1982; Cleveland and Devlin, 1988; Hastie and Tibshirani, 1990) established them and not local constant fitting as the standard approach for data analysis.

Of the two bandwidth specifications, fixed and nearest-neighbor, the latter has the attractive property that there are typically less radical swings in the variance of the smooth. Typically, neither method works well if there are radical changes in the smoothness of the function; in such a case, adaptive methods can be helpful.

We introduce the assessment of parametric localization as a general approach to adaptive fitting. We discuss methods of adaptive selection of mixing degree and for adaptive bandwidth selection. This leads to far better adaptive procedures than the pointwise bias estimation methods that have originated in the kernel literature.

Asymptotics for smoothing give only the roughest of indicators of what works when the number of observations is finite. One principal problem is that for finite samples, boundary regions are important because bandwidths can be large. Unfortunately, asymptotic results have sometimes been interpreted too much at face value. Two conclusions that have been drawn in the literature — (1) the fit of an odd order polynomial is better than the next lowest even order (2) fixed bandwidth smoothing is better than nearest-neighbor — do not hold up for finite samples.

REFERENCES

BRILLINGER, D. (1977). Discussion of a paper of Stone. *Ann. Statist.* **5**, 622-623.

CLEVELAND, W. S. (1979). Robust locally weighted regression and smoothing scatterplots. *J. Amer. Statist. Assn.* **74**, 829-836.

CLEVELAND, W. S. (1993). *Visualizing Data.* Hobart Press, Summit, NJ, books@hobart.com.

CLEVELAND, W. S. and DEVLIN, S. J. (1988). Locally weighted regression: an approach to regression analysis by local fitting. *J. Amer. Statist. Assn.* **83**, 596-610.

CLEVELAND, W. S., HASTIE, T., and LOADER, C. (1995). Adaptive local regression by automatic selection of the polynomial mixing degree. In preparation.

CLEVELAND, W. S. and GROSSE, E. H. (1991). Computational methods for local regression. *Statist. and Computing* **1**, 47-62.

CLEVELAND, W. S. and GROSSE, E. H. and SHYU, M. J. (1992). Local regression models. *Statistical Models in S*, J. M. Chambers and T. Hastie, editors, pages 309–376. Chapman and Hall, New York.

CLEVELAND, W. S. and LOADER, C. (1995). Computational methods for local regression from the 19th century to the present. In preparation.

COX, D. R. (1958). Some problems connected with statistical inference. *Ann. Math. Statist.* **29**, 357-372.

DANIEL, C. and WOOD, F. (1971). *Fitting Equations to Data.* Wiley, New York.

DE FOREST, E. L. (1873). On some methods of interpolation applicable to the graduation of irregular series. *Annual Report of the Board of Regents of the Smithsonian Institution for 1871*, 275-339.

DE FOREST, E. L. (1874). Additions to a memoir on methods of interpolation applicable to the graduation of irregular series. *Annual Report of the Board of Regents of the Smithsonian Institution for 1873*, 319-353.

DONOHO, D. and JOHNSTONE, I. (1994). Ideal spatial adaptation by wavelet shrinkage. *Biometrika* **81**, 425-455.

FAN, J. (1993). Local linear regression smoothers and their minimax efficiencies. *Ann. Statist.* **21**, 196-216.

FAN, J. and GIJBELS, I. (1992). Variable bandwidth and local linear regression smoothers. *Ann. Statist.* **20**, 2008-2036.

FAN, J. and GIJBELS, I. (1994a). Censored regression: local linear approximations and their applications. *J. Amer. Statist. Assn.* **89**, 560-570.

FAN, J. and GIJBELS, I. (1994b). Adaptive order polynomial fitting: bandwidth robustification and bias reduction. Unpublished manuscript.

FAN, J. and GIJBELS, I. (1995). Data-driven bandwidth selection in local polynomial fitting: variable bandwidth and spatial adaptation. *J. Royal Statist. Soc,* Ser. B. **57**, 371-394.

FRIEDMAN, J. H. Multivariate adaptive regression splines (1991). *Ann. Statist.* **19**, 1-141.

FRIEDMAN, J. H. and STUETZLE, W. (1981). Projection pursuit regression. *J. Amer. Statist. Assn.* **76**, 817-823.

FRIEDMAN, J. H. and STUETZLE, W. (1982). Smoothing of scatterplots. Technical Report Orion 3, Dept. Statistics, Stanford University.

GASSER, TH. and JENNEN-STEINMETZ, C. (1988). A unifying approach to nonparametric regression estimation. *J. Amer. Statist. Assn.* **83**, 1084-1089.

GRAM, J. P. (1883). Über Entwickelung reeller Functionen in Reihen mittelst der Methode der kleinsten Quadrate. *J. Math.* **94**, 41-73.

HALL, P., SHEATHER, S. J., JONES, M. C. and MARRON, J. S. (1991). On optimal data-based bandwidth selection in kernel density estimation. *Biometrika* **78**, 263-269.

HÄRDLE, W. (1990). *Applied Nonparametric Regression.* Oxford University Press, Oxford.

HASTIE, T. and LOADER, C. (1993). Local regression: automatic kernel carpentry (with discussion). *Statist. Science* **8**, 120-143.

HASTIE, T. J. and TIBSHIRANI, R. J. (1990). *Generalized Additive Models.* Chapman and Hall, London.

HENDERSON, R. (1916). Note on graduation by adjusted average. *Actuarial Soc. Amer.* **17**, 43-48.

HENDERSON, R. (1924). A new method of graduation. *Actuarial Soc. Amer.* **25**, 29-39.

HOEM, J. M. (1983). The reticent trio: some little-known early discoveries in life insurance mathematics by L. H. F. Oppermann, T. N. Thiele and J. P. Gram. *Inter. Stat. Rev.* **51**, 213-221.

HJORT, N. L. (1994). Local Bayesian regression. Unpublished manuscript.

HJORT, N. L. and JONES, M. C. (1994). Locally parametric nonparametric density estimation. Unpublished manuscript.

KATKOVNIK, V. YA. (1979). Linear and nonlinear methods of nonparametric regression analysis. *Soviet Automatic Control* **5**, 25-34.

KENDALL, M. G. (1973). *Time Series.* Oxford University Press, Oxford.

KENDALL, M. G. and STUART A. (1976). *The Advanced Theory of Statistics, Vol. 3.* Hafner, New York.

LANCASTER, P. and SALKAUSKAS, K. (1981). Surfaces generated by moving least squares methods. *Mathematics of Computation* **37**, 141-158.

LANCASTER, P. and SALKAUSKAS, K. (1986). *Curve and Surface Fitting: An Introduction.* Academic Press: London.

LOADER, C. (1993). Change point estimation using local regression. Unpublished manuscript, available by `ftp` from `netlib.att.com`.

LOADER, C. (1994). Computing nonparametric function estimates. *Computing Science and Statistics: Proceedings of the 26th Symposium on the interface*, 356-361.

LOADER, C. (1995). Local likelihood density estimation. *Ann. Statist.* , to appear.

MACAULAY, F. R. (1931). *The Smoothing of Time Series.* National Bureau of Economic Research, New York.

MALLOWS, C. (1973). Some comments on C_p. *Technometrics* **15**, 661-675.

MALLOWS, C. (1974). Discussion of a paper of Beaton and Tukey. *Technometrics* **16**, 187-188.

McDONALD, J. A. and OWEN, A. B. (1986). Smoothing with split linear fits. *Technometrics* **28**, 195-208.

MÜLLER, H.-G. (1984). Smooth optimum kernel estimators of densities, regression curves and modes. *Ann. Statist.* **12**, 766-774.

MÜLLER, H.-G. (1987). Weighted local regression and kernel methods for nonparametric curve fitting. *J. Amer. Statist. Assn.* **82**, 231-238.

NADARAYA, E. A. (1964). On estimating regression. *Theor. Probab. Appl.* **9**, 141-142.

PARZEN, E. (1962). On estimation of a probability density function and mode. *Ann. Math. Statist.* **33**, 1065-1076.

ROSENBLATT, M. (1956). Remarks on some nonparametric estimates of a density function. *Ann. Math. Statist.* **27**, 832-837.

ROSENBLATT, M. (1971). Curve estimates. *Ann. Math. Statist.* **42**, 1815-1842.

RUPPERT, D. and WAND, M. P. (1992). Multivariate locally weighted least squares regression. *Ann. Statist.* **22**, No. 3.

SHISHKIN, J, YOUNG, A. H., and MUSGRAVE, J. C. (1967). The X-11 variant of the Census Method II seasonal adjustment program. Technical Paper 15, U.S. Bureau of the Census.

SPECKMAN, P. (1995). Discussion of a paper of Donoho et al. *J. Royal Statist. Soc*, Ser. B. **57**, 337-338.

SPENCER, J. (1904a). On the graduation of the rates of sickness and mortality. *J. Inst. Act.* **38**, 334-347.

SPENCER, J. (1904b). Graduation of a sickness table by Makeham's hypothesis. *Biometrika* **3**, 52-57.

STANISWALIS, J. (1988). The kernel estimate of a regression function in likelihood-based models. *J. Amer. Statist. Assn.* **84**, 276-283.

STIGLER, S. M. (1978). Mathematical statistics in the early States. *Ann. Statist.* **6**, 239-265.

STOKER, T. M. (1993). Smoothing bias in density derivative estimation. *J. Amer. Statist. Assn.* **88**, 855-863.

STONE, C. J. (1977). Consistent nonparametric regression (with discussion). *Ann. Statist.* **5**, 595-620.

STONE, C. J. (1980). Optimal rates of convergence for nonparametric estimators. *Ann. Statist.* **8**, 1348-1360.

STONE, M. (1974). Cross-validatory choice of assessment of statistical predictions (with discussion). *J. R. Statist. Soc. B* **36**, 111-47.

TIBSHIRANI, R. and HASTIE, T. (1987). Local likelihood estimation. *J. Amer. Statist. Assn.* **82**, 559-567.

TSYBAKOV, A. B. (1986). Robust reconstruction of functions by the local-approximation method. *Prob. Inform. Trans.* **22**, 69-84.

WATSON, G. S. (1964). Smooth regression analysis. *Sankhya* Ser. A **26**, 359-372.

WAHBA, G. (1990). *Spline Functions for Observational Data.* SIAM, Philadelphia.

WHITTAKER, E. T. (1923). On a new method of graduation. *Proc. Edinburgh Math. Soc.* **41**, 63-75.

WOOLHOUSE, W. S. B. (1870). Explanation of a new method of adjusting mortality tables, with some observations upon Mr. Makeham's modification of Gompertz's theory. *J. Inst. Act.* **15**, 389-410.

Variance Properties of Local Polynomials and Ensuing Modifications

Burkhardt Seifert & Theo Gasser

Abteilung Biostatistik, ISPM, Universität Zürich, Sumatrastrasse 30, CH–8006 Zürich

Summary

Local polynomial regression estimation has a number of advantages and might become a "golden standard" for curve fitting. The attractive theoretical features are in partial contradiction to variance properties for random design and to practical experience. The conditional variance is unbounded. The unconditional variance is infinite when using optimal (compact) weights. A tutorial illustration of construction of weights for kernel and local polynomial estimators clarifies the mechanism of these problems. Properties are better for Gaussian weights, which are, however, computationally slow. We show the connection between numerical and statistical instabilities and corresponding solutions. The k–nearest–neighbour rule is shown to be an inadequate tool in this context. We propose a refined local modulation of bandwidth using a variance–bias compromise and local polynomial ridge regression.

Keywords: Kernel estimation; Local polynomials; Nonparametric estimation; Nonparametric regression; Smoothing.

1 Introduction

A number of different methods have been devised for estimating regression curves, their derivatives and other curves of relevance in applied statistics without the restrictive assumption of a parametric model. Common nonparametric regression estimators are Nadaraya–Watson and Gasser–Müller kernel estimators (Nadaraya 1964; Watson 1964; Gasser & Müller 1979, 1984),

[1]This work was part of the research program of project no. 21.–36042.92 of the swiss NSF

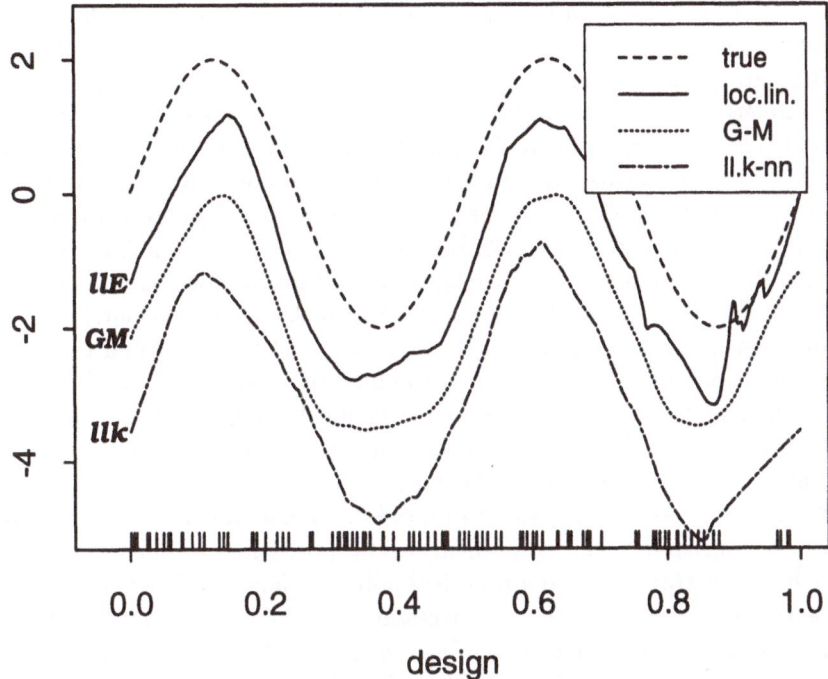

Figure 1: *True regression function* $r(x) = 2 \sin(4\pi x)$ *(small dashes) and 3 (shifted) estimators at their ISE-optimal bandwidths; using* $n = 100$ *random uniform design points and* $\sigma^2 = 1$. *Solid line is local linear fit with Epanechnikov weights (llE), dots are Gasser-Müller estimator with Epanechnikov kernel (GM), and dash-dots are local linear fit with Epanechnikov weights using k nearest neighbours (llk).*

smoothing splines (Schoenberg 1964; Reinsch 1967), and local polynomials (Stone 1977; Cleveland 1979; for historical notes see Müller 1988). Asymptotic properties of local polynomials are studied in Fan (1993), Ruppert & Wand (1994) and Fan et al. (1993). Much praise has been given to this method in Hastie & Loader (1993), who give a heuristic motivation.

There has been considerable debate about the pros and cons of various methods, and whether any method can be considered optimal (see e.g. Jennen-Steinmetz & Gasser 1988; Chu & Marron 1991). The proof that the local linear estimator achieves full asymptotic minimax efficiency among linear estimators (Fan 1993), and a number of desirable properties, seemed to decide the question definitely in favor of local polynomials. On the other hand, practical experience with local polynomials is often not as good as expected.

Figure 1 illustrates this with a simulated example. The comparison shows local linear fitting with (optimal) Epanechnikov weights (*llE*), the

Gasser–Müller Epanechnikov kernel estimator (*GM*), and local linear fit with Epanechnikov weights using a *k*–nearest–neighbour rule for the bandwidth (*llk*). In all cases, the bandwidth minimizing the finite sample integrated square error (ISE) on $[0.1, 0.9]$ has been used. Since such a bandwidth cannot be well estimated from the data (Hall & Marron 1987), this comparison shows the estimators, and in particular local linear fits, under unrealistically favorable conditions (see section 4, figure 10). The local linear fit (*llE*) looks rugged, in contrast to the kernel estimator. Using *k* nearest neighbours (with ISE–optimal $k = 16$) instead of a constant bandwidth reduced random peaks. It is interesting that the statistical problems have their equivalent in numerical problems for the local polynomial method. For figure 1 checks of ill–condition have been performed to guarantee numerical stability.

The somewhat rough appearance of local polynomial estimates has its equivalent in the fact that their variance is in general unbounded. For this reason, Fan (1993) modified the local linear estimator, while Stone (1977) trimmed the weight sequence induced by local linear fitting in order to achieve consistency. The modifications proposed for the theoretical analysis are hardly attractive from a practical point of view, and it would be nice if in practice no modifications were necessary. Asymptotic minimax–optimality for local polynomials has been derived at the theoretically optimal weight function and the optimal bandwidth. This leads to the further question, how sensitive results are to non–optimal choice of weights and bandwidth, and to what extent asymptotic results are valid for finite sample sizes. Our study concentrates on finite sample variance in theoretical and numerical terms.

The theoretical analysis leads to various conditions for achieving a finite, or infinite respectively, unconditional variance. A *k*–nearest–neighbour bandwidth rule ($k \geq 4$), strongly advocated by Bill Cleveland, is one modification leading to finite unconditional variance, in agreement with the practical improvement seen in figure 1. (The *k*–nn rule was originally derived to respond to systematic variation in design and not to solve variance problems intrinsic to some method.) In all cases no finite bound for the conditional variance exists (Seifert & Gasser 1996). Problems can be attributed to randomly occurring sparse regions in the design. This result leads to proposals for modifying local polynomials:

1. local polynomial ridge regression,

2. increasing locally the bandwidth where problems with variance arise.

These proposals combine good finite sample performance with asymptotic optimality.

2 Local polynomial and kernel regression estimation

Let $(X_1, Y_1), \ldots, (X_n, Y_n)$ be a set of independent and identically distributed pairs of random variables where the X_i are scalar predictors and the Y_i are scalar responses. In regression analysis, a functional relationship between predictor and response is assumed as

$$Y_i = r(X_i) + \varepsilon_i, \tag{1}$$

where the ε_i are independent and satisfy $\mathrm{E}(\varepsilon_i) = 0$ and $\mathrm{V}(\varepsilon_i) = \sigma^2(X_i)$. The predictors X_i are either "regularly spaced" $X_i = F_X^{-1}(i/n)$ for some distribution function F_X (fixed design) or distributed with density $f_X(x)$ (random design). They are assumed to be sorted $X_1 \leq \ldots \leq X_n$. The goal is to estimate $r(x_0)$ or its derivatives. For simplicity of notation we concentrate on estimation of $r(x_0)$ itself. Some popular (linear) estimators are the following:

1. Nadaraya–Watson (kernel) estimator (Nadaraya 1964; Watson 1964),

2. (convolution) kernel estimator (Gasser & Müller 1979),

3. k–nearest–neighbour estimator (Devroye 1978),

4. local polynomial estimator (Macauley 1931),

5. (cubic) smoothing splines (Schoenberg 1964).

Qualitative differences between estimators result mainly from:

(A) how weights are defined,

(B) dependence of bandwidth on design.

2.1 Construction of weights

In order to understand these differences let us first compare the geneses of weights:

The **Nadaraya–Watson estimator** (Nadaraya 1964; Watson 1964) is based on a local approximation of $r(x)$ by a constant:

$$r(x) \approx r(x_0) \tag{2}$$

provided that x is close to x_0. Representation (2) suggests minimizing

$$\sum_{i=1}^n (Y_i - \hat{r}_{\mathrm{NW}}(x_0))^2 \, K\left(\frac{X_i - x_0}{h}\right). \tag{3}$$

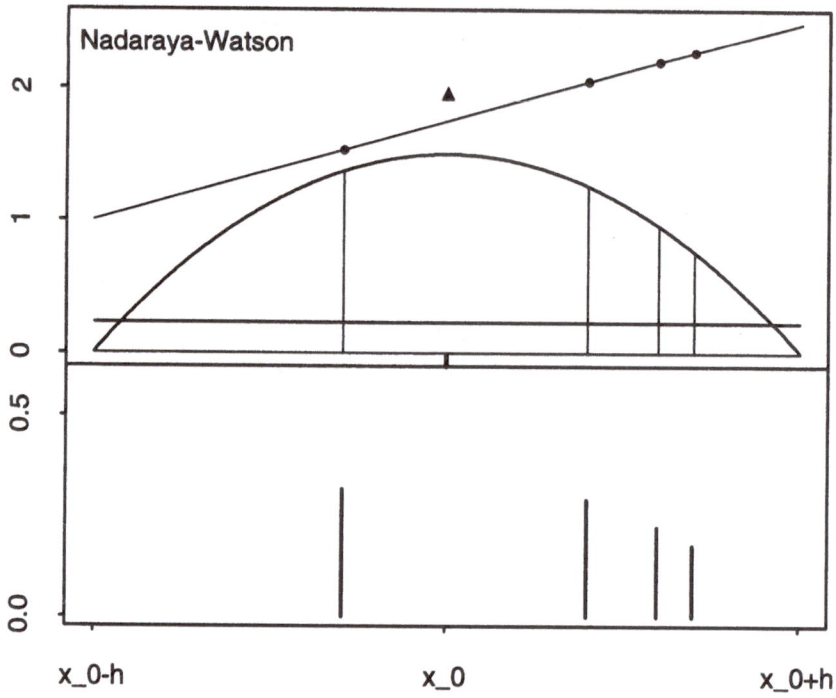

Figure 2: *Tutorial illustration of Nadaraya–Watson estimator. Above: 4 observations illustrating bias; middle: genesis of weights; below: resulting weights.*

Here K denotes a nonnegative kernel and h is a bandwidth, which may or may not depend on x_0. The Epanechnikov kernel $K(x) = (3/4)(1 - x^2)_+$ is asymptotically optimal. Formula (3) leads to an explicit representation of the Nadaraya–Watson estimator as weighted local mean

$$\hat{r}_{\mathrm{NW}}(x_0) = \sum_{i=1}^{n} \left\{ K\left(\frac{X_i - x_0}{h}\right) \bigg/ \sum_{j=1}^{n} K\left(\frac{X_j - x_0}{h}\right) \right\} Y_i. \qquad (4)$$

Figure 2 demonstrates the construction of weights for the Nadaraya–Watson estimator in a tutorial example with 4 design points in the smoothing interval. Above 4 observations (dots) without noise lying on a line are presented. The triangle is the resulting estimate demonstrating bias. In the middle the genesis of weights is demonstrated. The parabola is the Epanechnikov kernel $K\left(\frac{x - x_0}{h}\right)$. The weights then are proportional to this kernel evaluated at design points (vertical lines). Finally, the weights are multiplied by a constant ($\sum w_i = 1$). This constant is illustrated by the horizontal line at $1/\sum K\left(\frac{X_j - x_0}{h}\right)$. Below are the resulting weights. The figure shows rather smooth weights for the Nadaraya–Watson estimator, resulting in a small conditional variance. Also, the large conditional bias is seen in case of nonsymmetric position of design points. These properties are reflected by asymptotic theory: the asymptotic conditional bias and variance are obtained as

$$B(\widehat{r}_{NW}(x_0) \mid X) = E(\widehat{r}_{NW}(x_0) \mid X) - r(x_0)$$

$$= \frac{h^2}{2} \left\{ r''(x_0) + 2\, r'(x_0) \frac{f_X'(x_0)}{f_X(x_0)} \right\} \int t^2 K(t)\, dt\, (1 + o_P(1)) \tag{5}$$

$$V(\widehat{r}_{NW}(x_0) \mid X) = \frac{\sigma^2(x_0)}{n\, h\, f_X(x_0)} \int K^2(t)\, dt\, (1 + o_P(1)). \tag{6}$$

Here, the disturbing bias is seen in the term $r'(x_0)\, f_X'(x_0)/f_X(x_0)$, which in general results in a bias even for linear functions, and may shift peaks of the regression function. The conditional and unconditional measures are related by

$$V(\widehat{r}(x_0)) = E[V(\widehat{r}(x_0) \mid X)] + E[E(\widehat{r}(x_0) \mid X) - E(\widehat{r}(x_0))]^2, \tag{7}$$

$$B^2(\widehat{r}(x_0)) = E[B^2(\widehat{r}(x_0) \mid X)] - E[E(\widehat{r}(x_0) \mid X) - E(\widehat{r}(x_0))]^2, \tag{8}$$

$$MSE(\widehat{r}(x_0)) = E[MSE(\widehat{r}(x_0) \mid X)]. \tag{9}$$

Consequently, for finite samples, the unconditional variance is also increased due to bias. As will turn out in the later evaluation, the estimator converges slowly to its asymptotic performance.

Convolution type kernel estimators (Priestley & Chao 1972; Gasser & Müller 1979, 1984) are based on the approximation

$$r(x_0) \approx \frac{1}{h} \int_0^1 r(t)\, K\left(\frac{t - x_0}{h}\right) dt \tag{10}$$

for bandwidths $h \to 0$, where without loss of generality [0,1] is the range of design. Comparing (2) and (10), it is seen, that the affinity between the two estimators is rather vague, and calling both "kernel estimators" may be misleading. Again the Epanechnikov kernel is asymptotically optimal. Replacing $r(t)$ in (10) by a step function of observations, we obtain the **Gasser–Müller kernel estimator**

$$\widehat{r}_{GM}(x_0) = \sum_{i=1}^n \left\{ \frac{1}{h} \int_{s_{i-1}}^{s_i} K\left(\frac{t - x_0}{h}\right) dt \right\} Y_i, \tag{11}$$

where $s_i = (X_i + X_{i+1})/2$, $s_0 = 0$ and $s_n = 1$.

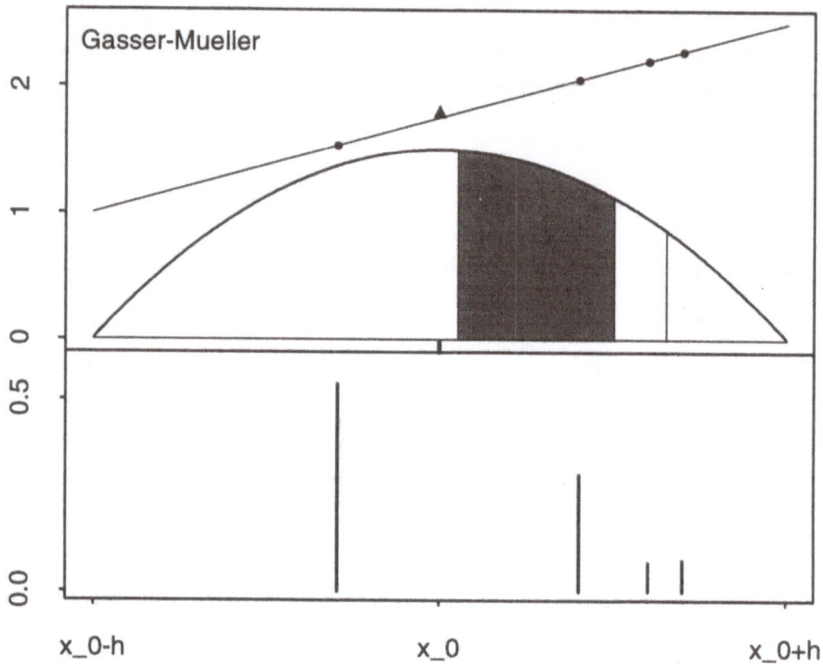

Figure 3: *Tutorial illustration of Gasser–Müller estimator. Above: 4 observations illustrating small bias; middle: genesis of weights; below: resulting weights.*

Figure 3 demonstrates the construction of weights for the Gasser–Müller estimator for the same tutorial example with 4 design points. Above 4 observations (dots) without noise on a line are presented. The triangle is the resulting estimate showing a small bias. In the middle the genesis of weights is demonstrated. The parabola is the Epanechnikov kernel $K\left(\frac{x-x_0}{h}\right)$. The weights then are the integrals of this kernel between averages of subsequent design points (vertical lines), divided by h. Below are the resulting weights. The figure demonstrates why the variance is increased for random design: If three design points are close together, the middle one is almost neglected due to its small weight. These properties are reflected by asymptotic theory: the asymptotic conditional bias and variance are obtained as

$$B(\widehat{r}_{GM}(x_0) \mid X) = \frac{h^2}{2} r''(x_0) \int t^2 K(t)\,dt \,(1+o_P(1)), \tag{12}$$

$$V(\widehat{r}_{GM}(x_0) \mid X) = c_{GM} \frac{\sigma^2(x_0)}{n\,h\,f_X(x_0)} \int K^2(t)\,dt \,(1+o_P(1)). \tag{13}$$

The disturbing term in the bias of the Nadaraya–Watson estimator no longer occurs. The factor c_{GM} in the variance is

$$c_{GM} = \begin{cases} 1 & \text{for fixed design} \\ 1.5 & \text{for random design} \end{cases}$$

Thus, the variance is inflated by a factor of 1.5 in case of random design.

Local polynomial estimators are based on the approximation

$$r(x) \approx \sum_{j=0}^{p} \beta_j (x - x_1)^j \tag{14}$$

provided that x is close to x_1, where x_1 is some point in the smoothing interval, and r is at least $(p+1)$ times differentiable with bounded $r^{(p+1)}$. The goal is still to estimate $r(x_0)$. The conventional choice is $x_1 = x_0$, but the polynomial in (14) is independent of x_1. The Nadaraya–Watson estimator is a local polynomial fit of order $p = 0$. Note that p should be odd according to asymptotic theory because of bias problems for even p similar to (5). Usually p is equal to 1 or at most 3 due to the local nature of the approximation. Representation (14) suggests minimizing

$$\sum_{i=1}^{n} \left(Y_i - \sum_{j=0}^{p} \beta_j (X_i - x_1)^j \right)^2 K \left(\frac{X_i - x_0}{h} \right) \tag{15}$$

with respect to $\beta = (\beta_0, \ldots, \beta_p)'$. Here K denotes a nonnegative weight function. Again, the Epanechnikov weight function is asymptotically optimal. Denote

$$X = \begin{pmatrix} 1 & (X_1 - x_1) & \cdots & (X_1 - x_1)^p \\ \vdots & \vdots & & \vdots \\ 1 & (X_n - x_1) & \cdots & (X_n - x_1)^p \end{pmatrix}, \quad Y = \begin{pmatrix} Y_1 \\ \vdots \\ Y_n \end{pmatrix}$$

and

$$W = \text{diag}(K \left(\frac{X_i - x_0}{h} \right)).$$

Then the solution of the least squares problem (15) is obtained as solution $\widehat{\beta}$ of the linear system

$$X'WX\widehat{\beta} = X'WY. \tag{16}$$

The resulting local polynomial

$$\sum_{j=0}^{p} \widehat{\beta}_j (x - x_1)^j \tag{17}$$

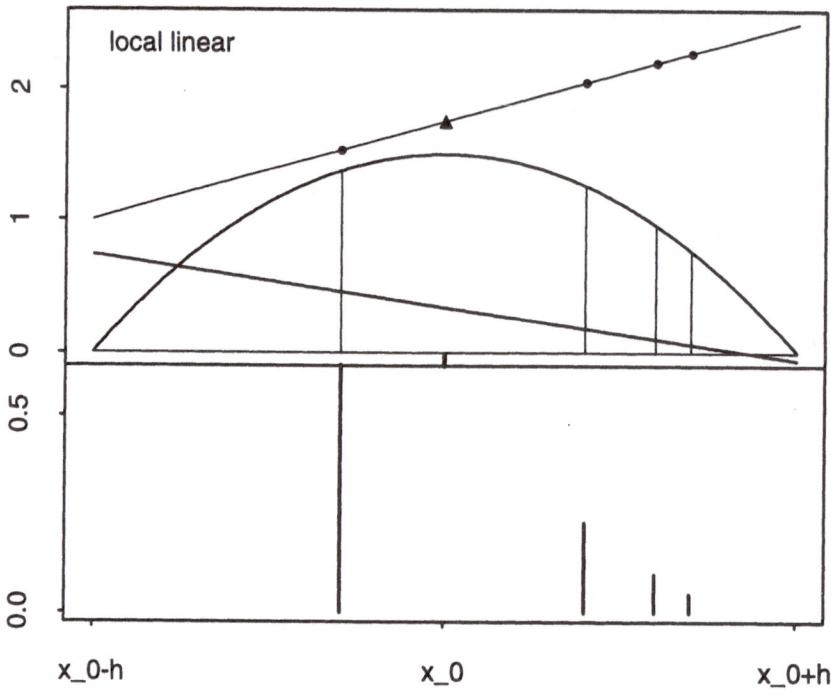

Figure 4: *Tutorial illustration of local linear estimator. Above: 4 observations illustrating unbiasedness; middle: genesis of weights; below: resulting weights.*

is independent of x_1. Hence x_1 can be chosen e.g. to improve numerical stability (Seifert, Brockmann, Engel & Gasser 1994) or to get more insight into the properties of the method. We estimate r at a point x_0 by the local polynomial (17) at x_0,

$$\widehat{r}_{lp}(x_0) \;\; = \;\; \sum_{j=0}^{p} (x_0 - x_1)^j \, \widehat{\beta}_j \, . \tag{18}$$

For $p = 1$ we get the local linear estimator, which for $x_1 = x_0$ has a simple explicit form:

$$\widehat{r}_{ll}(x_0) = \sum_{i=1}^{n} \left\{ \frac{S_2 - S_1 \, (X_i - x_0)}{S_0 \, S_2 - S_1^2} \, K\left(\frac{X_i - x_0}{h}\right) \right\} Y_i \, , \tag{19}$$

where

$$S_j = \sum_{i=1}^{n} (X_i - x_0)^j \, K\left(\frac{X_i - x_0}{h}\right) \, .$$

Figure 4 demonstrates the construction of weights for the local linear estimator in the tutorial example with 4 design points. Above 4 observations (dots) without noise on a line are presented. The triangle is the resulting estimator showing the unbiasedness. In the middle the genesis of weights is demonstrated for $x_1 = x_0$. The parabola is the Epanechnikov weight function $K\left(\frac{x-x_0}{h}\right)$. The weights then are constructed in two steps: First, the weight function at design points (vertical lines) is used. Then, these weights are multiplied by a line ($\sum w_i = 1$ and $\sum w_i(X_i - x_0) = 0$), which has a beneficial effect on the bias. This line $y = \frac{S_2 - S_1(x-x_0)}{S_0 S_2 - S_1^2}$ is, on the other hand, the intuitively plausible reason of variance problems of local linear estimators. Some of the resulting weights (below) are rather small, others may be very high. Even negative weights and weights larger than 1 are possible. Thus, the variance of local linear estimators is not bounded by σ^2, as is the case for the two other estimators $\hat{r}_{NW}$ and $\hat{r}_{GM}$. Asymptotic theory provides:

$$B(\hat{r}_{ll}(x_0) \mid X) = \frac{h^2}{2} r''(x_0) \int t^2 K(t) \, dt \, (1 + o_P(1)), \qquad (20)$$

$$V(\hat{r}_{ll}(x_0) \mid X) = \frac{\sigma^2(x_0)}{n \, h \, f_X(x_0)} \int K^2(t) \, dt \, (1 + o_P(1)) . \qquad (21)$$

The asymptotic bias is the same as that of the Gasser–Müller estimator. The conditional asymptotic variance is smaller in case of random design and the same as that of the Nadaraya–Watson estimator. The high variance of the estimator in figure 4 is not reflected in conditional asymptotic theory. For general local polynomial estimators the initial weights are multiplied by a polynomial of order p. This demonstrates the potential problems with variance for increasing polynomial order.

The construction of weights for **cubic smoothing splines** is quite different. We did not find a tutorial illustration similar to figures 2 to 4. The main reason is, that smoothing splines are not local, i.e. all n observations influence all weights. Asymptotically, smoothing splines can be considered as kernel estimators (Silverman 1984) with a kernel

$$K(x) = \frac{1}{2} \exp\left(-\frac{|x|}{\sqrt{2}}\right) \sin\left(\frac{|x|}{\sqrt{2}} + \frac{\pi}{4}\right)$$

with non–compact support, and a local, design–adaptive bandwidth

$$h(x_0) = \left(\frac{h}{n \, f_X(x_0)}\right)^{1/4} . \qquad (22)$$

Thus, smoothing splines behave like Gasser–Müller estimators of kernel order 4 and local cubic polynomial estimators with a design–adaptive bandwidth. This asymptotic behavior was confirmed by our finite sample evaluations, where the non–optimal kernel resulted in slightly worse MISE–performance for large sample size. For small sample size, however, MISE–performance and visual smoothness are good.

2.2 Dependence of bandwidth on design

The use of a variable bandwidth $h(x_0)$ has been advocated for various reasons. Such a local bandwidth should e.g. adapt to

(i) the local curvature of r (see (5), (12), (20)), to improve locally the bias–variance compromise,

(ii) the local variance in case of heteroscedasticity,

(iii) the local density of design points (see (6), (13), (21)), to stabilize locally the variance.

Since various methods differ how they deal with problem (iii) of design adaptation, we will concentrate on that point. Popular bandwidth schemes relied either on a constant bandwidth, a k–nearest–neighbours bandwidth or a bandwidth proportional to $f_X(x_0)^{-1/4}$ implicitly for smoothing splines.

A **constant bandwidth** $h(x_0) \equiv h$ relates naturally to a poor prior knowledge about the underlying design density, and also residual variance and curvature of regression function. Such a situation is common in exploratory data analysis.

A **k–nearest–neighbour** (k–nn) rule was designed to adapt to the density of design points in such a way that the variance becomes approximately constant (in case of homoscedasticity). This is born out by asymptotic theory since a k–nn rule leads to a bandwidth approximately proportional to $1/f_X(x_0)$. One should, however, note two things: firstly, variance stabilization is usually a natural procedure for unbiased, but not necessarily for biased estimators. Secondly, for finite samples — and in particular for random design — the variance depends not only on the number of observations but also on their locations in the smoothing interval. Therefore, a k–nn rule stabilizes variance only asymptotically.

Figure 5 demonstrates two local linear estimators with the same average bandwidth but either a constant or a k–nn bandwidth. Both estimators oversmooth since the ISE–optimal smoothing parameters gave rather wiggly results in the flat parts. Thus, a bandwidth leading to visually nice results was chosen. The local linear estimator with constant bandwidth ($h = 0.12$) flattens the peaks and the troughs, but still gives a qualitatively correct impression of the structure. The local linear estimator with k–nn rule ($k = 20$, which is in the average $h = 0.12$) does not get the structure at all. Instead of two peaks, we have only one. The impression of this figure is supported by a minimax result: Jennen–Steinmetz & Gasser (1988) studied asymptotic properties of the Gasser–Müller kernel estimator for different design dependent bandwidth schemes of the kind

$$h(x_0) = h_0 \, f_X^{-\alpha}(x_0) \quad (0 \le \alpha \le 1) \, .$$

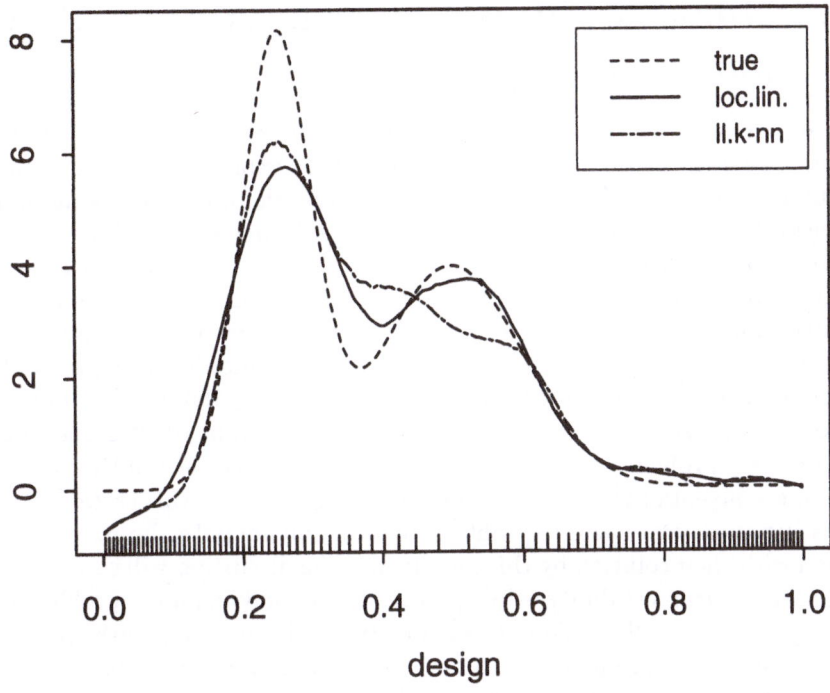

Figure 5: *True regression function* $r(x) = \varphi((x-0.25)/0.05) + \varphi((x-0.5)/0.1)$ *(dashes) and 2 local linear estimators with Epanechnikov weights; using* $n = 100$ *design points and* $\sigma^2 = 0.5$. *Solid line is constant bandwidth, dash dots are k–nn rule.*

These schemes include constant bandwidth ($\alpha = 0$) and k–nn rule ($\alpha = 1$) as extremes. Cubic smoothing splines are implicitly included in this consideration with $\alpha = 1/4$ (see (22)). The analysis showed, that $\alpha = 0$, i.e. a constant bandwidth, is asymptotically minimax optimal. Because of the qualitatively similar asymptotic behavior of Gasser–Müller kernel estimators and local polynomials ((20)–(12), (21)–(13)), the result holds for both estimators. None of the rules is uniformly optimal. Roughly speaking, the k–nn rule has its advantages, if data are dense in regions with high curvature, as in a well–planned study with high prior information. A constant bandwidth, however, is more robust against departures from such a desirable design. In figure 5, one would like to use a smaller bandwidth in the center and a larger bandwidth in the flat parts, just opposite to the k–nn rule. In this way the interesting structure can be recovered with low bias, while the variance can be fought with a small price in the flat parts. Such a data–adaptive local bandwidth selection has been proposed e.g. by Brockmann, Gasser & Herrmann (1993) with encouraging results (see Herrmann 1994a, for an application to "wavelet examples").

3 Numerically stable computation of local polynomials

Numerical and statistical stability of algorithms are closely related: If small differences in the data — replacing theoretical values by numbers in a computer — lead to differences in the estimator, the algorithm is numerically unstable. If the differences are larger — replacing values of the regression function by noisy data — the same mechanism leads to statistical instability of the estimator. These two related problems also come up for local polynomial fits. In addition to a fast and numerically stable algorithm, a numerically very stable algorithm for computation of local polynomials was proposed by Seifert, Brockmann, Engel & Gasser (1994). This paper proposed and discussed mainly a fast updating algorithm which leads to severe numerical problems. However, these problems arise in mild form also for a slow conventional algorithm, and this is the topic of this section. We will discuss numerical stability problems of local polynomial estimators in connection with their solution by this algorithm. The discussion will show, however, that numerical instability is inherent in the method and not a problem of bad programming. All considerations are strictly local; bandwidth, polynomial order and weight function are allowed to vary for different values of x_0. The present discussion is thus strictly independent from questions raised in the previous section.

3.1 Algorithms

The **conventional algorithm** for computation of local polynomial fits uses $x_1 = x_0$ in (14) for computation of

$$S = X'WX = \begin{pmatrix} S_0 & \cdots & S_p \\ \vdots & & \vdots \\ S_p & \cdots & S_{2p} \end{pmatrix} \quad \text{and} \quad T = X'WY = \begin{pmatrix} T_0 \\ \vdots \\ T_p \end{pmatrix} \quad (23)$$

with elements

$$S_j = \sum_{i=1}^{n} (X_i - x_1)^j K\left(\frac{X_i - x_0}{h}\right), \quad j = 0, \ldots, 2p, \quad (24)$$

$$T_j = \sum_{i=1}^{n} (X_i - x_1)^j K\left(\frac{X_i - x_0}{h}\right) Y_i, \quad j = 0, \ldots, p, \quad (25)$$

and then solve the linear equation (16). The main reason for numerical instability is the numerical condition of these normal equations, or more precisely the condition of the matrix of coefficients S. In this section we will describe a number of improvements of numerical stability, as proposed by Seifert et al. (1994). The resulting algorithm will be called **stable conventional**.

3.2 Cholesky decomposition

Theoretically, $p + 1$ different points with nonzero weights $K\left(\frac{X_i - x_0}{h}\right)$ are sufficient to ensure that S is positive definite. We use Cholesky decomposition to solve the normal equations and to decide, whether S is far enough from singularity.

The decomposition of $S = ((s_{jk}))$ is of the form $S = L D L'$. $L = ((\ell_{jk}))$ is a lower triangular matrix with diagonal elements $\ell_{jj} = 1$. $D = \text{diag}(d_j)$ is the diagonal matrix of Cholesky factors. The normal equations are then easily solved step by step. The well known formulae for the decomposition use only the four fundamental rules of arithmetic:

$$d_j = s_{jj} - \sum_{k < j} \ell_{jk}^2 d_k, \tag{26}$$

$$\ell_{jk} = \left(s_{jk} - \sum_{\ell < k} \ell_{j\ell} \ell_{k\ell} d_\ell \right) \bigg/ d_k. \tag{27}$$

Cholesky factors d_j should be sufficiently away from zero compared to s_{jj} to avoid the loss of significant digits in (26). For the local linear estimator these values are simply $d_1 = s_{11} = S_0$, $\ell_{21} = s_{21}/d_1 = S_1/S_0$, and $d_2 = s_{22} - \ell_{21}^2 d_1 = S_2 - S_1^2/S_0$. Large relative numerical errors may occur in the computation of d_2 which is proportional to $S_0 S_2 - S_1^2$, the determinant of S. Due to formula (19), this relative error may cause large absolute errors in $\widehat{r}_{11}(x_0)$.

3.3 Singularity

Following (26), the ratios d_j/s_{jj} are scale invariant. Due to its sensitivity the last ratio $d_{p+1}/s_{p+1,p+1}$ is used to assess stability and is henceforth called the "stability factor". Using some common assumptions, (24) leads to an asymptotic representation

$$
\begin{aligned}
s_{jk} &= S_{j+k-2}(x_0) \\
&= n \int (t - x_0)^{j+k-2} K\left(\frac{t - x_0}{h}\right) f_X(t)\,dt\,(1 + o(1)) \\
&= n\, f_X(x_0)\, h^{j+k-1} \int z^{j+k-2} K(z)\,dz\,(1 + o(1)). \tag{28}
\end{aligned}
$$

Formula (28) leads to theoretical values of the stability factor of S. The Cholesky factors are of order $d_j = O(n\, h^{2j-1})$ as is s_{jj}. Consequently, for an adequate number of design points which are well spaced, the stability factor of S is close to a value, which does not depend on n and h, but only on the weight function K. As to be expected a priori, singularity is only a problem in sparse regions of the design and for small bandwidths. Theoretically, $p+1$ points are sufficient to obtain a stable solution. Numerical problems may arise basically due to two reasons: First, the weight function decreases the influence of points close to $x_0 \pm h$. With only $p + 1$ points we switch to uniform weights, as a first step. This does not change the estimator, which is in this case just an interpolation, but makes its computation more stable. A second reason for stability problems is, that random design points may lie close together. S is defined to be singular, if

$$d_{p+1} / s_{p+1,p+1} < \text{sing} \times \text{"theoretical value"},$$

where the "theoretical value" is derived from (28) and "sing" can be chosen by the user. Its size is not critical and was set to sing $= 0.01$. Thus, the matrix is defined to be singular, if the numerical condition is less than 1% to be expected for equidistant design. If S is singular, the local smoothing interval is enlarged by one point. This local increase due to ill–condition is a true modification of local polynomials, and it is the only one. Thus, the algorithm can be used to evaluate later on the validity of asymptotic properties for finite samples.

Figure 6 demonstrates this modification for the conventional algorithm. Here we have $n = 100$ uniformly distributed design points, $m = 100$ equidistant output points, local polynomials of order $p = 3$, and a small bandwidth $h = 0.01$, just to provoke singularity. Observations are on a line $y(x) = 2\,x - 1$ without random noise. Using sing $= 0.01$, the two outstanding numerical errors of the order 10^{-10} are reduced to 10^{-13}. This modification becomes practically important for higher order polynomials and in particular for a fast updating algorithm (Seifert et al. 1994), where numerical errors accumulate. Note that the algorithm handles ill–conditioned realizations of the design fully automatically.

3.4   Further improvement of numerical stability

As was mentioned above, the resulting estimator $\widehat{r}_{\mathrm{lp}}(x_0)$ is independent of the choice of x_1 in formulae (15)–(18). Thus, x_1 can be chosen to improve numerical stability.

Take the case $p = 1$ as an example. If we choose

$$x_1 = \sum_{i=1}^{n} X_i\, K\left(\frac{X_i - x_0}{h}\right) \bigg/ \sum_{i=1}^{n} K\left(\frac{X_i - x_0}{h}\right)$$

the weighted mean of design points in the smoothing interval instead of $x_1 = x_0$, we get $S_1 = 0$ and thus per definition a stable matrix S. Following (18), (24) and (25), a numerically stable local linear estimator

$$\widehat{r}_{\mathrm{ll}}(x_0) = \frac{T_0}{S_0} + \frac{(x_0 - x_1)\,T_1}{S_2}$$

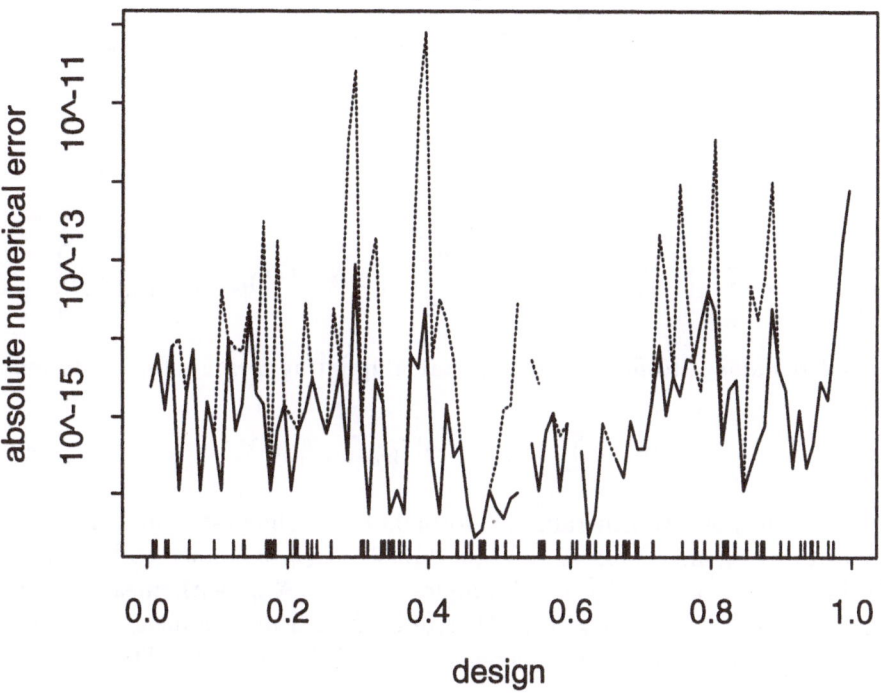

Figure 6: *Tutorial illustration of reduction of numerical errors by control of singularity for the conventional algorithm. Shown are absolute numerical errors (on logarithmic scale) for sing = 10^{-30} (dots) and sing = 0.01 (solid line). Missing lines indicate zero numerical error.*

is obtained, provided, the components of this formula are computed stably. In our algorithm (Seifert et al. 1994) we used

$$x_1 = \bar{X}_0 = \sum_{i=1}^{n} X_i \, \mathrm{I}_{[x_0-h,x_0+h]}(X_i) \Bigg/ \sum_{i=1}^{n} \mathrm{I}_{[x_0-h,x_0+h]}(X_i) \ ,$$

the unweighted mean of design points in the smoothing interval. This choice stabilizes the normal equations for every p. What remains is to find a fast and stable formula for computation of S_j in (24) and T_j in (25). Suppose K is a polynomial weight function of order a comprising in particular the optimal Epanechnikov, the minimum variance (uniform), biweight or triweight weights. Then

$$S_j \quad = \quad \sum_{i=1}^{n} (X_i - \bar{X}_0)^j \left(\sum_{k=0}^{a} a_k \left(\frac{X_i - x_0}{h} \right)^k \right) \mathrm{I}_{[x_0-h,x_0+h]}(X_i)$$

$$= \sum_{i=1}^{n} \sum_{k=0}^{a} h^{-k} a_k \sum_{\ell=0}^{k} \binom{k}{\ell} (X_i - \bar{X}_0)^{j+\ell} (\bar{X}_0 - x_0)^{k-\ell} I_{[x_0-h, x_0+h]}(X_i)$$

$$= \sum_{k=0}^{a} h^{-k} a_k$$

$$\times \sum_{\ell=0}^{k} \binom{k}{\ell} (\bar{X}_0 - x_0)^{k-\ell} \left\{ \sum_{i=1}^{n} (X_i - \bar{X}_0)^{j+\ell} I_{[x_0-h, x_0+h]}(X_i) \right\} . \quad (29)$$

What remains is to find a fast and stable updating formula for the terms in braces

$$m_j = \sum_{i=1}^{n} (X_i - \bar{X}_0)^j \, I_{[x_0-h, x_0+h]}(X_i) . \quad (30)$$

For this purpose we generalized a formula for pooling estimates of variance ($j = 2$) by Chan, Golub & LeVeque (1983) which is known to be fast and stable: Given two distinct subsamples $X_{\ell 1}, \ldots, X_{\ell n_\ell}$ with means $\bar{X}_\ell$ and central moments $m_{j,\ell}$, $\ell = 1, 2$. Denote by $\bar{X}$ and m_j the mean and central moments of the union of both subsamples and $d = \bar{X}_1 - \bar{X}$. Then

$$\bar{X} = \bar{X}_1 + n_2 (\bar{X}_2 - \bar{X}_1) / (n_1 + n_2) \quad (31)$$

and

$$m_j = \sum_{i=1}^{n_1} (X_{1i} - \bar{X})^j + \sum_{i=1}^{n_2} (X_{2i} - \bar{X})^j$$

$$= \sum_{k=0}^{j} \binom{j}{k} (\bar{X}_1 - \bar{X})^{j-k} m_{k,1} + \sum_{k=0}^{j} \binom{j}{k} (\bar{X}_2 - \bar{X})^{j-k} m_{k,2}$$

$$= \sum_{k=2}^{j} \binom{j}{k} d^{j-k} \left(m_{k,1} + \left(-\frac{n_1}{n_2} \right)^{j-k} m_{k,2} \right)$$

$$+ d^j n_1 \left(1 - \left(-\frac{n_1}{n_2} \right)^{j-1} \right) \quad (32)$$

for $j \geq 2$. Note, that $m_1 = 0$ and $m_0 = n_1 + n_2$. The computation of linear forms T_j is similar. The computation of central moments for a partition is more stable than the standard two–pass algorithm. Due to this and due to an appropriate centering we can expect an algorithm that is more stable than the conventional one.

Figure 7 demonstrates this for the local linear estimator. The maximal numerical error of the stable conventional algorithm is reduced by a factor

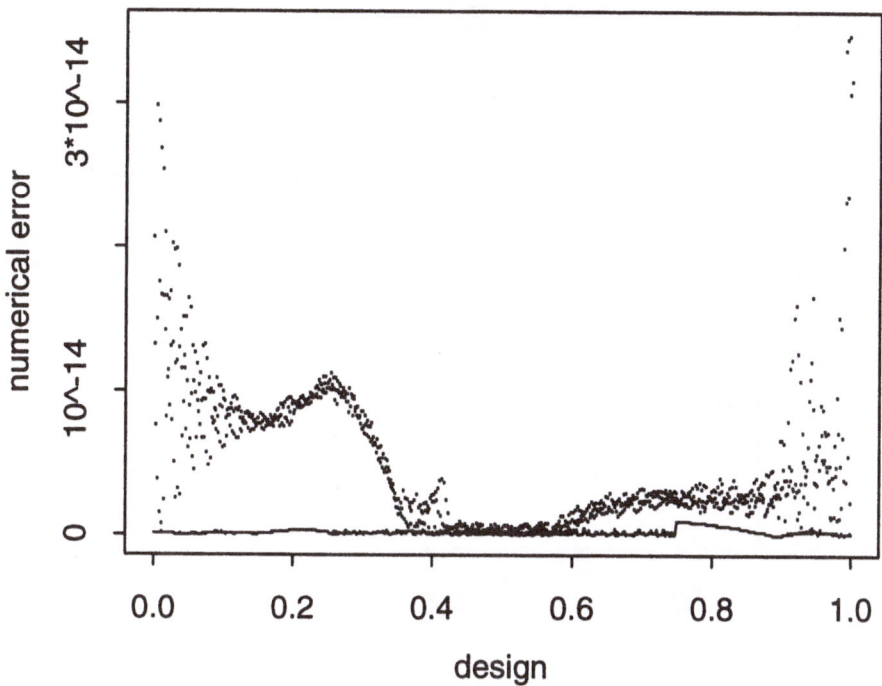

Figure 7: *Absolute numerical errors of conventional (dots) and stable conventional algorithm (solid line) of local linear estimator with Epanechnikov weights for $n = 1000$ random uniform design points, $r(x) = 2x - 1$, $\sigma^2 = 0$ and $h = 0.25$.*

35 compared to the conventional one, which has especially problems at the boundary. The matrix S becomes ill–conditioned for large p when centering at x_0 rather than $\bar{X}_0$. Let us note in passing that the stable conventional algorithm needs $O(n^{7/5})$ operations as compared to $O(n^{9/5})$ for the conventional one (for $p = 1$). Thus, it is faster and more stable.

4 Finite sample properties

In Seifert & Gasser (1996) we evaluated finite sample properties of several estimators (with emphasis on local polynomials), both at a theoretical and a numerical level. We will now summarize some relevant results in the light of the previous section.

Performance for prespecified bandwidth: First, we study the behavior of local linear fitting for a range of prespecified bandwidths and various weight functions, and compare it with (Epanechnikov) kernel estimators. Finite

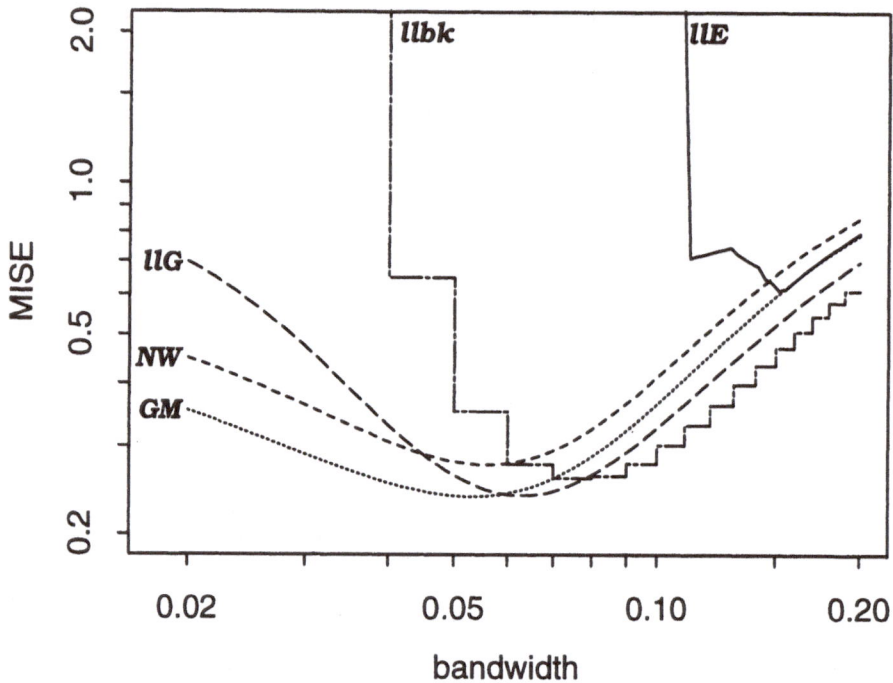

Figure 8: *MISE depending on bandwidth h (on logarithmic scales) of estimators of r in (33) for n = 50 uniformly distributed design points and $\sigma^2 = 0.5$. Solid line is local linear fit with Epanechnikov weights (llE), dash–dots are local linear fit with biweights and a k–nn rule (llbk), dashes are local linear fit with Gaussian weights (llG), dots are Gasser–Müller estimator (GM), and small dashes are Nadaraya–Watson estimator (NW), both with Epanechnikov kernel.*

sample mean ISE (MISE) for regression function

$$r(x) = 2 - 5x + 5\exp\left\{-\left(\frac{x - 0.5}{0.05}\right)^2\right\} \tag{33}$$

and $n = 50$ random uniform design points is given in figure 8. Local linear fitting with Gaussian weights (scaled to reach the same asymptotically optimal bandwidth as Epanechnikov weights) has an optimal MISE comparable to kernel estimators, and never breaks down as does local linear fitting with Epanechnikov weights for bandwidths smaller than 0.11. Its MISE increases steeply to values as large as 629 for $h = 0.02$. Since the bias is uniformly bounded for smooth regression functions, the problems arise due to an inflation of variance, which may become arbitrarily large. A large local variance arises, whenever the empirical distribution of design points in the smoothing interval is skew and the variance s_x^2 of these points becomes small.

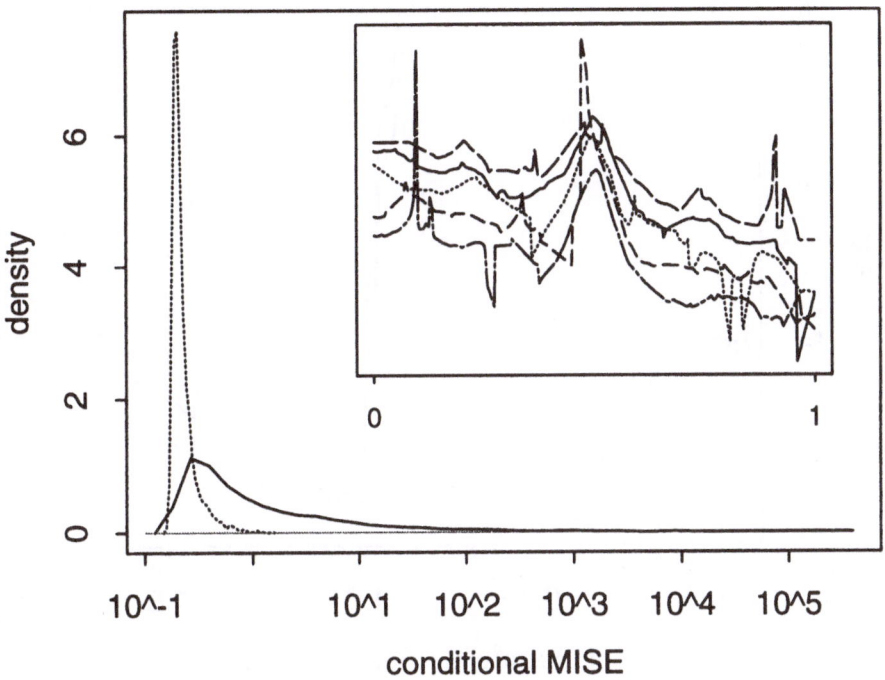

Figure 9: *Probability densities of logarithms of 4000 conditional MISEs of estimators of r in (33) for h = 0.05, n = 50 uniformly distributed design points and σ² = 0.5. Solid line is local linear fit with Epanechnikov weights and dots are Gasser–Müller estimator with Epanechnikov kernel. Above are 5 realizations of local linear fit with median ISE performance.*

This problem is known from parametric linear regression as difficulty of extrapolation. Heuristically, this is due to some weights becoming arbitrarily large, since negative weights may arise by construction (see figure 4). As a consequence, local linear estimators with compact weights (i.e. weights derived from a weight function with compact support such as Epanechnikov or biweight function) and non–adaptive bandwidth have infinite unconditional variance. Using a k–nearest–neighbour rule and biweights — close in spirit to Cleveland (1979) — leads to an improvement, but is still inferior to using Gaussian weights and a fixed bandwidth. The theoretical explanation is that unconditional variance of local linear fits with nonadaptive bandwidth is finite if and only if the weight function has non–compact support. Gaussian weights, on the other hand, do not allow a fast $O(n)$ computation mentioned in section 3, and are slightly inefficient asymptotically. The better behavior of a k–nn rule can be explained theoretically: requiring a minimum of 4 design points guarantees a finite unconditional variance, but not a bounded conditional variance. The MISE of the Nadaraya–Watson estimator was shifted

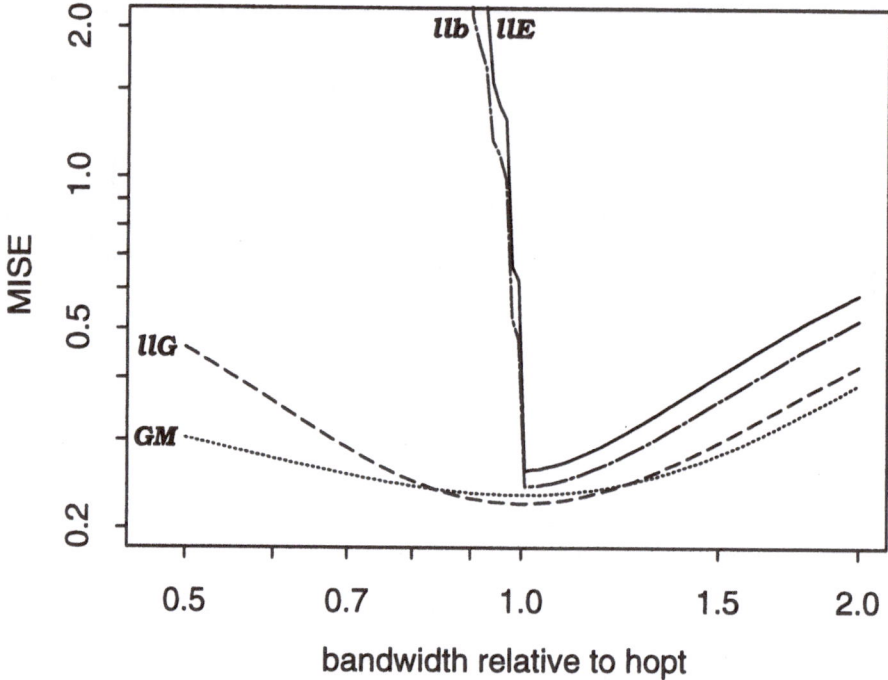

Figure 10: *MISE depending on deviations from conditional MISE–optimal bandwidths (on logarithmic scales) of estimators of r in (33) for n = 50 uniformly distributed design points and $\sigma^2 = 0.5$. Solid line is local linear fit with Epanechnikov weights (llE), dash–dots are local linear fit with biweights (llb), dashes are local linear fit with Gaussian weights (llG), and dots are Gasser–Müller estimator with Epanechnikov kernel (GM).*

upwards compared to the one of the Gasser–Müller estimator, in conflict to asymptotic theory. Similar results were obtained for other σ^2, and also for sine regression.

Figure 9 shows, that the breakdown of the local linear estimator is not a result of some outlying values: for $h = 0.05$, 90% of conditional MISEs of the Gasser–Müller estimator were smaller than 0.3, but 75% of local polynomials were larger (this is accompanied by an unattractive behavior of the median conditional MISE which is 0.2 for the Gasser–Müller estimator and 0.6 for local polynomials).

Sensitivity to bandwidth choice: Figure 10 allows a closer assessment of the sensitivity of MISE. In figure 8 it was demonstrated, that a design–independent bandwidth does not work for local polynomials. Now a conditional MISE–optimal bandwidth for each realization of the design is used. The x–axis is scaled to this bandwidth. Thus, MISE at $x = 1$ is the mean

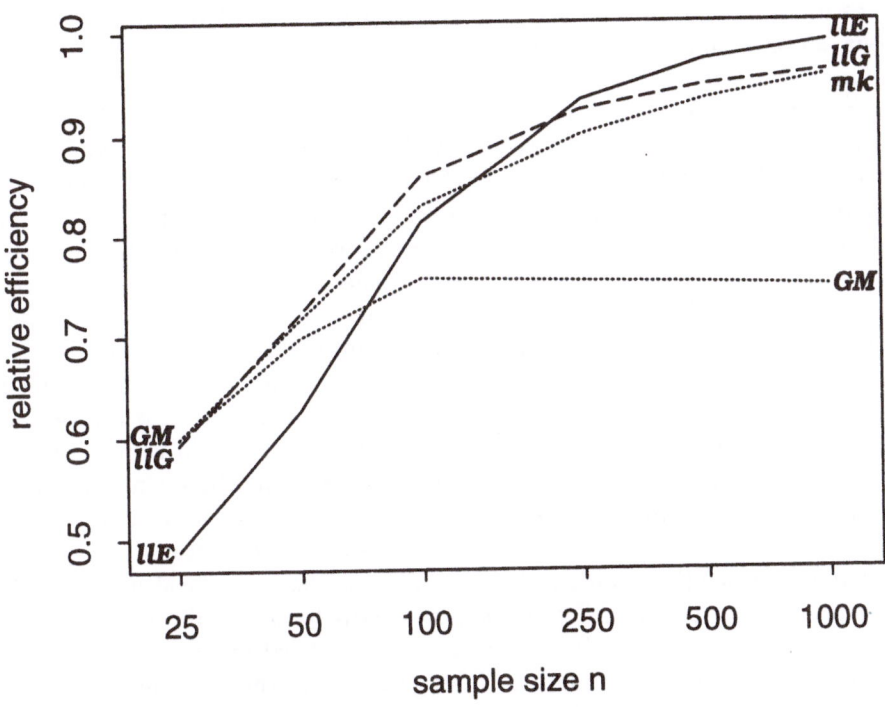

Figure 11: *Efficiency depending on sample size n (on logarithmic scale) of estimators of r in (33) at conditional MISE–optimal bandwidths for random uniform design and $\sigma^2 = 0.5$. Solid line is local linear fit with Epanechnikov weights (llE), dashes are local linear fit with Gaussian weights (llG), and dots are Gasser–Müller (GM) and modified kernel (mk) estimators.*

of optimal conditional MISEs. The most striking result is that local linear fitting with compact weights breaks down for bandwidths minimally smaller than the optimal ones. This can again be explained theoretically: Large conditional variances arise in regions where the design is locally sparse, and when the local distribution has low variance and high skewness. At which bandwidth this happens is dictated by random fluctuations in the design (and no such problems arise for fixed design). Local linear fits with Gaussian weights behaved best at the optimal bandwidths. They were still more sensitive to under– or oversmoothing compared to kernel estimators.

Performance with sample size: Figure 11 shows how fast asymptotic efficiency is approached. Relative efficiency here is the ratio of asymptotic MISE of the local linear estimator expected from (20) and (21) to the true finite sample MISE of an estimator. Since efficiency is studied at the conditional MISE–optimal bandwidths, this representation is somewhat optimistic for local polynomials given the fact that the optimal bandwidth is difficult to estimate and given the sensitivity seen in figures 8 and 10. All estimators behave worse than expected from asymptotic theory. The Gasser–Müller estimator starts relatively high and reaches its asymptotic performance with an efficiency of 0.72 fast. At $n = 25$, local linear fitting with Epanechnikov weights starts with less than 50% of asymptotic efficiency, while this is about 60% for Gaussian weights. Then a steep increase is seen, but optimal Epanechnikov weights reach the performance of Gaussian weights as late as $n \approx 200$. Here the chance for sparse regions in the design is greatly diminished (see below). For $n = 1000$, Epanechnikov weights behave best but still slightly below asymptotic performance. According to theory, the Nadaraya–Watson estimator (not shown) is fully efficient for uniform random design, but this is not borne out by finite sample results. A modified kernel estimator was developed by Herrmann (1994b), following a proposal by Chu & Marron (1991). The s_i in (11) are determined in a more stable way for random design by using a nonparametric quantile estimator. The modified kernel estimator combines a good small sample and large sample MISE. Again, similar conclusions can be obtained when studying e.g. the sine function or other random designs. The problems for local polynomials with compact weights came out even more clearly for non–uniform designs, as expected from theory. As is obvious, an extreme location of points becomes unlikely for an increasing number of design points in the smoothing interval. Thus, large conditional variances are mainly a problem of sparse regions in the realization of a random design.

Probability of sparse regions: Figure 12 shows that sparse regions in the realization of a random design may arise with considerable probability for sample sizes $n = 200$ or less. For nonuniform designs, this probability still increases.

Fixed designs: Differences between local polynomials and the Gasser–Müller estimator turned out to be small, the latter being somewhat better in MISE, in particular for undersmoothing bandwidths. All methods approached the asymptotic efficiency from above, in contrast to random design (figure 11). The loss in efficiency for Gaussian weights came out clearly even for moderate sample size. The asymptotic and the finite sample behavior agree well in this case.

5 Finite–sample modifications of local polynomials

The variance of local polynomials becomes infinite for compact weights, leading to a problematic finite sample behavior. Using non–compact Gaussian weights avoids these problems, but is not really an attractive alternative for

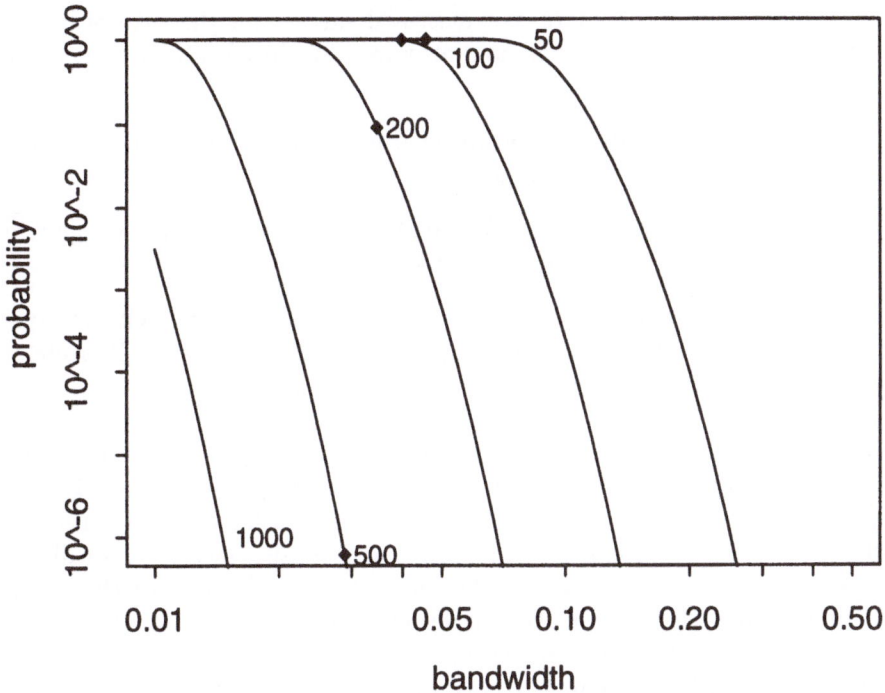

Figure 12: *Probability for a sparse region with 3 points in the realization of a random uniform design, depending on bandwidth and sample size. The points mark asymptotically optimal bandwidths for r in (33) and $\sigma^2 = 0.5$.*

reasons given earlier. Requiring at least 4 points in each smoothing interval for local linear fits also guarantees a finite variance (in general, one has to increase from a minimum of $p + 1$ to $p + 3$ points). Our evaluation of this approach has shown that it is too gross: optimal MISE is substantially higher compared to the two modifications described later in this section or to using Gaussian weights. Also, a k-nearest-neighbour bandwidth rule with $k \geq 4$ guarantees a finite unconditional variance. This rule is inadequate basically for two reasons:

- An unequal local distribution of design points may still arise leading to an unacceptable high conditional variance. Note that the first modification below could be used to achieve a finite sample variance stabilization, a goal achieved only asymptotically by a k-nn rule.

- A k-nn rule is basically an approach for design adaptation — irrespective of the estimator used — and not intended to solve a variance problem specific for local polynomials. As mentioned earlier, as a design-adaptive bandwidth rule it can be recommended only in case of well planned designs.

These methods are, therefore, not discussed further for the sake of brevity. Two modifications of local polynomials were introduced in Seifert & Gasser (1996):

a) Increase the bandwidth locally, specifically where problems with variance of the curve fit come up.

b) Replace local polynomial regression by local polynomial ridge regression to stabilize the variance of the estimator.

Both modifications lead to finite variance, are heuristically well motivated and share asymptotic properties with local polynomials. We will sketchily describe these two modifications for the local linear case.

Local increase of bandwidth: The strategy is to increase the bandwidth locally if a large gain in conditional variance is paid with a relatively small increase in conditional bias. Note the analogy with numerical problems, where in case of ill–condition, the bandwidth is also increased locally. A large variance is due to a local sparseness of the realization of the design, combined with a local realization of design points which is skew and/or has a small variance. Such a variance–bias compromise is reminiscent of data-driven bandwidth selection, but it is different in orientation. Denote by h_0 the preassigned bandwidth for estimation of $r(x_0)$. Then, we would like to find a local bandwidth h close to h_0 that guarantees an acceptable variance, such that MISE is also reduced. Asymptotically, the local linear fit is left unchanged. We solve this problem in two steps:

- Check whether finite sample variance is high:

$$V_{h_0}\left(\widehat{r}_{ll}(x_0) \mid X\right) \geq \text{cut}\ \widetilde{V}_{h_0}\ ,$$

where $V_h\left(\widehat{r}_{ll}(x_0) \mid X\right)$ is the finite sample variance of $\widehat{r}_{ll}(x_0)$ and $\widetilde{V}_{h_0}$ is the asymptotic variance to be expected from (21) for uniform design.

A cut–off at cut $= 1.0$, for example, behaves well. Note, that it is not sufficient to increase the bandwidth until the variance is small enough. There is no guarantee that an increase of bandwidth always decreases variance. We might buy an adequate variance at the price of an unacceptable large local bandwidth. Asymptotically an increase of h increases the squared bias proportional to $(h/h_0)^4$ (20). Then, taking h_0 as optimal, the following variance–bias compromise is sought:

- Find the minimum of

$$V_h\left(\widehat{r}_{ll}(x_0) \mid X\right) + \frac{1}{4}\widetilde{V}_{h_0}\left(\frac{h}{h_0}\right)^4$$

with respect to h.

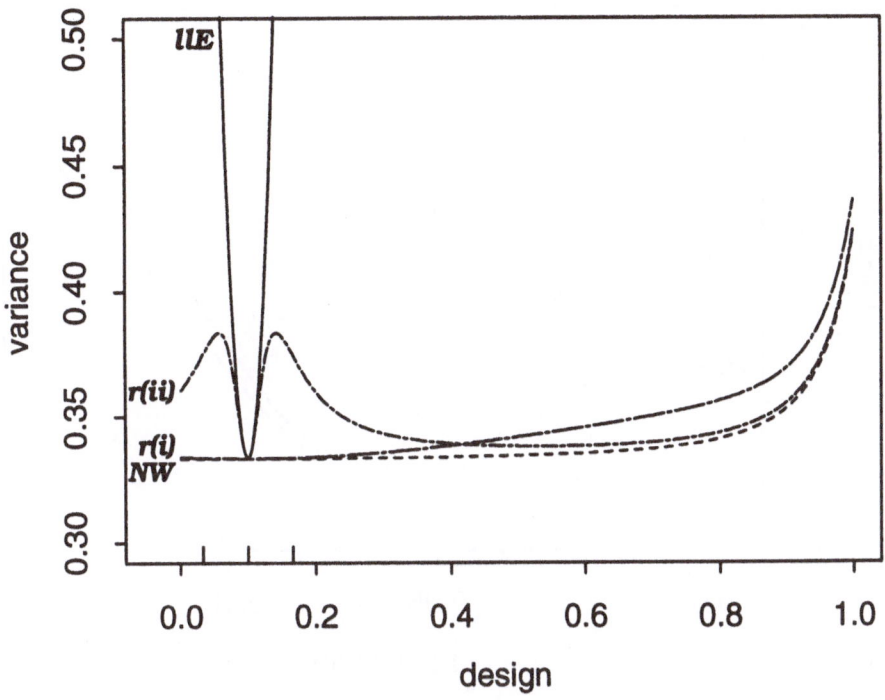

Figure 13: *Variance depending on design in a tutorial example with 3 design points, $\sigma^2 = 1$ and $h = 1$. Solid line is local linear fit (llE), dashes are Nadaraya–Watson estimator (NW), and dash–dots are local linear ridge regressions — (i) (r(i)-longer dashes) and (ii) (r(ii)-shorter dashes) — all with Epanechnikov weights.*

Local linear ridge regression: Ridge regression estimators are known to reduce MISE of linear estimators and to have minimax properties, if the regression parameters are assumed to lie in a bounded region. Our aim is to use ridge regression to cope with problems of variance. Note that these problems are associated with a local non–smooth behavior of the estimate. Thus, we will use the inherent smoothness assumption for the regression function for "ridging" the local linear estimator. If $\beta' H \beta \leq \sigma^2$ for some positive semidefinite matrix H, such that $H + X' W X$ is nonsingular, then

$$\widetilde{\beta} = (H + X' W X)^{-1} X' W Y$$

is the appropriate ridge estimator. Note, that the possibly ill–conditioned matrix $S = X' W X$ is replaced by a well–conditioned one. Consequently, ridge regression attacks — as a by–product — numerical instability at the price of a modification of the estimator.

We now propose two possible parameter restrictions, both specifying the smoothness assumption (using $\bar{X}_0$ from section 3.4).

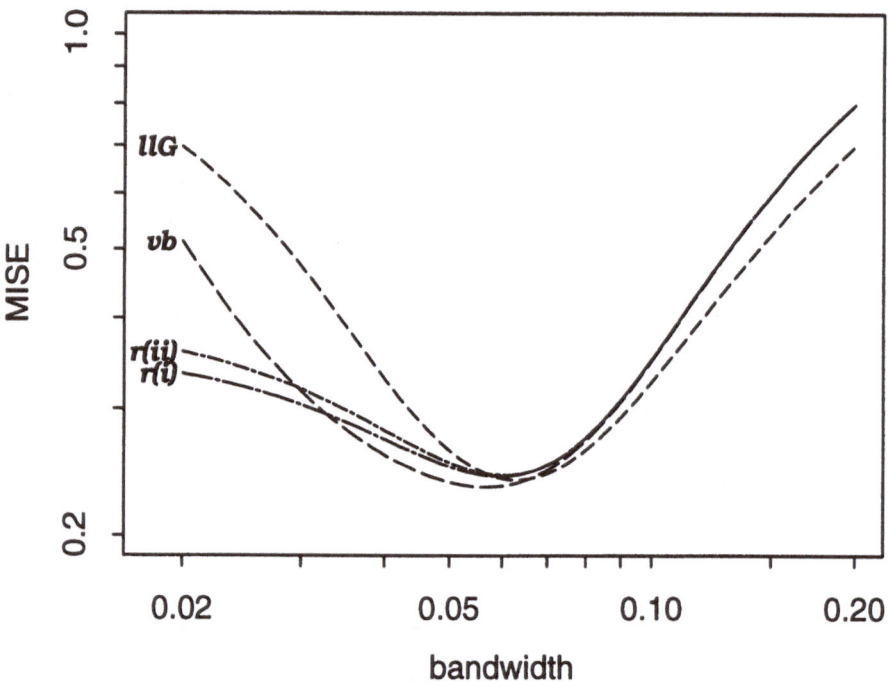

Figure 14: *MISE depending on bandwidth h (on logarithmic scales) of estimators of r in (33) for n = 50 uniformly distributed design points and $\sigma^2 = 0.5$. Dashes are local linear fit with Gaussian weights (llG), long dashes are local linear fit with a variance–bias compromise (vb) with cut = 0.5, and dash–dots are local linear ridge regression ((i) (r(i)-longer dashes) with $c_1 = 0.001$ and (ii) (r(ii)-shorter dashes) with c = 1).*

(i) "The derivatives r' of r at $x = \bar{X}_0$ are bounded."

and

(ii) "The difference of $r(\bar{X}_0)$ and $r(x_0)$ is bounded."

To see the differences between approaches (i) and (ii), compare the resulting ridging matrices H:

$$H_{(i)} = \begin{pmatrix} 0 & 0 \\ 0 & c_1 \end{pmatrix} \quad \text{and} \quad H_{(ii)} = \begin{pmatrix} 0 & 0 \\ 0 & c(x_0 - \bar{X}_0)^2 \end{pmatrix}.$$

Figure 13 demonstrates that the local behavior is much improved by both ridge variants. Note that the variance of a local linear fit with Epanechnikov weights goes up to a value of 113.5.

Comparison of different modifications: The modifications described entail the choice of further parameters (the cut–off point "cut" and the ridge parameters c_1 or c). In order to evaluate the potential benefits of the modifications, appropriate values are used.

Figure 14 compares MISE for various methods depending on bandwidth (related to figure 8). Modified local linear fits with Epanechnikov weights outperform the fit with Gaussian weights. The problems for unmodified fits for bandwidths below about 0.1 have completely disappeared. Judged from this example the variance–bias compromise seems to be particularly attractive around the optimal bandwidth. The cut–off and ridge parameters have been selected by prior experience, identical ones for all bandwidths and all realizations of the design.

We undertook an evaluation of efficiency depending on sample size, similar to figure 11. Here, the bandwidths, cut–off and ridge parameters were selected by minimizing the conditional MISE. The methods thus were evaluated at their best. The modified local linear methods with Epanechnikov weights dominated all other methods for small and large sample size. Both ridge methods were close together.

The results for the modifications are encouraging, but these methods require detailed further investigation. Whereas the problem of an adaptive optimal choice of cut–off and ridge parameters is still an open question, first evaluations show that any reasonable choice beats unmodified local polynomials with compact weights by far.

References

Brockmann, M., Gasser, T. & Herrmann, E. (1993). Locally adaptive bandwidth choice for kernel regression estimators. *J. Amer. Statist. Assoc.* **88**, 1302–1309.

Chan, T. F., Golub, G. H. & LeVeque, R. J. (1983). Algorithms for computing the sample variance: Analysis and recommendations. *The American Statistician* **37**, 242–247.

Chu, C. K. & Marron, J. S. (1991). Choosing a kernel regression estimator. *Statist. Sci.* **6**, 404–433.

Cleveland, W. S. (1979). Robust locally weighted regression and smoothing scatterplots. *J. Amer. Statist. Assoc.* **74**, 829–836.

Devroye, L. P. (1978). The uniform convergence of nearest neighbor regression function estimators and their application in optimization. *IEEE Trans. Inform. Theory* **IT–24**, 142–151.

Fan, J. (1993). Local linear regression smoothers and their minimax efficiencies. *Ann. Statist.* **21**, 196–216.

Fan, J., Gasser, Th., Gijbels, I., Brockmann, M. & Engel, J. (1993). Local polynomial fitting: A standard for nonparametric regression. Manuscript.

Fan, J. & Gijbels, I. (1992). Variable bandwidth and local linear regression smoothers. *Ann. Statist.* **20**, 2008–2036.

Gasser, Th. & Müller, H.-G. (1979). Kernel estimation of regression functions. In: *Smoothing techniques for curve estimation. Lecture Notes in Math.* **757**, 23–68. New York, Springer.

Gasser, Th. & Müller, H.-G. (1984). Estimating regression functions and their derivatives by the kernel method. *Scand. J. Statist.* **11**, 171–185.

Hall, P. & Marron, J. S. (1987). On the amount of noise inherent in bandwidth selection for a kernel density estimator. *Ann. Statist.* **15**, 163–181.

Hastie, T. & Loader, C. (1993). Local regression: Automatic kernel carpentry. *Statist. Sci.* **8**, 120–143.

Herrmann, E. (1994a). Local bandwidth choice in kernel regression estimation. Manuscript.

Herrmann, E. (1994b). A note on the convolution type kernel regression estimator. Manuscript.

Jennen–Steinmetz, Ch. & Gasser, Th. (1988). A unifying approach to nonparametric regression estimation. *J. Amer. Statist. Assoc.* **83**, 1084–1089.

Lejeune, M. (1985). Estimation non-paramétrique par noyaux: régression polynomiale mobile. *Revue de Statist. Appliq.* **33**, 43–68.

Macauley, F. R. (1931). *The smoothing of time series.* New York: National Bureau of Economic Research.

Müller, H.-G. (1988). *Nonparametric Regression Analysis of Longitudinal Data. Lecture Notes in Statistics* **46**. Berlin, Springer.

Nadaraya, E. A. (1964). On estimating regression. *Theory Probab. Appl.* **9**, 141–142.

Priestley, M. B. & Chao, M. T. (1972). Non–parametric function fitting. *J. Roy. Statist. Soc. Ser. B* **34**, 385–392.

Reinsch, C. H. (1967). Smoothing by spline functions. *Numer. Math.* **10**, 177–183.

Ruppert, D. & Wand, M. P. (1994). Multivariate locally weighted least squares regression. *Ann. Statist.* **22**, 1346–1370.

Schoenberg, I. J. (1964). Spline functions and the problem of graduation. *Proc. Nat. Acad. Sci. USA* **52**, 947–950.

Seifert, B., Brockmann, M., Engel, J. & Gasser, T. (1994). Fast algorithms for nonparametric curve estimation. *J. Comp. Graph. Statist.* **3**, 192–213.

Seifert, B. & Gasser, T. (1996). Finite sample variance of local polynomials: Analysis and solutions. *J. Amer. Statist. Assoc.*, to appear.

Silverman, B. W. (1984). Spline smoothing: the equivalent variable kernel method. *Ann. Statist.* **12**, 898–916.

Stone, C. J. (1977). Consistent Nonparametric Regression. *Ann. Statist.* **5**, 595–645.

Watson, G. S. (1964). Smooth regression analysis. *Sankhyā* Ser. A **26**, 359–372.

Comments

by Peter Hall and Berwin A. Turlach

Centre for Mathematics and its Applications, Australian National University, Canberra ACT 0200, Australia

An important issue common to these three articles is the matter of irregularly spaced design. The ridge regression methods addressed by Seifert and Gasser would not be necessary if design points were equally spaced; the rich variety of techniques considered by Marron is in part a response to the wide range of possible design configurations; and the nearest neighbour methods that underlie Cleveland and Loader's algorithms are designed to provide adaptation to variations in design density.

There are several approaches to dealing with design irregularities. One (e.g., that of Cleveland and Loader) is to develop an algorithm which reacts to regions of low design density by employing more smoothing there. While this is a laudable objective, it should be remembered that optimal selection of a smoothing parameter, particularly in a local sense, demands an appropriate response to variability of the target function as well as to changes in design. Thus, standard k-nearest neighbour methods for smoothing fall short of being fully adaptive, in that they do not effect an optimal trade-off between systematic and stochastic sources of variability.

A second remedy for design irregularities is that suggested by Seifert and Gasser. Those authors use essentially a constant amount of smoothing across the support of the design, and employ a secondary modification—their ridging method—to deal with regions suffering from sparse design. Of course, selection of the ridge parameter must influence to some extent both bias and variance, but from at least a theoretical viewpoint these affects will be negligible. Ridge parameters are not smoothing parameters, to first order.

It should be noted, however, that selection of a ridge parameter does not absolve the statistician from adjusting the smoothing parameter. The ridge matrices in Seifert and Gasser's paper are singular, and as a result the smoothing parameter may need to be increased in regions where so few data values fall that the estimator is not well-defined. In the context of smoothing by locally fitting a polynomial of degree p, at least p data values should fall within the support of the scaled kernel function if proper definition is to be ensured. In practice, at least $p + 1$ data would be recommended.

To appreciate the importance of this point it is helpful to calculate the probability that each interval of width $2h$, centred between 0 and 1, contains at least one data value when design points are uniformly distributed. Using the Poisson approximation of Seifert and Gasser (1994) this probability may be shown to equal 0.742 when $n = 50$ and $h = 0.05$. It can fall to extremely low levels; for example, to 0.002 when $n = 50$ and $h = 0.02$. That is, when $(n, h) = (50, 0.02)$ the statistician would expect to have to intervene and adjust the bandwidth for about 99.8% of realizations, even when using a ridge parameter.

Nevertheless, there are advantages to using fixed bandwidth rules. In particular, they are often faster to compute, and provide more predictable responses to sampling fluctuations. Indeed, many fixed bandwidth rules (excluding cross-validation) are relatively robust against stochastic variability, since they use the entire data set to compute the smoothing parameter. Of their very nature, locally adaptive methods rely heavily on only a small fraction of the data to make each smoothing parameter decision, with the result that they are more liable to be affected by stochastic aberrations.

In this context we would suggest that Marron's six factors for assessing methods of smoothing parameter choice be augmented by a seventh:

F7 robustness (is it reasonably resistant against stochastic aberrations?)

Fixed bandwidth rules conform to this recommendation particularly well. While they might not enjoy all the "efficiency" advantages (Marron's factor **F3**) of a good local rule, they will outperform a suboptimally efficient rule, such as k-nearest neighbours, in terms of **F7**; and will be neither superior or inferior, in general, in terms of **F3**. We agree with Marron's comments about the importance and usefulness of fixed bandwidth rules, noting at the same time (as we are sure he would too) that they are outperformed in some respects and on some occasions by other, more locally adaptive approaches. As Marron notes, no single bandwidth selector is uniformly superior to any other.

In conclusion, we would like to mention an approach alternative to that of Seifert and Gasser, suitable for overcoming the problem of design fluctuations in fixed-bandwidth kernel-type methods. It is based on adding interpolated "pseudodata" to the original data set, so as to overcome problems arising from sparse design. The simulations that we shall report are based on the following rule, but alternative methods are feasible. In the notation of Seifert and Gasser, suppose $x_1 \leq \ldots \leq x_n$, and the x_i's come from a distribution supported on the interval $\mathcal{I} \equiv [0, 1]$; and consider the spacings $s_i \equiv x_{i+1} - x_i$, for $0 \leq i \leq n$. (We define $x_0 = 0$ and $x_{n+1} = 1$.) Suppose the kernel K is supported on the interval $(-1, 1)$, and fix a real number $\nu > 1$. If the spacing

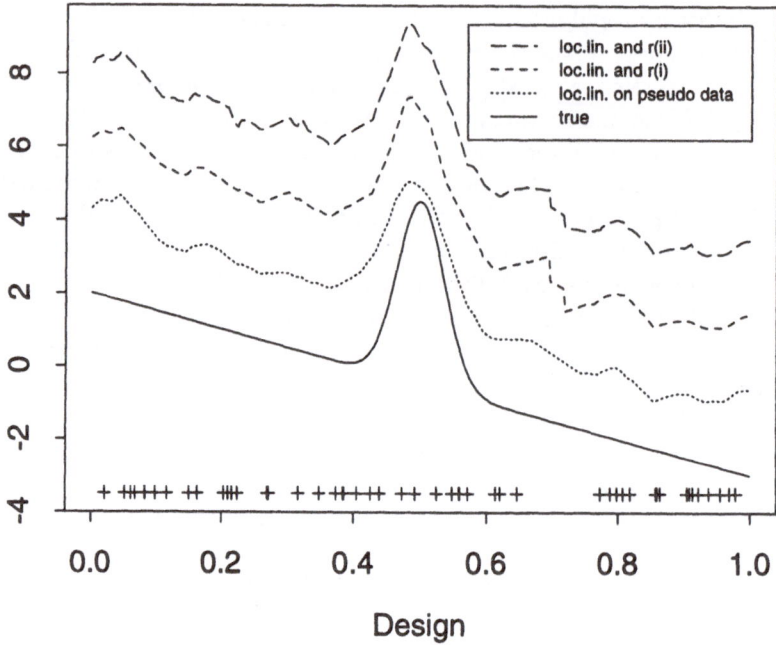

Figure 1: Estimates of the function r. The unbroken line depicts the true regression function r, defined at (1), the dotted line represents the local linear fit of the pseudo data method with $\nu = 3$, shifted by 2. Dashed lines show local linear ridge regression. The short dashed line represents method (i) with $c_1 = 0.001$, shifted by 4, and the long dashed line represents method (ii) with $c = 1$, shifted by 6. There were $n = 50$ uniformly distributed design points indicated by crosses, and we took $\sigma^2 = 0.5$. All fits were done using the Epanechnikov kernel and the (asymptotically optimal) bandwidth of $h = 0.0457$.

s_i exceeds $2h/\nu$, add new design points to the interval $\mathcal{I}_i \equiv (x_i, x_{i+1})$, the number added being the smallest necessary for the largest spacing among the revised set of design points in that interval to not exceed $2h/\nu$. Thus, the number added will equal the smallest integer not less than $\nu(x_{i+1} - x_i)/2h$. With each design point added to $\mathcal{I}_i$ we associate a pseudo value of Y, derived by linear interpolation between (x_i, Y_i) and (x_{i+1}, Y_{i+1}).

Within a bandwidth of either end of the design interval $\mathcal{I}$ it is appropriate to modify this rule by replacing $2h/\nu$ by h/ν. At the very ends of $\mathcal{I}$ we use horizontal extrapolation from the first or last data pair in place of linear

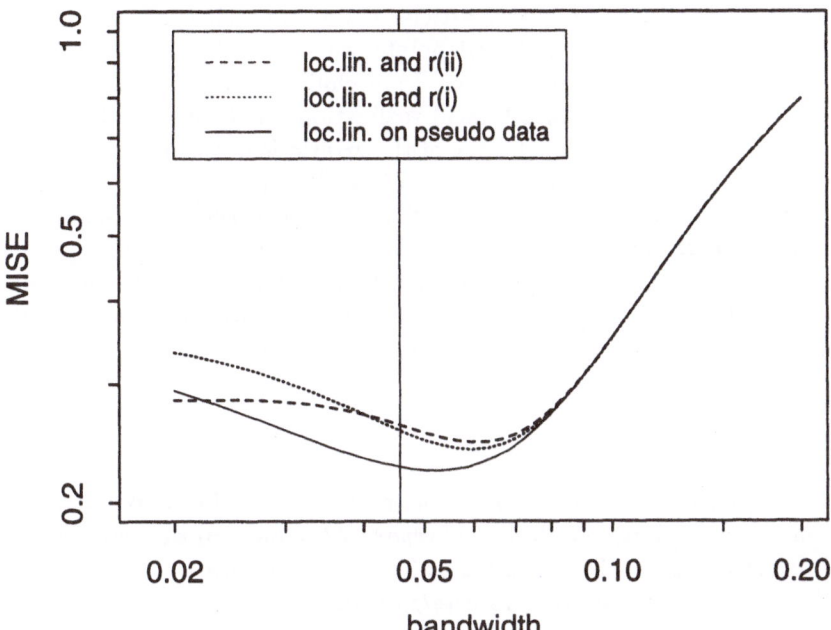

Figure 2: MISE as a function of bandwidth h (on logarithmic scales) for estimators of r, defined at (1), with $n = 50$. Each curve is based on 4000 simulations, with uniformly distributed design points and $\sigma^2 = 0.5$. The unbroken line represents the local linear fit to the pseudo data method with $\nu = 3$, and the broken lines show local linear ridge regression. The dotted line represents method (i) with $c_1 = 0.001$, while the dashed line represents method (ii) with $c = 1$. The vertical line indicates the asymptotically optimal bandwidth.

interpolation. It turns out that the procedure is rather insensitive to the value of ν, provided ν is not too small.

There are sound theoretical reasons for taking $\nu = 3$ or 4. These issues will be discussed elsewhere. There we shall also show that the above choice has good numerical properties. Additionally, it preserves the traditional first-order properties of the curve estimators, just as Seifert and Gasser's technique does.

To illustrate our method we used as our target the function

$$r(x) = 2 - 5x + 5\exp\{-400(x - 0.5)^2\}, \quad x \in [0, 1], \tag{1}$$

borrowed from Seifert and Gasser's article. Following their lead we employed a design of 50 uniformly distributed points, and generated additional data outside the interval $[0, 1]$. We used the Epanechnikov kernel and took $\nu = 3$. Figure 1 depicts the result of a typical realization. It illustrates the way in which interpolation has avoided some of the erratic features of a ridged estimator, without needing any intervention. Figure 2 compares MISE properties of our approach with those of Seifert and Gasser's two ridged estimators. It confirms that performance in terms of Marron's factor **F3** is, if anything, slightly superior for the interpolation method. The bandwidth providing optimal performance is somewhat smaller, and closer to the asymptotically optimal bandwidth, in the case of the interpolation rule.

References

Seifert, B. and Gasser, T. (1994). Finite sample variance of local polynomials: Analysis and solutions, *Technical report*, Abteilung Biostatistik, ISPM, Universität Zürich, Sumatrastrasse 30, CH-8006 Zürich.
URL: *http://www.unizh.ch/biostat/fin_SG.ps*

Comments

by M.C. Jones

The Open University, Milton Keynes, United Kingdom

I am very grateful for the opportunity to contribute to the discussion of these three fine papers on smoothing.

First, I would like to endorse Marron's "personal view" of smoothing and statistics. It is very helpful to see such a balanced view explicitly written out. What always strikes me about the different smoothing methods is just how *similar* they are rather than how different. I guess this stems from them (almost) all being linear in the data: there are only a few sensible principles for choosing the weights. Remaining differences are to do with localisation of bandwidths. I often feel that whatever goes on within one's favourite method could be mimicked very closely indeed (if one were clever enough to figure out how) by appropriate choices within another approach.

While Cleveland and Loader are right to remind people like me that we shouldn't get too carried away with asymptotics, Marron's view that they have a role in conjunction with more immediately practical investigations is also to be endorsed. A good example is the bandwidth selection problem for kernel density estimation which was stalled at poor methods before insights based in asymptotics "opened a flood gate of improvements". I should add that while some asymptotic analysis can be useful, much isn't; in particular, the provision of asymptotic biases and variances in smooth situations can yield insights (if taken with appropriate pinches of salt) but much other stuff seems, to me, rather pointless.

Cleveland and Loader give a very useful update on what's new in "local regression" from the Bell Labs camp. While some of these ideas have parallels in others' work, others were new to me. The idea of embedding polynomials in larger, more continuously indexed, families is an attractive one. But the linear combinations of fitted polynomials that make up the authors' polynomial mixing isn't quite this. I wonder if there are (even) more elegant approaches. One possibility that I haven't thought through *might* be to localise the "fractional polynomials" of Royston and Altman (1994).

By choosing different parametric families for local fitting, one gets, as Cleveland and Loader are well aware, different semiparametric approaches to regression. These are attractive as a one–parameter bridge between para- metric fits (large bandwidths) and nonparametric fits (small bandwidths). If the authors will forgive me for saying so, there are intriguing asymptotic consequences for the biases in the small bandwidth case. In particular, it is then the *number* of parameters fitted locally that matters — in direct anal- ogy with the degree of polynomial, for obvious Taylor series reasons — much more than the precise family used (or of other details in local likelihood fit- ting). This transfers work of Hjort and Jones (1994) in density estimation back to regression (Jones and Hjort, 1995).

A few other things on Cleveland and Loader's paper: (i) what do the authors do about bias in constructing "t intervals"? (ii) another historical paper in which local polynomial fitting is suggested, but in a spectral density estimation context, is Daniels (1962); (iii) wouldn't it be better in compar- ative figures such as Figures 2, 3, 7 and 8 to have in each column smooths with equal equivalent degrees–of–freedom rather than equal, but not compa- rable, bandwidths? (iv) the convenience of $C(h)$ in Section 9.2 would seem to be based on using a (considerably?) sub–optimal bandwidth, h itself, for estimating the local goodness–of–fit criterion, a disadvantage that other ap- proaches try to rectify (likewise for Scott and Terrell's, 1987, BCV method for (global) kernel density estimation); and indeed (v) without asymptotics, I wonder how the authors justify their faith in such as C_p.

Some days I think the deep study (apparently, studies) of Seifert and Gasser of problems due to sparse data is very valuable, some days less so. Perhaps in practice it is something like local bandwidths and a degree of common sense that is called for.

What seems to me to be a nice asymptotic explanation for Figures 2 to 4 is given in Jones, Davies and Park (1994). There, we observe that the estimators concerned differ in their treatment of the design density f as a multiple of the kernel weights: the Nadaraya–Watson estimator asymptotically includes $1/f(x)$, which at given x is no differential weighting at all, while Gasser- Müller and local linear estimators asymptotically include the much more appropriate, and changing with datapoints, $1/f(X_i)$.

Finally, aside perhaps from boundary differences and ignoring sparseness problems (and not including local bandwidth variation — perhaps this is too long a list of exceptions!), I am struck by how non–uniform (indeed locally non–uniform) one's design has to become to observe serious qualitative differences between estimators.

ADDITIONAL REFERENCES

Daniels, H.E. (1962) The estimation of spectral densities. *J. Roy. Statist. Soc.* Ser. B, **24**, 185–98.

Jones, M.C., Davies, S.J. and Park, B.U. (1994) Versions of kernel–type regression estimators. *J. Amer. Statist. Assoc.*, **89**, 825–32.

Jones, M.C. and Hjort, N.L. (1995) Local fitting of regression models by likelihood: what's important? Submitted.

Royston, P. and Altman, D.G. (1994) Regression using fractional polynomials of continuous covariates: parsimonious parametric modelling (with discussion). *Appl. Statist.*, **43**, 429–67.

Comments

by C. Thomas-Agnan

Université Toulouse 1, Toulouse, France

The debate on which smoothing method is "best", discussed for example in Chu and Marron (1991), Breiman and Peters (1992), or Sheather (1992), had to be extended to other estimators, due to renewed interest for local polynomial fitting starting in the early 1990's . Let us congratulate Marron, Cleveland and Loader, Seifert and Gasser for providing pertinent arguments on this question and valuable contributions to the study of this problem.

1 State of the art in smoothing

Following Marron's thorough account of the present state of the art in smoothing, I would first like to add a few items to the list of available methods and to the list of factors used to select them. To be more exhaustive, I would like to include in the list of methods:

M7 Orthogonal series estimators

M8 Convolution type kernel smoothers

More significantly, a number of additional items could be added to the list of factors:

F7 adaptability to dimensions greater than one

F8 possibility of incorporating shape constraints

F9 easy estimation of derivatives

F10 existence of asymptotic normality results

F11 good boundary behaviour

F12 robustness to moderate heteroscedasticity or error correlation.

Let us now comment on some of these factors.

About factor **F1**, the problem of availability for kernel methods may be a wrong question to ask: if you are not using a ready-to-use software package but rather an environment which allows some programming, it does not take more than a few minutes to write your own program to compute a kernel estimator. Splines or wavelets are another story, and the available versions in ready-to-use packages are unfortunately rather poor. For an environment like MATLAB, even though the Statistics Toolbox does not yet have smoothing macros, with the help of other toolboxes, one has all the building blocks for programming all these estimators very quickly, and this is why I recently chose this software to teach introductory smoothing.

I think that one aim in smoothing is not only to provide a fit at the data points but also in the interval between data points. In order to save time when implementing a smoother, it is common that a software package only computes the fit at the data points (or a subset thereof) and then does some kind of interpolation. If this fit is thus obtained, what one use in practice does not correspond to the estimator which has been investigated theoretically. The problem of speed may be relevant for simulation studies for example if you have to evaluate the smoother a large number of times, but in the end, for actual applications, the software should allow you to compute the real estimator more slowly but more exactly if you wish.

About factor **F2**, it is a very frequent statement that fixed bandwidth local constant kernel methods are more easily interpretable than others by their nature of simple local averages. This idea may come from overrating closed form formulas and can be debated as far as regression is concerned. It seems to me that the same is true for all linear smoothers (those for which the weights may depend on the design but do not depend on y_i). As soon as your software is able to compute the smoother, it is easy to make it generate the "equivalent kernel" as defined in Hastie and Tibshirani (1990), also called "effective kernel" in Hastie and Loader (1993), and from there to interpret the weights in the same fashion.

About factor **F8**, Delecroix and Thomas-Agnan (1994) point out that more methods up to now have been developped with splines than with kernel smoothers, and very little with other types of smoothers.

Finally, the great number of factors is responsible for there being so many divergent views on the "right" choice of smoothing method. Assume for a moment that to avoid the problem of non comparable factors, we try to rank the smoothers on the basis of their mathematical properties. Let us even restrict attention to the asymptotic minimax rates of convergence. We still face quite a difficult task because the various results available in the literature differ by

    a) regularity conditions imposed to the true regression

    b) loss function

    c) local or global nature of the performance criteria

    d) random design or fixed design.

If we thus fail to get to a conclusion, is there a hope to settle the question by simulation studies? The problem of finding by simulations which smoother performs the best for its own best choice of smoothing parameter is interesting but may not be of much practical interest for data analysts if the methods for estimating the best smoothing parameter are not efficient enough. This is why the two issues: choice of smoother, choice of smoothing parameter are difficult to separate. Nevertheless, one should try to avoid misinterpreting the conclusions of such studies. Take for example the following sentence in Breiman and Peters (1992) : delete-knot regression splines give "overall good performance at all sample sizes and X-distributions, is the clear winner in the smoothness category, is very accurate on straight line data, and fairly efficient computationally". One may easily be tempted to only memorize: regression splines are best.

My last point is that we should not forget that most smoothing methods have their counterpart as approximation methods in the field of numerical analysis. The two fields interact to the point that some papers are difficult to classify in one or the other category. It is clearly true for orthogonal series, smoothing splines, regression splines and wavelets. One should give thought to the fact that, aside from the Lancaster and Salkauskas (1981, 1986) reference mentioned by Cleveland and Loader, kernel smooths and local polynomials are pretty much ignored by our numerical colleagues.

2   Assessing the performance of local polynomials

Local polynomials share with splines the nice property of defining a continuous path from interpolation to linear regression.

Seifert and Gasser provide us with a very good tutorial on the comparative construction of weights for different smoothers and an interesting discussion of numerical stability of algorithms for the computation of local polynomials. Their simulations with finite sample MISE clearly underline with non-asymptotic arguments, the problem of large conditional variance of local polynomials due to sparse regions in the realization of the design.

In this paper I found particularly interesting the study of how performance varies with sample size and approaches asymptotic efficiency. It yields quantitative information on Cleveland and Loader's suspicions on the limitations of asymptotics.

I would like to question the statement that "splines are not local", i.e. all n observations influence all weights. First of all, the same can be said for kernels with unbounded support. Moreover, it is clear from the study of "equivalent kernels" (Hastie and Tibshirani, 1990) that the influence of the design points on the weights at a given point rapidly decays with the distance to this point.

I was very surprised to hear that smoothing splines displayed "slightly worse MISE performance for large sample size", because it was contradictory with my own experience with simulations. Moreover, it is very unclear whether this could be attributed to the asymptotic kernel being non-optimal.

Is it admissible to combine the asymptotic equivalence result with the MISE expansions?

Cleveland and Loader give us a good supply of modeling strategies and guidance on what works in practice when using local polynomials, illustrated by a number of examples, but no simulation. It is troublesome that no clear assumptions are stated on the function f so that it is "well approximated" (to be defined !) by a member of a parametric class like polynomials of given degree. The authors claim in paragraph 1.4 that the "good properties" of the method do not require smoothness and regularity assumptions like classical smoothers. Surprisingly later, when they start some theory with "minimal assumptions", it turns out that f has to have at least a "continuous second derivative". Even if we agree with their rejection of asymptotic guidance, should one believe that a good alternative be accumulation of empirical prescriptions?

It is interesting to note that the authors of these two papers, although not quite with the same motivations, propose a similar kind of variant to local polynomial regression: local linear ridge regression for Seifert and Gasser, and "polynomial mixing" for Cleveland and Loader.

3 Conclusions

Coexistence of divergent views on the "right" choice of smoothing method can be perceived as a positive fact: a testimony of the field's richness and activity. Such debates are healthful and generate progress. I think that they will not discredit the field for outsiders or give the impression that it is split into schools, as long as the arguments remain on scientific grounds.

The importance of **F1** may recede with the fast development of software. But recent history has seen for example the emergence of new questions like local adaptivity: the ability of a smoother to perceive and respond to spatially inhomogeneous smoothness. It is hard to predict the field's future evolutions. Will the entropy keep rising with an increasing number of new estimators together with a growing number of selection factors ?

References

[1] Breiman L. and Peters S.(1992) Comparing automatic smoothers (A public service enterprise), *International Satistical Review*, 60, 3, pp. 271-290.

[2] Chu C.K. and Marron J.S. (1991) Choosing a kernel regression estimator, *Statistical Science*, 6, 4, pp. 404-436.

[3] Delecroix M. and Thomas-Agnan C. (1995) Kernel and spline smoothing under shape restrictions. Submitted.

[4] Hastie T.J. and Tibshirani J.R. (1990) Generalized additive models, Chapmann and Hall

Comments

by S.J. Sheather, M.P. Wand, M.S. Smith and R. Kohn

Australian Graduate School of Management, University of New South Wales, Sydney, 2052, Australia

1 Introduction

We welcome the opportunity to comment on a set of papers seeking to advance nonparametric regression techniques. Especially as the authors of all three papers have already made substantial and influential contributions in this field. We agree with the authors on many points, both in principle and in substance. However, we have chosen to concentrate our comments on those aspects of the papers that we disagree with or that we find ambiguous. We take this critical role as we think it will be more interesting for readers and elicit further clarification from the authors. Most importantly, it points out that there are enormous opportunities for research in this field. Despite the large volume of work in the 1980's and 1990's we believe that the subject is still in its infancy.

2 Comments on Cleveland and Loader

Cleveland and Loader provide a useful outline of the history of local regression, introduce several innovative ideas and give their views on the state of the field. They have chosen to discuss their local regression approach mainly for the one dimensional case but it is clear from their other work that the great strength of their approach is its ability to generalize to higher dimensions. Despite this generality we believe that some fundamental aspects of their approach can be improved. Our views on their procedures are outlined in the following sections.

2.1 Graphical diagnostic procedures

The examination of residuals is now widely accepted as an important step in parametric regression analysis. Cleveland and Loader demonstrate the importance of graphical diagnostic methods in the nonparametric regression setting and recommend that these tools be routinely used. We endorse this practice and the authors' views that "residual plots provide an exceedingly powerful diagnostic that nicely complements a (model) selection criterion" (Section 8) and that "when we have a final adaptive fit in hand, it is critical to subject it to graphical diagnostics to study the performance of the fit" (Section 1).

Both Cleveland (1993) and the current article use a loess fit to plots of residuals against fitted values to detect lack of fit. Though looking at the residuals from a nonparametric fit is an extremely useful process, we question the validity of judging the appropriateness of a nonparametric regression estimate by fitting another such curve to the residuals.

The problem is how to choose the fit for the residual plot. The same local polynomial fit is not appropriate for both the original data and the residuals because if this fit allowed structure to go undetected in the original data, then it is very unlikely to capture structure in the residuals. Cleveland (1993, Section 3.6) and Section 8.1 of the current paper suggest a local linear fit to the residuals using an arbitrary smoothing parameter such as $\alpha = 1/3$. The inadequacy of this recommendation as a general principle is illustrated using the authors simulated example.

We generated 100 observations from the model given in section 8.2 and used loess to obtain locally quadratic fits with smoothing parameters increasing in equal percentage steps as suggested by the authors. These are given on the left hand side of Figure 1, along with local linear fits to each set of residuals on the right using $\alpha = 1/3$. It can be seen that the fits to the data improve as α increases, whereas the fits to the residuals deteriorate. If such residual fits were used as a guide, the roughest fit (where $\alpha = 0.2$) would be regarded as the best, while the best fit (where $\alpha = 0.675$) would be classified as the very worst. This effect is more disturbing when one considers that in higher dimensions simple scatter plots of the data cannot be used to resolve this apparent contradiction.

An alternative way of comparing the local fits based on different values of α is based on F-tests. Cleveland, Grosse and Shu (1992, page 331) provide an example of the use of this procedure. What concerns us is that these F-tests are based on the assumption of no bias and in general this is not the case.

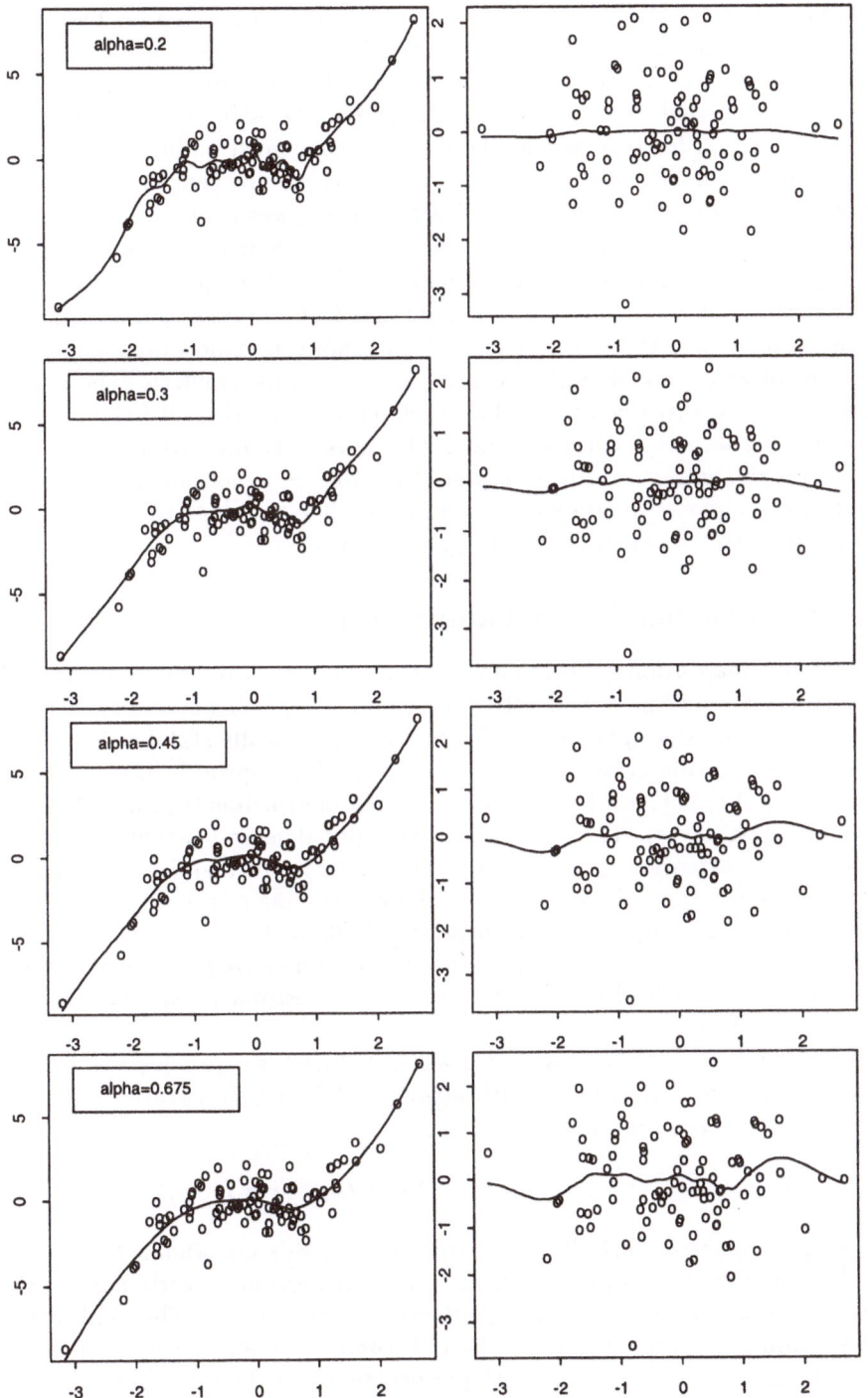

Figure 1. Local quadratic fits to data for various values of α (left) and local linear fits to the resulting residuals with α fixed at 1/3

We agree with Cleveland and Loader's practice of displaying fits based on different values of the smoothing parameter. The point at which we differ is that we would typically recommend as the default bandwidth a fixed plug-in bandwidth, h_{PI} that minimises an error criterion like estimated mean square error. The reason for this, as Cleveland and Loader mention in Section 10.3, is that plug-in bandwidths work quite well in the sense that they produce bandwidths quite close to the value that minimizes mean square error. Thus values of h smaller than h_{PI} will produce curve estimates with smaller bias but larger variance than the estimate based on h_{PI}. The opposite is true for values of h larger than h_{PI}. As Cleveland and Devlin (1988) point out, low bias is critical when the estimate is used for graphical exploration, since the eyes can tolerate some noise but cannot recover a missed effect. The curve estimate with $h = h_{PI}$ can be used as a reference point, since it balances the bias and variance of the curve estimate. This makes the fits with the different values of h straightforward to interpret. In addition, it is usually clear from a plot of the curve estimate with $h = h_{PI}$ whether an adaptive rather than a global bandwidth is needed for the given data set.

2.2 Nearest-neighbour bandwidths

The implicit adaptation of the bandwidth through the use of nearest-neighbour distances is a convenient and often successful means of addressing design sparseness. Nearest-neighbour bandwidths asymptotically stabilise the variance of the resulting curve estimate. However, this approach has a serious shortcoming that detracts from its use as a general principle (e.g. as a default in computer packages). The problem is that a global nearest-neighbour bandwidth pays no attention to bias, that is, curvature in the regression function. Since the adaptation is based solely on the positioning of the design points, for a given α and design the global nearest-neighbour bandwidth is the same whether the underlying curve is a straight line or a high frequency sine curve. This can sometimes lead to an unattractive curve estimate, especially when the design is skewed.

An example of such behaviour is given in Figure 2 where a loess estimate is plotted against a fixed bandwidth estimate. The (X_i, Y_i) regression data were generated according to

$$X_i = 1 - U_i^{0.7}; \quad Y_i = \sin(2\pi X_i^3) + 0.2Z_i, \quad i = 1, \ldots, 200,$$

where the U_i's are uniform $[0, 1]$ variates and the Z_i's are standard normal. The dashed curve is a local linear fixed bandwidth estimate, with bandwidth chosen using the rule of Ruppert, Sheather and Wand (1995). The solid curve is the estimate obtained using the S-PLUS loess() function with a span of 0.12, chosen so that the each smooth performed about the same in the valley on the right. Default values were used for the other loess() parameters. The cost of the reasonably good performance in the valley is the aberrant behaviour of loess on the left hand side, despite the fact that there are more data here and less structure to resolve. Increasing the span value doesn't help since it leads to the valley being smoothed away. Such disconcerting behaviour is due simply to the fact that there are more data where there is less curvature, and vice versa.

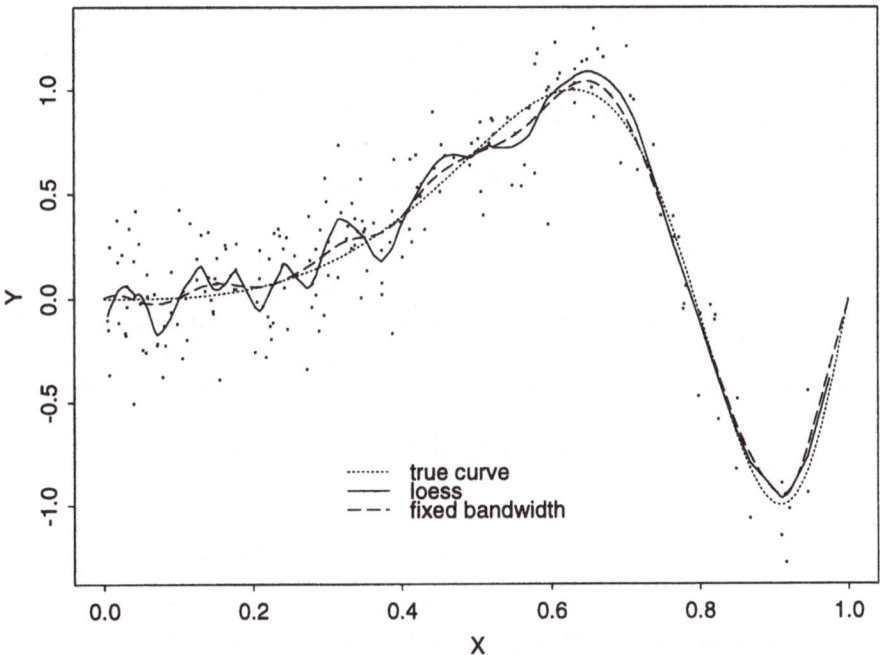

Figure 2. Nearest-neighbour (loess) and fixed bandwidth estimates for a simulated example.

The overall performance of fixed bandwidth estimate in this case is certainly more pleasing and leaves us somewhat skeptical about the authors' claim that "nearest neighbors appear to perform better overall in applications" compared to fixed bandwidths. There is also room for improvement in the fixed bandwidth estimate since it does not properly resolve the peak and valley and has some slight spurious bumps on the left hand side. Clearly some sort of local adaptation of the bandwidth is required for data such as these. We look forward to seeing research into further methodology for doing this that addresses both the sparsity and curvature issues.

2.3 Equivalent quadratic smoothers

In Section 7, Cleveland and Loader argue that local quadratic fitting does better than local linear and local constant fitting "when there is a rapid change in slope, for example, a local minimum or maximum". Then in Section 10 they recommend that one compare the fit of a local constant or linear fit with that of a local quadratic regression based on a bandwidth with the same variance as the local constant or linear fit.

We undertook the suggested comparison using a local linear fit based on the plug-in bandwidth rule of Ruppert, Sheather and Wand (1995). The regression functions considered were $f_1(x) = 1 - 48x + 218x^2 - 315x^3 + 145x^4$ and $f_2(x) = \sin(10\pi x)$. These were chosen since they each have at least one minimum and maximum. The x's were generated as uniform $(0,1)$ random variables and the errors as normal random variables with mean 0 and standard deviation equal to one quarter of the range of f. The goodness of fit criterion used was the mean square error summed over the x's, which Cleveland and Loader denote by $R(\hat{f}, f)$ (equation (1) in Section 8.2).

In order to compare the local linear fit $(\hat{f}^L)$ and the local quadratic fit $(\hat{f}^Q)$ we computed their relative efficiency, that is, the ratio of their mean square errors, $RE = R(\hat{f}^L, f)/R(\hat{f}^Q, f)$ for 100 samples of size $n = 100$ and 10,000. Estimates of the mean relative efficiency and associated 95% confidence limits are given in Table 1. The confidence limits were calculated using a normal approximation to the distribution of mean relative efficiency.

Function	Sample Size	Mean RE	95% CI for Mean RE
f_1	100	0.9729	(0.9707, 0.9752)
f_1	10000	0.9833	(0.9832, 0.9834)
f_2	100	1.1216	(1.1140, 1.1292)
f_2	10000	1.5428	(1.4862, 1.5993)

Table 1. Relative efficiency of local quadratic to local linear fits

In terms of mean square error, the local quadratic fit of the sine curve (f_2) provides an improvement over the local linear fit, especially for the larger sample size. However, this improvement is not evident for the quartic (f_1). On the other hand, the loss of efficiency due to using a local quadratic fit is small and thus the idea of converting a local linear bandwidth into a local quadratic one is worthy of further investigation.

The results for f_1 in Table 1 illustrate a relatively well-known phenomenon that in some cases very large sample sizes are needed in order to see the benefits of higher order methods, which improve rates of convergence from $O_p(n^{-2/5})$ to $O_p(n^{-4/9})$.

2.4 Local adaptability

Methodologies that employ a single global smoothing parameter are not appropriate for use on functions that take on highly dissimilar forms on different sections of the domain of the independent variable. What is required is a degree of *local adaptability* – something that can be achieved by varying the bandwidth over the domain, as the authors outline in Section 9.2 using a local version of Mallows C_p. Nevertheless, it can be computationally difficult to locally estimate such bandwidths at each point in the design. Instead, some authors (e.g., Härdle and Marron, 1995) split up the design domain into a series of regions and a different bandwidth is used in each of these regions. The problem then becomes one of choosing these regions.

An alternative strategy that is computationally tractable and avoids ad hoc domain splitting rules is to link regression splines with Bayesian variable selection within the computational framework of the Gibbs sampler as outlined in Smith and Kohn (1994). Here, a large number of potential knots are introduced along the domain of the independent variable (for example, 1 knot every 5 design points) and a significant subset determined, resulting in a regression spline that is both locally adaptive and smooth. The following example clearly illustrates the importance of local adaptability in nonparametric regression.

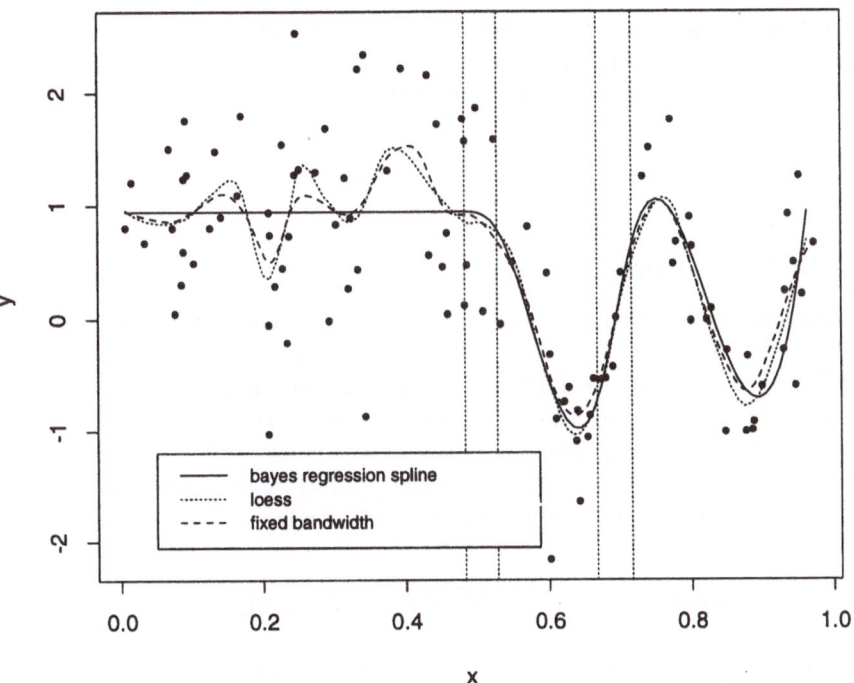

Figure 3. Nearest-neighbour (loess), fixed bandwidth and Bayes estimates for a simulated example.

One hundred observations were generated from a regression model with the x's uniformly distributed on (0,1) and $f(x) = 1$ if $x < 0.5$ and $f(x) = \cos(8\pi x)$ if $x > 0.5$. The errors were distributed normally with mean zero and standard deviation 0.75. A kernel based local linear smooth with automatic bandwidth selection (Ruppert, Sheather and Wand, 1995) was fit to the data, along with a local quadratic fit given by loess with $\alpha = 0.2$. Figure 3 shows that both fits capture the cosine curve well, but are not at all smooth in on the rest of the domain. No global bandwidth value will successfully produce a fit that is both smooth and captures the oscillations of the cosine curve well. However, the locally adaptive Bayesian regression spline (the solid curve) is both smooth and possesses low bias. The four significant knots found for these data are marked as vertical lines in Figure 3. We look forward to the opportunity to compare the performance of this Bayesian regression spline with the locally adaptive procedure of Cleveland and Loader.

3   Comments on Marron

Marron stresses the importance of reliable automatic smoothing parameter selection. It seems to us that it is time that the current, often arbitrary, default choices for smoothing parameters common in software packages were replaced by what Marron describes as the "second generation of bandwidth selection methods". Other situations which we would add to the authors' list in Section 3 are:

- analysis of discrete regression data sets, such as those with binary responses, and

- multivariate smoothers.

In these cases it is not as straightforward to use graphical techniques to assess the fit.

We agree with the author's view regarding the importance of evidence for performance of smoothing parameter selection rooted in E1, E2 and E3. High quality smoothing parameter choice is a very subtle and challenging problem. It cannot be properly developed and evaluated without the use of a sensible combination of all of these research tools.

4   Comments on Seifert and Gasser

It is clear that fixed bandwidth local polynomial smoothers require modifications to guard against degeneracies in the local fitting process. Seifert and Gasser have made an important first step into the development of such modifications. The results from their bandwidth inflation and ridge regression ideas are very encouraging.

Another lesson that is apparent from their work (e.g. Figure 8) is that the Gaussian kernel weight can go a long way to alleviating degeneracy problems.

We are not so much concerned about claims by the authors that Gaussian weights are "computationally slow" because of the existence of fast computational methods other than theirs (see e.g. Fan and Marron, 1994) that do not impose restrictions on the kernel type.

An important question that arises from Section 5 of this article is: at what point is a design become so sparse that a smoothing technique should not be used at all? Figure 2 shows some contrived designs over $[0, 1]$ generated from the beta mixture density $\frac{1}{2}\text{Beta}(1, s) + \frac{1}{2}\text{Beta}(s, 1)$ for increasing values of s. The regression curve is $m(x) = \sin(8\pi x)$, shown without noise for clarity.

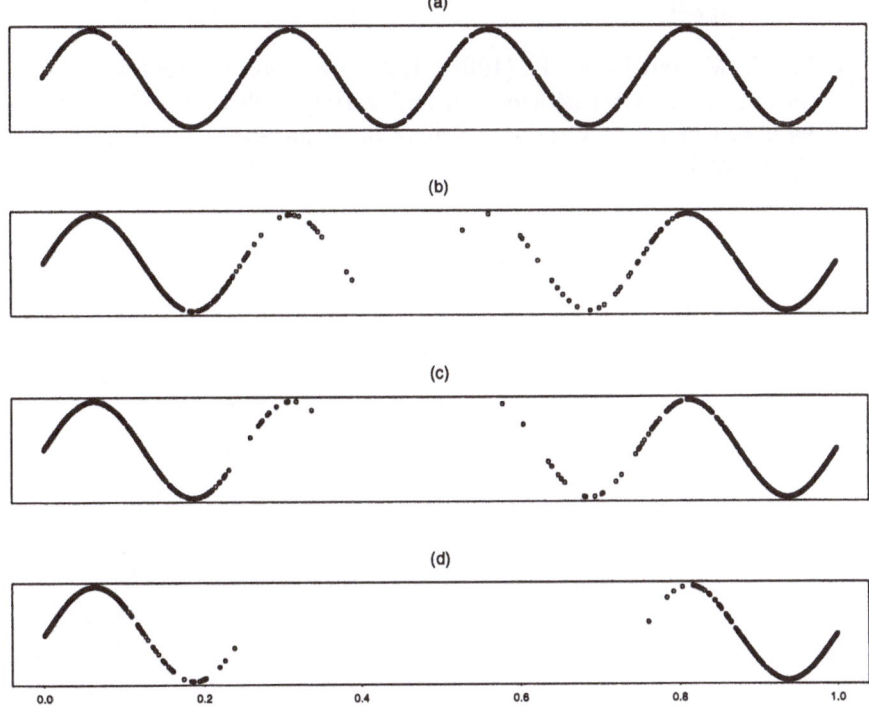

Figure 4. Regression designs with increasing sparsity in the centre.

For design (a) there is clearly no problem with application of a smooth over the whole of $[0, 1]$. At the other extreme most would agree that for design (d) the data is non-existent in the middle part of $[0, 1]$ and that it is pointless to try to use local regression here. Designs (b) and (c) are not as clear-cut. There are some data in the middle, but are there enough for one to expect reasonable recovery of the underlying regression function? It would seem that methodology for estimating when the design is thick enough to apply a smoothing technique be required. Density estimation could lead to effective solutions. Such technology would be even more important for two- and three-dimensional designs where there are more obscure ways in which a design can "peter out".

References

[1] Fan, J. and Marron, J.S. (1994). Fast implementations of nonparametric curve estimates. *J. Comput. Graph. Statist.*, **3**, 35–56.

[2] Härdle, W. and Marron, J.S. (1995). Fast and simple scatterplot smoothing. *Comput. Statist. Data Analysis*, **18**, to appear.

[3] Ruppert, D., Sheather, S.J. and Wand, M.P. (1995). An effective bandwidth selector for local least squares regression, *J. Amer. Statist. Assoc.*, **90**, to appear.

[4] Smith, M. and Kohn, R. (1994). Robust nonparametric regression with automatic data transformation and variable selection. Working Paper, 94-026, Australian Graduate School of Management, University of New South Wales.

Rejoinder

J.S. Marron

Department of Statistics, University of North Carolina, Chapel Hill, NC 27599-3260, USA

1 Comments on "the debate"

Many people are aware of some substantial recent controversy in smoothing. The current discussion is partly a consequence of this. I have observed a number of differing ideas as to "what the debate is all about". Of these various ideas, I view the main ones as being:

I1 Should one use local polynomial (degree higher than 0) or local constant (i.e. conventional kernel methods) smoothers?

I2 Should one choose "smoothing windows" using the same width everywhere, or a width based on nearest neighbor considerations?

I3 Is LO(W)ESS the best possible way to do smoothing?

I4 Is mathematical analysis a useful tool in statistics?

I5 Are there effective methods of choosing the bandwidth from the data? Are these good enough to become defaults in software packages?

Here are some personal comments on each of these:

1.1 Kernels vs. local polynomials

There is less disagreement here than some people seem to expect. My personal current default, e.g. what I use when faced with a new data set to smooth, is the local linear. This choice is based on its excellent handling of boundary problems and random design issues, as made clear in a number of ways by several authors.

However, I understand and respect those who select any of local constant, quadratic or cubic (each has its reasons for being used, and situations where it is "best"). Since local constant is the one that seems to be a favorite target, I will comment further on that.

While it has some serious weaknesses (which is why I currently don't use it myself as a general rule), there are also situations where its use is appropriate, and even "best". In particular, when there is an equally spaced design, and boundary issues are not important (e.g. a "circular design", or the regression is flat at the boundaries, or boundaries are not relevant to the problem being addressed), then such an estimator can be "best" in terms of: simplicity, in terms of interpretability (i.e. "what is it doing to the data"), and in terms of being "close to the true regression" (e.g. will have the best performance in a simulation, or best theoretical properties).

Another context in which such estimators are sensible is when doing mathematical analysis. This is because simple moving averages are very tractable to analyze, with the benefit that one can analyze more deeply (e.g. tackle harder problems). A first class analysis would go on to show that the lessons learned hold also for the more useful higher degree local polynomials. But in many contexts it is sensible to first look at the simplest case: degree 0, and in some instances such work by itself would be publishable in my view.

1.2 Width of smoothing windows

Actually there are at least three general classes of ideas here:

1. fixed width,

2. nearest neighbors,

3. location adaptive bandwidths that use more of the data than simply the design.

Each of these has its strengths, its weaknesses, and its advocates.

The appeal of the fixed width approach comes from its simplicity and interpretability. Its weaknesses are that some data sets can show structure at different amounts of smoothing (usually in different locations), and it can behave erratically (even to the point of numerical instability) in regions where the data are sparse.

I view nearest neighbor methods as a quick and dirty fix to the data sparsity problem. However, they are very weak in terms of interpretability (just how does the resulting curve relate to the data?), and adapt only to the design (the X's, not the Y's) , while ignoring other reasons for varying the amount of smoothing, such as curvature in the regression function and heteroscedasticity. This point is discussed by Hall and Turlach, and made quite clear in section 2.2 of the discussion of Sheather, Wand, Smith and

Kohn, where the nearest neighbor based method LO(W)ESS has difficulties because it ignores curvature in the regression function.

Higher tech methods of location adaptive smoothing have more potential for "getting the amount of smoothing right", than either of the others. The example in section 2.4 of the discussion of Sheather, Wand, Smith and Kohn makes this point very clear. This has motivated useful research on this topic (and I am involved in this). However, higher tech methods have their attendant costs. The obvious costs are in terms of complexity, and interpretability. A less well understood drawback is the ease with which spurious features can be introduced, and important ones be eliminated. See Marron and Udina (1995) for discussion of this issue, and an environment which allows user choice of a location adaptive bandwidth function in kernel density estimation.

When I analyze data, I place a very high priority on interpretability, so my personal default method is a fixed width approach.

Any problems with data sparsity are readily apparent when one tries several bandwidths. When the sparsity is so severe it causes numerical problems, my current personal preference is for a simple ridge regression approach, which can often work well for surprisingly small ridge values, see Hall and Marron (1995), However, I plan to rethink this in view of the ideas in the contribution of Gasser and Seifert and the discussion of Hall and Turlach.

In my view, the problem of different important features being visible at different smoothing levels is more important than data sparsity. I do not believe that any location adaptive method will be able to completely replace the data analyst who looks interactively at different amounts of smoothing. However, insights gained from this process are also available from the "family approach" discussed in Marron (1995a). The main idea is to present a "one plot summary", which shows different smoothing levels, through overlaying smoothers with several bandwidths. Nearest neighbor methods are hopeless in this context as "continuity in the smoothing direction" is crucial to gaining visual insight.

While I do not believe that good location adaptive methods have much future in exploratory data analysis, I consider them worth more research, for the goal of display of final conclusions. An important open problem in this area is to find a device to show that the bandwidth function used is not hiding important features or highlighting spurious ones.

1.3 Is LO(W)ESS best?

LO(W)ESS includes a number of really excellent ideas, many of which were implemented well before many "experts in smoothing" even understood the issues. These ideas include local polynomials instead of local constants, really clever computational approaches, and adding robustness in smoothing. For these reasons, LO(W)ESS is recommended in the popular software package

S, and its variant Splus. This has resulted in this method being very broadly used for effective data analysis.

However, the fact that this is a useful method does not imply that it is the only useful method, or that it is the best method. As noted in my original contribution, there are many other effective approaches to data analysis by smoothing (M1 - M6). Furthermore, there are many different criteria by which smoothers can be judged (F1 - F12, including those of Thomas-Agnan). A commonly agreed upon answer to this question is impossible, because different people weight these criteria differently (all with good reason).

A more important issue is that I have (and hope others will gain) respect for:

1. the many clever ideas that went into LO(W)ESS, many of which were far ahead of their time.

2. the idea that many different viewpoints on how to choose a smoother are allowable, acceptable and rational.

1.4   Mathematics in statistics

Cleveland and Loader take issue with the use of mathematical methods in curve estimation. Their basic premise seems to be that "there are a lot of not very informative papers out there". I agree with this, but am not willing to accept a blanket condemnation of the usefulness of mathematical approaches to understanding smoothing. As noted in section 4 of my original contribution mathematics has played, and will continue to play an important role in methodological statistics. Moreover, the ability to bring to bear an array of methods (including real data examples, mathematics and simulation) is becoming increasingly important as we tackle harder and harder methodological problems.

1.5   Does bandwidth selection "work"?

In my original contribution it was made quite clear, from all of the important viewpoints of real data examples, asymptotic analysis, and simulations, that "modern" bandwidth selection methods do work quite well. As detailed in the specific comments section below, I take issue with Cleveland and Loader on this point, and several variants of it.

In fact, modern bandwidth selection is at a point that it is worth implementing as the default in software packages. This would be much more useful than an arbitrary choice based on experience with a few examples.

The issue of defaults for smoothing parameters is an important one. Cleveland and Loader convey the impression that it doesn't matter so much, because a proper analysis will involve looking at several bandwidths. While I agree with the latter in an abstract theoretical sense (e.g. I always try several myself, despite that fact that a modern bandwidth selection method will be effective), I find it contrary to much of the actual practice of smoothing. I have attended many public talks by, and have had much private discussion with, people who have used smoothing. A large proportion of these use LOWESS because of its strong recommendation in S. When I ask what smoothing parameter they use, a too common answer is "the default in S". While it is easy (and theoretically correct) to say "they shouldn't do that", I submit that software developers should take into account how users choose to use their product, as opposed to attempting to dictate how it should be used. From this point of view, the use of modern data based methods as defaults would make smoothing a much more useful tool for the very large number of non-expert user who do not choose to (or think they have no time to) try out several bandwidths.

An interesting approach to the problem of users attaching too much significance to defaults is implemented in the Software package R-code discussed in Cook and Weisberg (1994). In this package the bandwidth is controlled by a slide bar, which is manipulated by a mouse. This is not a new idea, but the clever contribution is that there is no initial default smooth presented. To get a smooth, the user must start moving the bandwidth slider. This is useful because then the user automatically scans through several bandwidths immediately, and sees how important it is to try several values of this parameter. With this user interface, problems with the user thinking "the default is good enough for me" are avoided.

2   Specific comments

Here are some additional comments on the original papers and the resulting discussion

2.1   Cleveland and Loader

These authors are to be congratulated: some excellent contributions are made here.

My favorite among these is the fascinating historical discussion in Section 2. The level of scholarly work done in this section, with detailed attention to early references of which I (and surely many others) had been unaware, is truly outstanding. This is an important contribution that deserves to be frequently referenced.

Another important contribution is the interesting and convincing data analysis. This well written section makes it clear that the methods advocated by the authors can be effective tools for data analysis. It shows that in the hands of an expert, who has a lot of time to devote to the problem, methods that I personally find less intuitive and easy to work with, such as those based on nearest neighbors, and higher degree local polynomials, can give good results. However, I have not yet been convinced of how these tools will work in the hands of a user with less expertise, and less time to devote to the project.

Section 10.1 is also very useful. It is a simple and elegant mathematical way of seeing why local polynomials are better than local constants.

I wish to take issue with some of the other aspects of the paper.

Fix and Hodges (1951) (a preprint that was finally published as Fix and Hodges (1989)!), and Akaike (1954) predate the famous work of Rosenblatt and Parzen as the origins of the "modern view of density estimation" discussed at the beginning of Section 2.2.

The remarks in section 5 are generally correct, but other preferences for choice of kernel function are reasonable as well. My personal preference is for the Gaussian, because it gives very smooth pictures (smoother than piecewise polynomials with three derivatives, such as the tricube, as noted in Marron and Nolan (1989)). It also has some useful "continuity in smoothness properties" when the bandwidth changes. For example this seems to be the only kernel where the number of bumps in the smooth is a monotone function of the bandwidth, see Silverman (1981), and the kernel with the nicest "mode tree" which provides another way of understanding how smooth the Gaussian kernel is, see Minnotte and Scott (1993). Cleveland and Loader's objection to the Gaussian on speed grounds is not important when a fast "binned implementation" is used, for example as described in Fan and Marron (1994).

The examples in Section 8.2 were interesting and informative, but not convincing. The main reason is doubts about Mean Squared Error (and its many variants) as an error criterion. As noted in Marron and Tsybakov (1995), such error criteria too often "feel aspects of the problem" that are quite different from those that are relevant for data analysis. In my view it is more sensible to assess smoothers from the viewpoint of "how well do they convey the important lesson in the data?", than from the viewpoint of "how close to the true curve is the estimate in the sense of MSE?". From this viewpoint, the very interpretable global bandwidth smoother is more attractive than suggested here. Another important point to keep in mind is that this type of presentation is necessarily anecdotal in nature, with lessons depending on the example chosen. Examples are useful for making points, but in my view it would have been more useful to show examples where both were in turn better, thus illustrating the respective strengths and weaknesses of each method. Cleveland and Loader's examples together with those of Sheather, Wand, Smith and Kohn make a reasonable collection of this type.

The last paragraph of Section 8.2 doesn't say much, other than the obvious "don't take asymptotics too literally". It appears intended to suggest that asymptotics are useless, but is not convincing. It carries no more weight than an assertion that simple linear regression is useless on the basis of an example with silly results from extrapolation far outside the range of the data.

Lejeune (1984, 1985) probably deserves explicit mention at the beginning of section 10.2.

The issue of when bias approximations work and when they don't, discussed in the ninth paragraph of Section 10.2 is certainly an interesting and important one. However the suggestion that mathematical statisticians are unaware of this is incorrect. For example, see Marron and Wand (1992) and Marron (1995b) for theoretical work and insights on this issue.

The last paragraph of section 10.2 is not a fair representation of how competent mathematical statisticians work and think. Such workers do not draw "sweeping conclusions", they report what they find. Mathematical results are of the form "if A, then B". Accusations of this type do not apply as long as A is made clear, and discussed in a relevant statistical light. I agree that the term "optimal" gets overused, and is often used for things which don't turn out as well as one might hope. However, when the term is used in a paper about asymptotics, it is a short form of "asymptotically optimal". In this form, the objections of Cleveland and Loader do not apply.

In my view, the most dissappointing part of the paper comes near the beginning of section 10.3, with the statement "much effort has been expended, ... with little regard for the possibility that the optimal bandwidth may be very inadequate". As noted in Section 3 of my original contribution, there has been quite a lot of work (much of it already published), on assessing how bandwidth selectors actually work (as opposed to merely their asymptotic properties) in terms of both simulation and real data examples.

The criticisms of plug-in bandwidth selection methods in the third and fourth paragraphs of Section 10.3 are well known, and have had an important effect on how the field has developed. It has been interesting to think about why such methods work as well as they do when they are actually used, in view of this type of consideration. Some personal thoughts on precisely this "paradox" are given in the appendix of my original contribution.

Cleveland and Loader's attitude on bandwidth selection is mystifying. Through most of the paper, they are pragmatic, with effective data analysis as their primary goal. However they diverge dramatically from this principle at this point in their discussion, and "sweeping conclusions are drawn solely from asymptotics, ... without any real justification". Also the asymptotics are not quite right, in that selection of a global bandwidth depends on things more like the integral $\int (f'')^2$ than like the second derivative $f''(x)$ itself. The asymptotics are different for estimation of the integral (e.g. estimators with rate of convergence $n^{-1/2}$ are available under strong enough assumptions for the former, but not the latter). But this is a minor issue. The important point is that there is a convincing case for the effectiveness of modern bandwidth selectors . See Section 3 of my original contribution for references.

I agree with Cleveland and Loader's position in Section 10.4, and have also not been convinced by the suggestions that odd order local polynomials are "better". However, I do not agree with the suggestion that this is an example where asymptotics give the "wrong answer".

2.2 Gasser and Seifert

This is a good exposition of how local polynomial methods work, including important numerical issues. The analysis of data sparsity is deep and interesting. The probability of having three points in some neighborhood, displayed in Figure 12, turns out to be important in the asymptotic analysis of ridge regression done by Hall and Marron (1995).

2.3 Hall and Turlach

The alternative approach to data sparsity looks interesting. I agree with the addition of the factor F7, for the choice of the smoothing parameter, although my F1 - F6 were intended for the choice of smoother, e.g. choice among M1 - M6.

2.4 Jones

I like the point about many good methods being rather similar in many cases. Much of the controversy has been about only some of the cases that one encounters. E.g. most smoothers work quite well (and similarly) in the important and simple case of a fixed equally spaced design.

2.5 Kohn, Sheather, Smith and Wand

This discussion provides some thought provoking examples. I also liked the additional reasons given as to why modern bandwidth selection methods should be considered for implementation in software packages.

The point of Figure 4 is a thought provoking one.

2.6 Thomas-Agnan

I agree with the addition of F8 - F12 to the list of factors that go into the choice of a smoother. I am not so sure about adding M7 and M8 to the list of methods, on the grounds that I view M7 as essentially the same as M4, and M8 as a special case of M1. However, this is personal, and there are certainly other ways of classifying estimators.

Thomas-Agnan raises an interesting point about effective kernels and my factor F2. I agree that one can make a plot of the *large number of different* effective kernels resulting from say a nearest neighbor estimator (or many others as noted). But for what I call "less interpretable methods", there will be different effective kernels for each point where the estimate is constructed. Because it is hard to keep all these different kernel functions in my mind, (note the contrast to the *single* effective kernel for a fixed bandwidth method), it is hard to understand what the estimator tells about the data. However, this whole discussion underlines one of my main points: there are many different viewpoints on smoothing. Factors that are important to some (e.g. F2 is quite important to me), are less important to others.

The points on comparability are well taken. Comparisons with analogs in numerical analysis need to be interpreted with caution for two reasons. First they tend to deny the presence of variance, which is crucial in statistical smoothing. Second a "good smooth" is much harder to define in statistics. I am not prepared to accept "the smooth should be as close as possible to the true regression curve" as a criterion, because it is not always the same as the more important "the curve should tell us as much as possible about the data".

I am not sure that the importance of F1 will recede in time, because no software can make everything easily available. Choices have to be made about what is easy, and what takes more effort to use (e.g. menus can only be so long, and things that are on submenus are anticipated to get less use).

The "entropy" analogy is thought provoking.

3   Acknowledgement

The author is grateful to many people for input on these issues. N. I. Fisher, R. Kohn and M. P. Wand have made especially many helpful comments, and spent substantial time discussing these issues.

References

[1] Akaike, H. (1954) An approximation to the density function, *Annals of the Institute of Statistical Mathematics*, 6, 127-132.

[2] Cook, R. D. and Weisberg, S. (1994) *Regression Graphics*, Wiley, New York.

[3] Fix, E. and Hodges, J. L. (1951) Discriminatory analysis - nonparametric discrimination: consistency properties. *Report No. 4, Project No. 21-29-004*, USAF School of Aviation Medicine, Randolph Field, Texas.

[4] Fix, E. and Hodges, J. L. (1989) Discriminatory analysis - nonparametric discrimination: consistency properties, *International Statistical Review*, 57, 238-247.

[5] Fan. J. and Marron, J. S. (1994) Fast implementations of nonparametric curve estimators, *Journal of Computational and Graphical Statistics*, **3**, 35-56.

[6] Hall, P. and Marron, J. S. (1995) On the role of the ridge parameter in local linear smoothing, unpublished manuscript.

[7] Lejeune, M. (1984) Optimization in nonparametric regression, *Compstat 1984 (Proceedings in Computational Statistics)*, eds. T Havranek, Z. Sidak, and M. Novak, Physica Verlag, Vienna, 421-426.

[8] Lejeune, M. (1984) Estimation non-paramétrique par noyaux: régression polynomiale mobile, Revue de Statistiques Appliquées, 33, 43-67.

[9] Marron, J. S. (1995a) Presentation of smoothers: the family approach, unpublished manuscript.

[10] Marron, J. S. (1995b) Visual understanding of higher order kernels, to appear in *Journal of Computational and Graphical Statistics*.

[11] Marron, J. S. and Nolan, D. (1989) Canonical kernels for density estimation, *Statistics and Probability Letters*, 7, 195-199.

[12] Marron, J. S. and Udina, F. (1995) Interactive local bandwidth choice, unpublished manuscript.

[13] Marron, J. S. and Wand, M. P. (1992) Exact mean integrated squared error, *Annals of Statistics*, 20, 712-736.

[14] Minnotte, M.C. and Scott, D. W. (1993). The Mode Tree: A Tool for Visualization of Nonparametric Density Features, *Computational and Graphical Statistics*, 2, 51-68.

[15] Silverman, B. W. (1981) Using kernel estimates to investigate modality, *Journal of the Royal Statistical Society, Series B*, 43, 97-99.

Rejoinder

William S. Cleveland and Clive Loader

AT&T Bell Laboratories, 600 Mountain Avenue, Murray Hill, NJ 07974, USA

1 Sound Premises

Theoretical work in any area of statistics can have a substantial impact on the statistical methods that we use to analyze data in that area. But to do so, the premises on which the theory is based must be sound. The premises must sensibly model the sources of variation in the data. And the premises must address the methodology as it is used in practice, and set criteria that are of genuine importance for that usage. Having set the premises, the investigator must then derive results. This requires command of the necessary technical tools.

The theoretical work by actuaries in the early smoothing literature from the 1880s to the 1920s led to important advances such as local polynomial fitting. The strength of the work derived, in part, from elegant and insightful application of the technical tools, which were based on the algebra of finite difference operators. But far more important was an incisive setting of premises which came from a deep understanding of the behavior of the data under study, mortality and sickness data. These theoreticians were practitioners as well. They did far more than simply plug in data from some remote source to test an already developed tool. They started with the data. They shaped the premises of their theory from the data. *They were as responsible for the subject matter conclusions that emanated from their tools as they were for the scientific validity of the tools.* The consequence of this grounding in the data was the construction of the major pieces of intuitive insight that still guide smoothing today:

1. The trade-off between bias and variance.

2. The need for smooth weight functions to produce smooth fits.

3. Local polynomial fitting.

4. The poor performance of local constant fits compared with higher order fits.

5. Optimal weight functions.

6. Penalty functions and smoothing splines.

7. Smoothing in likelihood models.

The research community in smoothing today, or any other statistics research community, cannot expect to develop sound premises for theoretical work without a similar grounding in data.

2   Very Small Bandwidths

A major premise of the Seifert and Gasser paper is that we can learn about smoothers with bandwidths that are exceedingly small. In some cases the bandwidths get so small that the definitions of the smoothers are indeterminate without further rules about what to do with when there are no data or just a single point in a neighborhood. For example, in Figure 8, they investigate properties of smoothers by artificially generated data with a sample size of $n = 50$. The explanatory variable is uniform over $[0, 1]$. Their smoothing neighborhoods range from 0.04 to 0.4 so that the *expected* number of points in these neighborhoods ranges from 2 to 20. They point out that for local linear fitting with a neighborhood size of 0.04, the MISE "increases steeply to values as large as 629". With a neighborhood this small the results depend heavily on what one decides to do when 0 or 1 point appears inside the neighborhood.

These neighborhoods are far too small to give us an understanding of the relative performance of smoothers in practice. Except for studies where the data have little little or no noise and interpolation is the goal rather than smoothing, the fits with such small bandwidths would be far too noisy. A single attempt to use such small bandwidths on a real set of noisy data where we were responsible for the conclusions would make this obvious. Jones puts it well: "Some days I think the deep study (apparently, studies) of Seifert & Gasser of problems due to sparse data is very valuable, some days less so. Perhaps in practice it is something like local bandwidths and a degree of common sense that is called for."

A theoretical treatise is not wrong when its premises are at odds with the data and with the sensible practice of the methodology. It is simply uninformative.

3   Pseudodata

We find the Hall and Turlach suggestion of pseudodata interesting, although we reserve final judgment until we see their full account. There is a precedent for this idea, although in a much more limited form. For smoothers targeting the case of one explanatory variable, some, such as Tukey (1977), have faced the boundary problem by predicting the data beyond the boundaries and then smoothing the data and the predictions.

4   The Golden Standard: Past and Present

In a number of cases, there has been mention of local polynomial fitting as an emerging new discovery that holds great promise for the future, a coming golden standard. But what is emerging is awareness, not discovery. Local polynomial fitting, as we have emphasized in earlier papers, in our paper in this collection, and just now in this rejoinder, began in the actuarial research community and has been under development for over a century (e.g., Woolhouse 1880, Spencer 1904a, Henderson 1916, Macaulay 1931, Stone 1977, Cleveland 1979, Hastie and Loader 1993, Fan 1993.)

5   Plug-In Estimates

It makes little sense to use local quadratic or cubic estimates solely to fine-tune less efficient local constant and local linear methods. This statement is one consequence of our Section 10.3 and of the simulations of Sheather, Wand, Smith and Kohn in their Section 2.3. If we need local quadratic or local cubic fitting to adequately describe the characteristics of a regression curve or surface, then we should use the local quadratic or cubic smoother to fit the data. (We find ourselves somewhat surprised that it has fallen to us to make a statement that is nearly a tautology.)

Plug-in methods should be considered an idea that failed, and allowed to die a natural death. For those attempting life support, consider the following. The pilot estimates used to tune plug-in methods can beat the plug-in methods. Table 1 of SWSK confirms our point nicely; in seven out of eight examples local quadratic has won, despite the tuning of the amount of smoothing to the local linear estimate. (N. B. We do not suggest doing this in practice; we have simply pointed this out to increase our understanding of the issues.)

6   Local Constant Fitting

Our skepticism about local constant fitting (over all x) is simply a matter of not having found cases where it convincingly models the data more satisfactorily than higher-degree fitting. We can imagine cases where local constant fitting might do better, for example, when the underlying pattern is constant at all boundaries and locally linear elsewhere. But we believe that in practice, cases of better modeling by degree zero at all x are at best rare. We hasten to emphasize that this was the conclusion of the early actuarial research community discussed above; in their work, cubic fitting became a standard.

Of course, if we use adaptive methods, then it is possible that at some special values of x where the surface is flat, we might well chose degree zero locally. But we do not expect to choose it at all x because that would imply an uninteresting selection of the explanatory variables.

Marron seems intent on holding on to local constant fitting. He writes: "Fixed bandwidth local constant kernel methods put the interested analyst in closest possible intuitive contact with the data, because they are simple, understandable, local averages. Note that I am not advocating this estimator as the solution to all problems ... but instead am merely pointing out it cannot be dismissed out of hand."

To the contrary, we do have every right to dismiss it out of hand until someone, perhaps Marron himself, provides data sets where local-constant smoothing with fixed bandwidths at all x does a better job of modeling the data than higher order fits. We have demonstrated the reverse, in our paper in this collection, and elsewhere.

And we disagree with the claim of greater intuition for local constant fits. It is no more acute than for higher order fits. In practice, our intuition about how well a smoother models the data is most acute when it is based on diagnostic and on our understanding of the broad performance characteristics of the smoother such as the class of functions it reproduces, the frequencies that is passes, and the frequencies that it suppresses.

Thomas-Agnan also disagrees with Marron's statement about intuition, pointing out that it overrates closed-form formulas and that the popular linear smoothers end up as linear combinations of the y_i. Clearly we agree with the discussant.

7 Residual Plots

Sheather, Wand, Smith, and Kohn have raised questions about the validity of smoothing residual plots as an aid to judging the performance of a smoother. They also implicate Cleveland (1993), but they should also implicate Tukey (1977), who introduced the systematic smoothing of residuals.

Unfortunately, the discussants have missed an important property of smoothers, and they have misinterpreted the process that is used to judge a fit from a smooth of the residuals.

They state: "The same local polynomial fit is not appropriate for both the original data and the residuals because if this fit allowed structure to go undetected in the original data, then it is very unlikely to capture structure in the residuals." But this is not so because smoother operators, unlike least-squares operators, are not idempotent. In fact, the use of the same smoother on the residuals that was used to produce the fit is given the name *twicing* by Tukey (1977).

The process is not to judge a fit to be adequate if a smooth curve on its residual plot is flat. A flat curve means simply that no systematic, reproducible lack of fit has been detected. The fit may well be too noisy. As Cleveland (1994) points out: "This method of graphing and smoothing residuals is a one-sided test; it can show us when [the smoothing parameter] is too large but sets off no alarm when [the smoothing parameter] is too small. One way to keep [the smoothing parameter] from being too small is to increase it to the point where the residual graph just begins to show a pattern and then use a slightly smaller value ... .

In Figure 1 of the discussion of Sheather, Wand, Smith and Kohn, one sees lack of fit only marginally at the largest bandwidth, and so one would usually select a parameter close to 0.675. A very similar pattern is shown in the residual plots of our paper — especially Figure 8 — and we chose the larger bandwidths.

One always looks at residual plots in conjunction with looking at plots of the fit. When a prominent feature in a residual plot corresponds to a feature in the fit that has a rapidly changing derivative, lack of fit is typically the cause. For example, in Figure 11 of our paper, the large residuals are seen to line up with the breaks in the fitted curve, indicating lack of fit rather than a heavy-tailed residual distribution.

8 What is Loess ?

We do not understand Marron's M-classification of smoothers. Lowess and loess are local polynomial methods and therefore belong to category M1.

9 Fixed vs. Nearest-Neighbor Bandwidth Selection

We have argued for the value of nearest-neighbor bandwidth selection as a reasonable default method.

In providing smoothing tools to data analysts it makes sense to have available, perhaps along with a reliable adaptive method, a simple bandwidth selection procedure that is based only on the x_i and that has one easy-to-understand parameter. Both fixed and nearest-neighbor selection would provide this. But of the two, we have found that nearest-neighbor typically does as well or better than fixed. Fixed selection can result in dramatic swings in variance; the greatest drama occurs when there are no data in the interior of a fixed neighborhood. Deciding how to protect users from a fit that is not well defined at some places where an evaluation is requested is a thorny problem, one that does not need solving for nearest-neighbor selection.

But this defense of nearest-neighbor selection is not meant to imply that fixed-bandwidth selection can never perform better than nearest-neighbor. In their Section 2.2, Sheather, Wand, Smith and Kohn provide one example, an artificially generated one, where fixed does better. Seifert and Gasser do the same in their Section 2.2, again with artificially data. Both examples have the same phenomenon that makes fixed selection do better. The curvature is greatest where the data are the sparsest. But this should not be construed as a statement that fixed selection in some sense reacts to changes in curvature in any way. Neither fixed nor nearest-neighbor do so. Note that in the example of Sheather, Wand, Smith and Kohn, if we alter the example and take X_i to be $U_i^{0.7}$ instead of $1 - U_i^{0.7}$, then nearest neighbor will perform better than fixed. But the curvature is the same in both cases. To react to changing curvature we need adaptive methods such as those we discuss in Section 9 of our paper.

There are claims that fixed selection is optimal and that nearest-neighbor is not. These claims result from an asymptotic approximation to the nearest neighbor bandwidth; In Section 8.2 of our paper we showed this approximation does not work. A claim of optimality is a strong assertion because a single a single counterexample suffices to disprove it. In the papers in this collection, and in the discussions, there are examples where fixed is best, examples where nearest neighbor is best, and examples where both are inadequate.

In summary, we use nearest neighbor methods as a default in software for smoothing because (1) there is direct control over the number of points being smoothed, thus avoiding problems that arise from windows with a very small number of observations; and (2) it usually results in larger bandwidths at boundary regions, which is often desirable due to the one-sided nature of smoothing at boundaries compounded by the tendency for data to be sparse at the boundaries.

10   Comparing Smoothers

Seifert and Gasser compare the performance of smoothers at the same values of the bandwidth. This is inappropriate because the degrees of freedom of two different smoothers can be radically different for the same bandwidth values. In fact, by a rescaling of the kernel of a smoother, we can effect a large change in its performance relative to others if we match bandwidth. For example, the performance of the smoothers in Figure 8 of Seifert and Gasser is almost wholly due to differing degrees of freedom. They scale the local linear fitting with Gauss weights to "reach the same asymptotically optimal bandwidth as Epanechnikov weights." Because the comparison is carried out as a function of h, the behavior of the this estimate could be made to look quite different by scaling by some other method. Clearly, in a study such as this we need to compare smoothers by matching degrees of freedom.

But do we fail to practice what we preach? Jones asks: "wouldn't it be better in comparative figures such as Figures 2, 3, 7 and 8 to have in each column smooths with equal equivalent degrees-of-freedom rather than equal, but not comparable, bandwidths?" The answer is "no", since our aim was not to make a blanket statement as to which smoother was best for the problem, but rather model selection, for which we want to consider a range of models with differing amounts of smoothing. If we fit all degrees with equal degrees of freedom, one would often be choosing between all undersmoothed or all oversmoothed fits, and the mixing would not work.

11   Software

We agree with the view expressed by Sheather, Wand, Smith and Kohn and some other papers and discussants — most notably Marron — that software should incorporate automated bandwidth selectors. But we strongly disagree with the views as to what, or how, to implement. No existing bandwidth selectors come close to being suitable for use as a default in general purpose software such as loess, either in terms of general applicability or reliability of performance. An adaptive method such as that used in Figure 9 is appropriate when there is plenty of data with low noise; it would be quite inappropriate for smoothing residual plots.

As with any sensible model choice criterion, any bandwidth selector will fit several models, and attempt to decide which will be best. As such, a bandwidth selector is not part of a basic smoothing algorithm, but something that can be built on top. The bandwidth selection should not be confounded as part of one's basic smoothing algorithm as Sheather, Wand, Smith and Kohn suggest. Rather, the basic smoothing algorithm should return sufficient quantities of diagnostic information to assess the fit. The user can then decide whether to automatically minimize a criterion, or to actually look at the fit using diagnostics as a guide.

Consider for example the C_p type statistics used by Cleveland and Devlin (1988) as part of loess. The use of C_p is not forced upon users. Rather, the loess method returns sufficient diagnostic information about each fit to enable the C_p statistics to be readily computed. Locfit software (Loader, 1995) is a recent development built on the loess model, and provides a wealth of additional diagnostic information that enables many different bandwidth selectors with short S scripts. The basic point here is a software design issue: different computations, such as bandwidth selection and the underlying local fitting, need to be kept separate.

12 Testing

Sheather, Wand, Smith and Kohn as well as Jones raise the issue of F-tests and t-intervals being contaminated by bias. But this issue arises in parametric fitting where these methods of inference are used routinely. In both parametric fitting and smoothing there must be at least one fit for which we have a reasonable assurance that no bias is present. For t-intervals we need this fit to estimate the variation in the errors. Bias in t-intervals has similarities to other bias problems and is discussed in Sun and Loader (1994) and Loader (1993). For an F-test we need such a fit to provide an alternative model. The null model will have larger bandwidth; if the F-test reveals substantial lack of fit, then clearly the null model is biased and inadequate.

13 Computation

Sheather, Wand, Smith and Kohn criticize our adaptive procedure on the grounds it is "computationally difficult to locally estimate such bandwidths at each point in the design." Their statement is correct, but this is not what we do. Rather, the bandwidth is computed at an adaptive sequence of knots, with the greatest knot density in regions of lowest bandwidth. For example, in the wavelet example of Figure 11 of our paper, the bandwidth is computed at 130 points, far less than the 2048 design points. Moreover, most of these knots are in regions where small bandwidths are used, and the evaluation is relatively cheap. Jones also seems to misinterpret our procedure; the local $C(h)$ statistics are computed over a wide range of bandwidths.

New References

Cleveland, W. S. (1994) *The Elements of Graphing Data.* Hobart Press, books@hobart.com.

Loader, C. R. (1993). Nonparametric regression, confidence bands and bias correction. *Proc. 25th Symposium on the Interface*, 131-136.

Loader, C. R. (1995). LOCFIT: A program for local fitting.
http://netlib.att.com/netlib/att/stat/prog/lfhome/home.html

Silverman, B. W. (1985). Some aspects of the spline smoothing approach to nonparametric regression curve fitting (with discussion). *Journal of the Royal Statistical Society, B*, **47**, 1-52.

Stoker, T. M. (1994). Smoothing bias in density derivative estimation. *Journal of the American Statistical Association* **88**, 855-871.

Sun, J. and Loader, C. R. (1994). Simultaneous confidence bands in linear regression and smoothing. *Ann. Statist.* **22**, 1328-1345.

Tukey, J. W. (1977). *Exploratory Data Analysis.* Addison-Wesley, Reading, Massachusetts.

Rejoinder

Burkhardt Seifert & Theo Gasser

Abteilung Biostatistik, ISPM, Universität Zürich,
Sumatrastrasse 30, CH-8006 Zürich

Let us first thank all contributors of this discussion for the stimulating comments concerning various aspects of nonparametric regression estimation. We indeed appreciate the open scientific atmosphere seen in all contributions. Some major topics have emerged from papers and discussions, and for a number of problems a consensus is in reach, within some margin of tolerance.

Choice of estimator (or: weighting scheme)

The prevalent opinion is that most methods considered are rather close together in practical performance, and we fully agree. Differences are negligible for equally spaced design, and quite marked deviations from this design are necessary to make differences appreciable. This point has been well made by Jones. Thus, there is no need to try to persuade adherents of smoothing splines, local polynomials or convolution kernel estimators to use another method (the asymptotic loss in efficiency of convolution kernel estimators for random design can be eliminated by an idea of Chu & Marron (1991), which was further explored by Herrmann (1994b)). What is needed is good, flexible software which should have an option to estimate the MISE–optimal bandwidth. We support the opinion of Thomas–Agnan, that the available versions often are rather poor. Our Fortran routines for local polynomials — still without automatic bandwidth choice — are available by anonymous ftp (ftp biostat1.unizh.ch, cd pub) and www (http://www.unizh.ch/biostat). We expect to offer kernel estimators with bandwidth choice soon, including also derivative estimation.

We like the discussion of asymptotic versus finite sample analysis. The observation, that the outstanding asymptotic properties of local polynomials do not hold for small and medium sample sizes, was the main reason for us to analyze finite sample properties and to propose modifications (Seifert & Gasser 1996). On the other hand, convergence to asymptotic performance

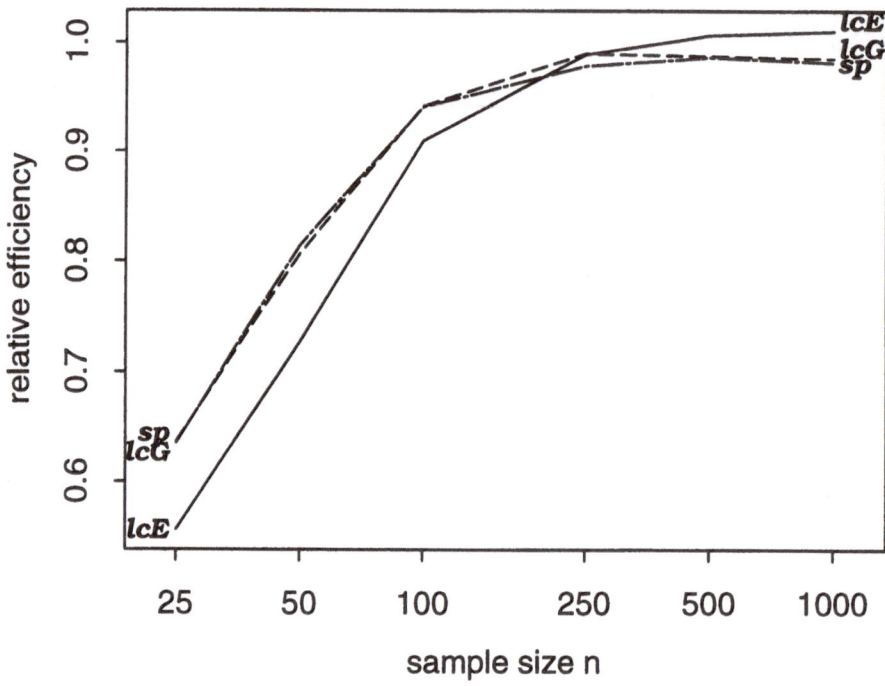

Figure 1: *Efficiency depending on sample size n (on logarithmic scale) of estimators of r in (33) at conditional MISE-optimal bandwidths for random uniform design and $\sigma^2 = 0.5$. Solid line is local cubic fit with Epanechnikov weights (lcE), dashes are local cubic fit with Gaussian weights (lcG), and dash-dots are smoothing splines (sp).*

is faster for the Gasser–Müller kernel estimator and for smoothing splines. Thus, we get the paradoxical situation, that the Gasser–Müller kernel estimator may behave better than the local linear estimator for small sample size even for random design, where it has an asymptotic efficiency of only 0.72. It should also be noted that convergence to asymptotic performance is even slower for higher order polynomials or kernels (compare also Marron & Wand 1992). As a consequence, the distrust expressed by Cleveland & Loader in the validity of an asymptotic analysis is much more justified for local polynomials compared to smoothing splines or Gasser–Müller kernel estimators.

Thomas–Agnan raised the question of finite sample behavior of smoothing splines. As pointed out in our paper, smoothing splines are related to kernel estimators of order 4 and local cubic polynomial estimators. Figure 1 shows the asymptotic behavior of several estimators for a random uniform design. For this case, the asymptotic, design–adaptive bandwidth in (22) of the equivalent kernel estimator is just a fixed one. To avoid bound-

ary effects, additional observations were generated outside. The behavior
was studied at the conditional MISE–optimal bandwidths. Efficiency here
is considered relative to the asymptotically optimal MISE for local cubic estimators. Interestingly, the behavior of smoothing splines and local cubic
estimators with Gaussian weights was nearly the same for all sample sizes.
To see whether the slightly worse MISE performance of smoothing splines
could be attributed to the asymptotic non–optimality of the induced kernel,
we computed asymptotic efficiency for some kernels. Following e.g. Fan et al.
(1993), the equivalent kernel K^* of a local cubic estimator with symmetric
weight function K is

$$K^*(t) = \frac{s_4 - s_2\,t^2}{s_4 - s_2^2}\,K(t)\,, \tag{R.1}$$

where $s_j = \int t^j K(t)\,\mathrm{d}t$. The asymptotic relative efficiency of two estimators
then is

$$\mathrm{eff}(1/2) = \left(\frac{\int (K_2^*)^2}{\int (K_1^*)^2}\right)^{8/9} \left|\frac{\int t^4 K_2^*}{\int t^4 K_1^*}\right|^{2/9}.$$

We get the following efficiencies relative to the asymptotically optimal
Epanechnikov weights (lc is local cubic, weights are $K(t) \propto (1 - t^2)_+^j$, $j =
0, \ldots, 3$, and tricube $K(t) \propto (1 - |t|^3)_+^3$):

estimator	s_2	s_4	$\int (K^*)^2$	$\int t^4 K^*$	efficiency
lc, uniform	0.3333	0.2	1.125	-0.08571	0.964
lc, Epanechnikov	0.2	0.08571	1.25	-0.04762	1
lc, biweight	0.1429	0.04762	1.4073	-0.03030	0.995
lc, triweight	0.1111	0.03030	1.5549	-0.02098	0.988
lc, tricube	0.1440	0.04545	1.4252	-0.02848	0.998
lc, Gaussian	1	3	0.4760	-3	0.939
smoothing spline			0.2652	-24	0.995

Thus, the non–optimal kernel is not the reason for the behavior of smoothing
splines in our analysis. One reason of the slight inefficiency might be the
regression function, which is not well approximated by cubic splines.

Choice of polynomial order

To start with, we would like to propose to avoid usage of the term "kernel
estimator" both for the Nadaraya–Watson and the Gasser–Müller estimator,
since they are fundamentally different. This habit could lead to misunder-
standings for non–specialists in the statistical community. The Nadaraya–
Watson estimator corresponds to local constant fitting which is undesirable,
and we agree in this respect fully with Cleveland & Loader. A Gasser–Müller
estimator based on one of the common positive and symmetric kernels corresponds closely to a local linear fit in case of fixed design, and grossly in case
of random design.

The question is now, whether even order fits should be used, given their additional large sample bias term. Sheather, Wand, Smith & Kohn discussed the behavior of a local quadratic fit versus a local linear one. As they show, a local quadratic estimator can gain relative to the local linear one. Of interest also is the comparison between local quadratic and local cubic fits. Both estimators asymptotically have the same variance, whereas the local quadratic fit has an additional bias, depending on the design. Note, that Sheather et al. used a uniform design, where this additional bias term disappears. We would like to contribute to this discussion with a small simulation study similar to that by Sheather et al., using their regression function f_1. Instead of mean square errors, we used the integrated square error (ISE), which also measures behavior in sparse regions of the design. We used the true ISE–optimal bandwidth to see the potential benefits inherent in the different methods. But then, to our surprise, among others, we obtained the following result for the comparison of local quadratic and local cubic fits and vice versa (ll, lq, and lc are local linear, quadratic, and cubic respectively):

Function	Sample Size	σ	Mean RE(lc/lq)	RE(lq/lc)
f_1	100	0.93	1.23	1.08

It should be noted that the distribution of ratios of ISEs is rather skew, and thus the mean always favors the estimator in the denominator. We therefore, present results for medians of these ratios, for f_1, $\sigma = 0.93$, and for uniform and a skew design:

n	Design	ISE(ll)	ISE(lq)	ISE(lc)	RE(ll/lq)	RE(lq/lc)
100	$X_i = U_i$	0.0763	0.0557	0.0541	1.35	1.02
100	$X_i = U_i^{0.7}$	0.0900	0.0743	0.0681	1.36	1.02
1000	$X_i = U_i$	0.0120	0.0072	0.0057	1.60	1.17
1000	$X_i = U_i^{0.7}$	0.0149	0.0098	0.0081	1.52	1.14

In this example, local quadratic and local cubic are superior to local linear estimators, whereas local quadratic and local cubic fits are similar for the small sample size, and local cubic superior for the large sample size. Thus, we concur with Sheather et al., that one might gain by using higher order polynomials, but there is no reason to use local quadratic instead of local cubic as default. This observation is supported by figure 1 of Cleveland & Loader, where the C–V score is plotted versus mixing degree. While often a degree 2 might be appropriate, there is never a local minimum there. We also support the observation by Cleveland & Loader, that the bias of local quadratic estimators usually is not so dramatic as expected from asymptotic theory. In fact, because of an increased variability of local cubics, the finite sample relative efficiency is here not influenced for the skew design.

It should be noted in this context, that for equidistant design — using symmetric weight functions K and estimating at the design points — local polynomials of even order coincide with those of the next higher order. For symmetric weights w_i, $\sum i^{(2k+1)} w_i = 0$ holds for all k. Thus, Spencer's fifteen and 21 point rules as well as Henderson's Ideal formula exactly reproduce cubics, not only quadratics, as mentioned by Cleveland & Loader.

Boundary

It is an advantage of local polynomials that they can automatically adapt to the boundary, without the use of boundary kernels as for the Gasser–Müller estimator (Gasser et al. 1985). However, this holds only for odd order polynomials, and this fact is a further argument to be cautious in the use of even orders, in particular the Nadaraya–Watson estimator. This automatic boundary adjustment does not imply, that the performance at the boundary is superior to other estimators with appropriate boundary treatment. We are grateful to Cleveland & Loader (section 10.4) for pointing out this fact: "However, this observation alone is not an argument in favor of odd order fitting, and does not imply odd orders have no boundary effects. Variance has so far been ignored: local polynomial fitting is much more variable at boundary regions ... What we can conclude is that boundary regions dominate the comparison." When comparing the performance at the boundary for odd order local polynomials with our boundary kernels (as implemented), the results were quite comparable. Let us add that problems with variance at the boundary inevitably increase with increasing polynomial or kernel order, and in particular for derivatives.

Modifications for local polynomials

We are glad that our suggestion that there is room for improving the behavior of local polynomials (Seifert & Gasser 1996) found such a positive echo. While Cleveland & Loader also considered ridging a good modification, Hall & Turlach propose a new small sample modification of local polynomials. Their method is effective and very easy to implement. When starting work on modifying local polynomials to better adapt to sparse regions in 1992 we also considered interpolation. For us, the main reason for not pursuing the idea was that problems might arise later on when estimating the bandwidth from the data. It is of interest to study interpolation in more detail and relate it to our proposals: From a mathematical point of view, their method is a local increase of the bandwidth, combined with a modification of the weight function. Thus, the method is related to our variance–bias compromise (Seifert & Gasser 1996). But while our method was restricted to a symmetric increase of bandwidth, Hall & Turlach's method allows an asymmetric increase.

To see how the method works, consider a linear polynomial interpolation rule of order p. Given $p + 1$ distinct points $(X_{i_1}, \ldots, X_{i_{p+1}})$ and values $(Y_{i_1}, \ldots, Y_{i_{p+1}})$, say, there exists a unique polynomial of order p connecting these points. The aim of an interpolation rule is to find the value $\tilde{y} = (Y_{i_1}, \ldots, Y_{i_{p+1}}) w$ of this polynomial at $\tilde{x} = (X_{i_1}, \ldots, X_{i_{p+1}}) v$, given v. It can be seen that w is the unique solution of the equation

$$(1, (\tilde{x} - x_0), \ldots, (\tilde{x} - x_0)^p) = w' \begin{pmatrix} 1 & (X_{i_1} - x_0) & \ldots & (X_{i_1} - x_0)^p \\ \vdots & \vdots & & \vdots \\ 1 & (X_{i_{p+1}} - x_0) & \ldots & (X_{i_{p+1}} - x_0)^p \end{pmatrix}.$$

Straightforward calculation shows that a local polynomial estimator, based on the original data including m interpolated data points $\tilde{x}_j$, $j = 1, \ldots, m$, is a local polynomial, based only on the original data, but with modified weights

$$\widetilde{W} = \text{diag}(K\left(\frac{X_i - x_0}{h}\right)) + \sum_{j=1}^{m} K\left(\frac{\tilde{x}_j - x_0}{h}\right) w_j w_j',$$

where the vectors w_j are blown up with zeros outside $(i_1, \ldots, i_{p+1})$. This formula allows exact finite sample computation of variance and bias as for the unmodified estimator. The formulae above do not use that $\tilde{x}$ is in the range of $(X_{i_1}, \ldots, X_{i_{p+1}})$. Thus, the results apply to extrapolation as well as to interpolation. The choice of a nondiagonal weight is new in local polynomial estimation, and we expect further discussion of this topic.

Short remarks

As a side remark we might point out that the nonparametric residual variance estimator used by Cleveland & Loader is a special case of an estimator defined in Gasser et al. (1986) for one dimension and in Herrmann et al. (1994) for two dimensions. An alternative variance estimator has been derived by Hall et al. (1990).

Besides smoothing data in a given region, as pointed out by Sheather et al., detection of empty regions — and also of edges, outliers, dependencies, heteroscedasticity, etc. — have to be considered.

Additional references

Gasser, Th., Müller, H.-G. & Mammitzsch, V. (1985). Kernels for nonparametric curve estimation. *J. Roy. Statist. Soc. Ser. B* **47**, 238–252.

Gasser, Th., Sroka, L. & Jennen–Steinmetz, Ch. (1986). Residual variance and residual pattern in nonlinear regression. *Biometrika* **73**, 625–633.

Hall, P., Kay, J. W. & Titterington, D. M. (1990). Asymptotically optimal difference–based estimation of variance in nonparametric regression. *Biometrika* **77**, 521–528.

Herrmann, E., Wand, M. P., Engel, J. & Gasser, Th. (1994). A bandwidth selector for bivariate kernel regression. *J. Roy. Statist. Soc. Ser. B* **57**, 171–180.

Marron, J. S. & Wand, M. P. (1992). Exact mean integrated squared error. *Ann. Statist.* **20**, 712–736.

Robust Bayesian Nonparametric Regression

C. K. Carter and R. Kohn

Australian Graduate School of Management, UNSW,
Kensington, NSW 2033, Australia

Summary

We discuss a Bayesian approach to nonparametric regression which is robust against outliers and discontinuities in the underlying function. Our approach uses Markov chain Monte Carlo methods to perform a Bayesian analysis of conditionally Gaussian state space models. In these models, the observation and state transition errors are assumed to be mixtures of normals, so the model is Gaussian conditionally on the mixture indicator variables. We present several examples of conditionally Gaussian state space models, and, for each example, we discuss several possible Markov chain Monte Carlo sampling schemes. We show empirically that our approach (i) provides a good estimate of the smooth part of the regression curve; (ii) discriminates between real and spurious jumps; and (iii) allows for outliers in the observation errors. We also show empirically that our sampling schemes converge rapidly to the posterior distribution.

Keywords: Gaussian mixture; Kalman filter, Markov chain Monte Carlo, spline smoothing, state space model.

1 Introduction

In the nonparametric regression literature, the case where the regression function is smooth throughout its range has been studied extensively; see, for example, Hastie and Tibshirani (1990) and the references therein. McDonald and Owen (1986) point out that there are a number of applications where the regression function is smooth except for a small number of discontinuities in the function or its derivatives, with the discontinuity points often unknown. Examples include Sweazy's kinked demand curve in microeconomics (Lipsey, Sparks and Steiner 1976), the transition of air resistance from a quadratic function to a linear function at high velocities (Marion 1970, Section 2.4), and the smoothing of sea surface temperature data where the discontinuities arise from changes in ocean currents (McDonald and Owen 1986). Nonparametric regression methods for estimating such functions should (i) provide a good estimate of the smooth part of the regression curve; (ii) provide some means of discriminating between real and spurious jumps; and (iii) allow for outliers in the observation errors and not confound outliers in the data with jumps in the regression function. McDonald and Owen (1986) and Müller (1992) propose nonparametric (and non Bayesian) methods for fitting curves with discontinuities which to some extent satisfy the first goal, but not the other two. We believe that the Bayesian approach discussed in this paper satisfies all three goals.

Our approach uses Markov chain Monte Carlo methods to perform a Bayesian analysis of conditionally Gaussian state space models. The use of Gaussian state space models for nonparametric regression using spline smoothing is well known; see, for example, Wecker and Ansley (1983) and the references therein. In this approach, the smoothing parameter is estimated either by generalised cross-validation or by marginal likelihood and the regression function is estimated using the Kalman filter and a state space smoothing algorithm. However, it seems computationally difficult to extend the approach in Wecker and Ansley (1983) to allow for outliers in the observations or discontinuities in the regression function.

Recent developments in Markov chain Monte Carlo methods have made it possible to perform a Bayesian analysis of conditionally Gaussian state space models; see, for example, Carter and Kohn (1994a and b) and Shephard (1994). In these models, the observation and state transition errors are assumed to be mixtures of normals, so the model is Gaussian conditionally on the mixture indicator variables. In this paper, we present several examples of the use of conditionally Gaussian state space models for robust nonparametric regression. For each example, we discuss the possible Markov chain Monte Carlo sampling schemes and show empirically that there exist sampling schemes which converge rapidly to the posterior distribution.

The paper has two aims. The first is to acquaint the reader with the Bayesian approach to spline smoothing and its implementation by Markov chain Monte Carlo. The second aim is to show that sampling schemes to carry out Markov chain Monte Carlo can have very different rates of convergence. The best schemes converge rapidly, whilst others can exhibit unacceptably slow convergence. Section 2 presents a Markov chain Monte Carlo approach to Bayesian spline smoothing. Section 3 uses mixture of normal errors to make spline smoothing robust against outliers and discontinuities in the underlying function. Section 4 contains our conclusions.

2 Bayesian spline smoothing

In this section we discuss the Gaussian state space model for spline smoothing given in Kohn and Ansley (1987). We put priors on the scale factors of the state space model to obtain a Bayesian model. We compare the performance of this Bayesian model with marginal likelihood — a commonly used method for fitting splines that has been shown to give good performance; see Kohn, Ansley and Tharm (1991) and the references therein. We then discuss several Markov chain Monte Carlo methods to generate from the posterior distribution of the Bayesian model and show that there can be large differences in performance between the methods.

2.1 Description of the state space model

We consider the following signal plus noise observations

$$y(i) = f(t_i) + e(i) \qquad i = 1, \ldots, n \qquad (2.1)$$

where the errors $e(i), i = 1, \ldots, n$, are independent and identically distributed $N(0, \sigma^2)$. We assume that the signal $f(t)$ is a smooth function, and choose the prior for the signal to be the solution to the stochastic differential equation

$$\frac{d^m f(t)}{dt^m} = \tau \frac{dW(t)}{dt}, \qquad (2.2)$$

where $W(t)$ is a Wiener process with $W(0) = 0$ and $\mathrm{var}\{W(t)\} = t$. We choose a diffuse prior for the initial conditions at $t = 0$ in (2.2), that is,

$$\{f(0), \ldots, f^{(m-1)}(0)\}' \sim N(0, cI_m) \qquad \text{with} \qquad c \to \infty. \qquad (2.3)$$

The motivation for the priors given in (2.2) and (2.3) comes from Wahba (1978) who showed that if $\lambda = \tau^2/\sigma^2$ then $\lim_{c \to \infty} E\{f(t)|Y, \lambda, c\}$ is the mth order spline smoothing estimate of $f(t)$. We assume, without loss of generality, that $0 \leq t_1 \leq \ldots \leq t_n$ and we let $\delta_i = t_i - t_{i-1}$ with $t_0 = 0$. Then, as in Kohn and Ansley (1987), we can write (2.1) and (2.2) as the linear state space model

$$y(i) = h'x(i) + e(i) \qquad (2.4)$$
$$x(i) = F(i)x(i-1) + u(i), \qquad (2.5)$$

where $x(i) = \{f(t_i), \ldots, f^{(m-1)}(t_i)\}'$, $h = (1, 0, \ldots, 0)'$,

$$F(i)_{jk} = \begin{cases} \delta_i^{k-j}/(k-j)! & \text{if } k \geq j \\ 0 & \text{otherwise,} \end{cases}$$

$u(i) \sim N\{0, \tau^2 U(i)\}$ with $U(i)_{jk} = \delta_i^{2m-j-k+1}/\{(m-j)!(m-k)!(2m-j-k+1)\}$, and, from (2.3), the initial state $x(0)$ is diffuse. We complete the Bayesian specification of the model by imposing the improper priors $p(\sigma^2) \propto 1/\sigma^2 \exp(-\beta_\sigma/\sigma^2)$, with β_σ small, and $p(\tau^2) \propto 1$. In the simulations reported below we take $\beta_\sigma = 10^{-10}$. The reason we impose a flat prior for τ^2 rather than the prior $p(\tau^2) \propto 1/\tau^2$, is that the latter prior results in an improper posterior distribution. The local mode at $\tau^2 = 0$ has nothing to do with the data and is due solely to the prior. We denote the Bayesian model given by (2.4) and (2.5), with the above priors by M.2.1.

2.2 Comparison with marginal likelihood

In this section, we study empirically the differences between the Bayesian model M.2.1 given in Section 2.1 and marginal likelihood methods. Our results suggest that the two methods give almost identical results so that the empirical studies on the performance of marginal likelihood spline smoothing should also apply to Bayesian spline smoothing. We take $m = 3$ which gives quintic splines. By marginal likelihood methods we mean choosing σ^2 and τ^2 to maximize the normalized marginal likelihood $\bar{p}(Y|\sigma^2, \tau^2)$, where

$$\bar{p}(Y|\sigma^2, \tau^2) = \lim_{c \to \infty} c^{m/2} p(Y|\sigma^2, \tau^2, c).$$

The marginal likelihood estimate of $f(t)$ is then $E\{f(t)|Y, \sigma^2, \tau^2\}$. For details see Kohn and Ansley (1987) and the references therein.

We use data generated by (2.1) with the function

$$f_1(t) = \frac{1}{3}\beta_{10,5}(t) + \frac{1}{3}\beta_{7,7}(t) + \frac{1}{3}\beta_{5,10}(t), \tag{2.6}$$

where $\beta_{p,q}(\cdot)$ is a beta density with parameters p and q, and $0 \leq t \leq 1$. This function was used by Wahba (1983) in her simulations. The error standard deviation is $\sigma = 0.2$, the sample size is $n = 50$, and the design is equally spaced over $[0, 1]$. Figure 1 shows the function $f_1(t)$ and the generated data, together with the marginal likelihood, maximum posterior, and posterior mean estimates of $f_1(t)$. The posterior mean estimate is calculated using the Markov chain Monte Carlo sampling scheme 2.3 as described in the next section. From Figure 1 the various estimation methods give almost identical results.

Figure 2 shows the marginal likelihood $\bar{p}(Y|\sigma^2, \tau^2)$ and the marginal posterior $p(\sigma^2, \tau^2|Y)$ as functions of σ^2 and τ^2. Table 1 shows the marginal likelihood and maximum posterior estimates of σ^2 and τ^2. From Figure 2 and Table 1 we see that the choice of priors given in Section 2.1 has only a small affect on the posterior surface.

2.3 Markov chain Monte Carlo sampling schemes

Let $Y = \{y(1), \ldots, y(n)\}'$ be the vector of observations, let $X = \{x(1)', \ldots, x(n)'\}'$ be the total state vector, and let $Z = (X', \sigma^2, \tau^2)'$.

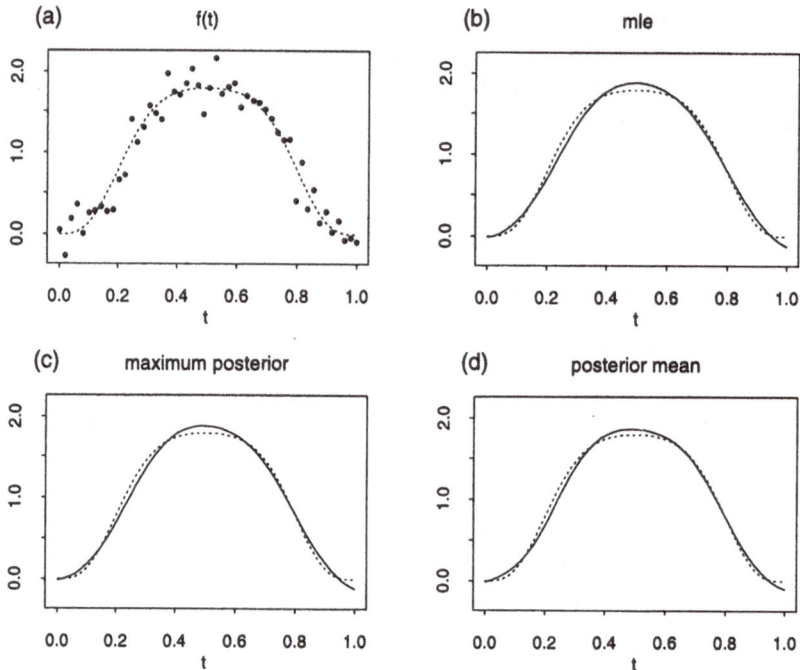

Figure 1: Part (a) shows the signal $f_1(t)$ with the generated data. Part (b) shows the signal $f_1(t)$ (dashes) with the marginal likelihood estimate of $f_1(t)$ (solid). Part (c) shows the signal $f_1(t)$ (dashes) with the maximum posterior estimate of $f_1(t)$ (solid). Part (d) shows the signal $f_1(t)$ (dashes) with the Markov chain Monte Carlo estimate of $f_1(t)$ (solid).

The Markov chain Monte Carlo approach for the Bayesian analysis of model M.2.1 constructs a Markov chain $Z^{[0]}, Z^{[1]}, \ldots$ such that the distribution of $Z^{[t]}$ converges to $p(Z|Y)$ as $t \to \infty$. A sufficient condition for convergence is invariance, aperiodicity, and irreducibility with respect to $p(Z|Y)$; see, for example, Tierney (1994). All the sampling schemes considered in this paper generate variables from conditional distributions with strictly positive densities, and hence are irreducible and aperiodic with respect to $p(Z|Y)$. In this section, we discuss several Markov chain Monte Carlo sampling schemes; in section 2.4, we discuss the efficiency of these Markov chains.

Carlin, Polson and Stoffer (1992) show how to generate X, σ^2 and τ^2 from the following Gibbs sampling scheme.

Sampling scheme 2.1 [Gibbs sampler without grouping] Generate from

(i)$p(\tau^2|Y, X)$; (ii)$p\{x(t)|Y, x(s), s \neq t, \sigma^2, \tau^2\}$ for $t = 1, \ldots, n$; (iii)$p(\sigma^2|Y, X)$.

variable	maximum likelihood	maximum posterior
$\hat{\sigma}^2$	0.034	0.033
$\hat{\tau}^2$	11,000	12,600

Table 1: Maximum likelihood and maximum posterior estimates of σ^2 and τ^2.

(a)

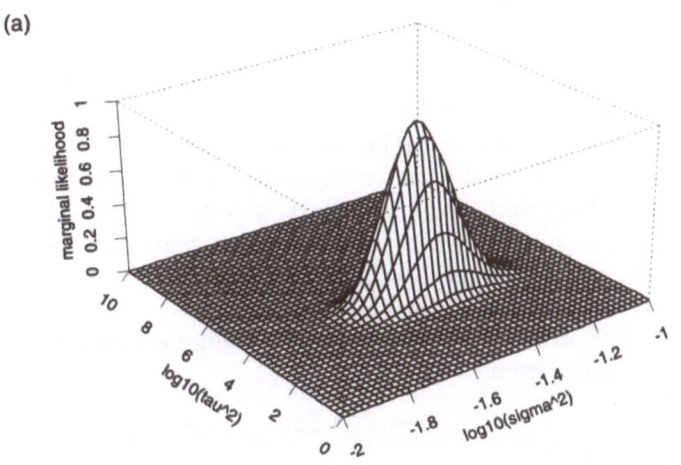

(b)

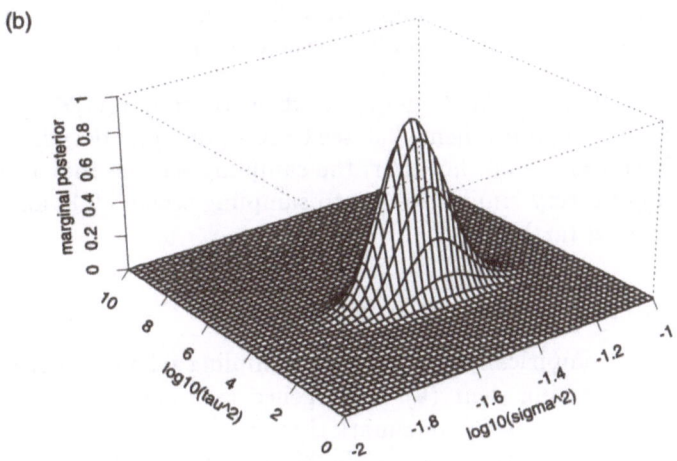

Figure 2: Part (a) shows the marginal likelihood as a function of σ^2 and τ^2. Part (b) shows the marginal posterior as a function of σ^2 and τ^2.

Invariance with respect to $p(Z|Y)$ follows from the construction of the Gibbs sampler.

Carter and Kohn (1994a) show how to generate X, σ^2 and τ^2 from the following Gibbs sampling scheme; similar results are also in Frühwirth-Schnatter (1994).

Sampling scheme 2.2 [Gibbs sampler with grouping but without using reduced conditionals] Generate from

$$(i)p(\tau^2|Y, X); (ii)p(X|Y, \sigma^2, \tau^2); (iii)p(\sigma^2|Y, X).$$

We show in Section 2.4 that sampling scheme 2.2 can be much more efficient than sampling scheme 2.1.

Let $g(t) = h(t)'x(t)$ and $G = \{g(1), \ldots, g(n)\}'$. Carter and Kohn (1994a) show how to generate σ^2 and τ^2 from the reduced conditional distribution $p(\sigma^2, \tau^2|Y, G)$; de Jong and Shephard (1994) give an efficient algorithm to generate from $p(G|Y, \sigma^2, \tau^2)$. These results suggest generating G, σ^2 and τ^2 from the following Gibbs sampling scheme.

Sampling scheme 2.3 [Gibbs sampler with grouping and using reduced conditionals] Generate from

$$(i)p(\tau^2|Y, G); (ii)p(G|Y, \sigma^2, \tau^2); (iii)p(\sigma^2|Y, G).$$

We show in the next section that sampling scheme 2.3 is more efficient than sampling scheme 2.2. Accordingly, we will use an extension of sampling scheme 2.3 for the more complex models considered in Section 3.

Remark 2.1 It is also possible to generate efficiently from $p(G|Y, \lambda)$ instead of $p(G|Y, \sigma^2, \tau^2)$ in sampling scheme 2.3; see Carter (1993) for details. For the examples we have considered, however, the sampling scheme that generates from $p(G|Y, \lambda)$ gives very similar results to sampling scheme 2.3, and hence will not be discussed further.

2.4 Comparison of sampling schemes

In this section, we empirically compare the sampling schemes discussed in Section 2.3 by examining their (a) convergence rate and (b) efficiency in estimating posterior moments. We remark that the convergence rate depends on the choice of starting distribution for the Markov chain and that, even with a fixed starting distribution, it is difficult to accurately determine when the Markov chain has converged. Accordingly, we have compared convergence rates using several different starting values and we only draw conclusions when the differences are large. For simplicity, we only show results for one starting value.

We first compare sampling schemes 2.1 and 2.2. Using the same model and data as in Section 2.2, we ran both sampling schemes for 20,000 iterations with the starting values $(\sigma^2)^{[0]} = 1$ and $X^{[0]} = E(X|\sigma^2 = 1, \tau^2 = 1)$. Figures 3(a) and 3(b) are plots of the iterates of σ^2 and τ^2 respectively for sampling scheme 2.1. Figures 3(c) and 3(d) are plots of the iterates of σ^2 and τ^2 respectively for sampling scheme 2.2. Figures 3(e) and 3(f) show the corresponding plots for the first 2,000 iterates of sampling scheme 2.2. The same starting values were used. Inspection of the marginal likelihood and the posterior distribution of σ^2 and τ^2 indicates that sampling scheme 2.2 converged in about 1,000 iterations, whereas sampling scheme 2.1 did not converge even after 20,000 iterations. We remark that inspection of Figures 3(a) and 3(b), without further investigation, could give the misleading impression that sampling scheme 2.1 has converged. We obtained similar results for other arbitrary starting values.

Next, we show that sampling scheme 2.2 traverses the posterior distribution much faster than sampling scheme 2.1 once they have both converged. This implies that estimates of posterior moments and densities based on sampling scheme 2.2 will be more efficient than those based on sampling scheme 2.1. To avoid problems of convergence associated with initializing far from the posterior mode, we initialized both sampling schemes at the marginal likelihood estimates of σ^2 and τ^2. We ran both sampling schemes using a warmup period of 1,000 iterations and a sampling period of 10,000 iterations. Figures 4(a) and 4(b) are plots of σ^2 and τ^2 for sampling scheme 2.1, and Figures 4(c) and 4(d) are the corresponding plots for sampling scheme 2.2. By comparing Figure 4(a) with Figure 4(c), and Figure 4(b) with Figure 4(d), it is clear that sampling scheme 2.2 traverses the posterior parameter space much faster than sampling scheme 2.1. This is obvious in particular for the plot in Figure 4(b) where the spread of the iterates is far smaller than for Figure 4(d).

We now empirically compare sampling schemes 2.2 and 2.3 using the same model as in Section 2.2, except with $m = 2$. We use data generated by (2.1) with the function $f_1(t)$ given in (2.6). The observation error standard deviation was $\sigma = 0.15$, the sample size was $n = 100$, and the design was equally spaced over $[0, 1]$. We ran both sampling schemes for a warmup period of 1000 iterations followed by a sampling period of 5000 iterations. Both sampling schemes seemed to converge within a few hundred iterations for arbitrary choices of starting values. Thus the convergence rates are similar and we will not present a direct comparison.

We compare the efficiency of both sampling schemes as follows. Figure 5 shows the function estimates and the sample autocorrelation functions of $\log(\tau^2)$ for both sampling schemes for the sampling period of 5000 iterations. The starting values were $\sigma^2 = 1$ and $x(i) = E\{x(i)|Y, \sigma^2, \tau^2 = 1\}$, $i = 1, \ldots, n$.

Figure 5 confirms that the function estimates are almost identical, as

136

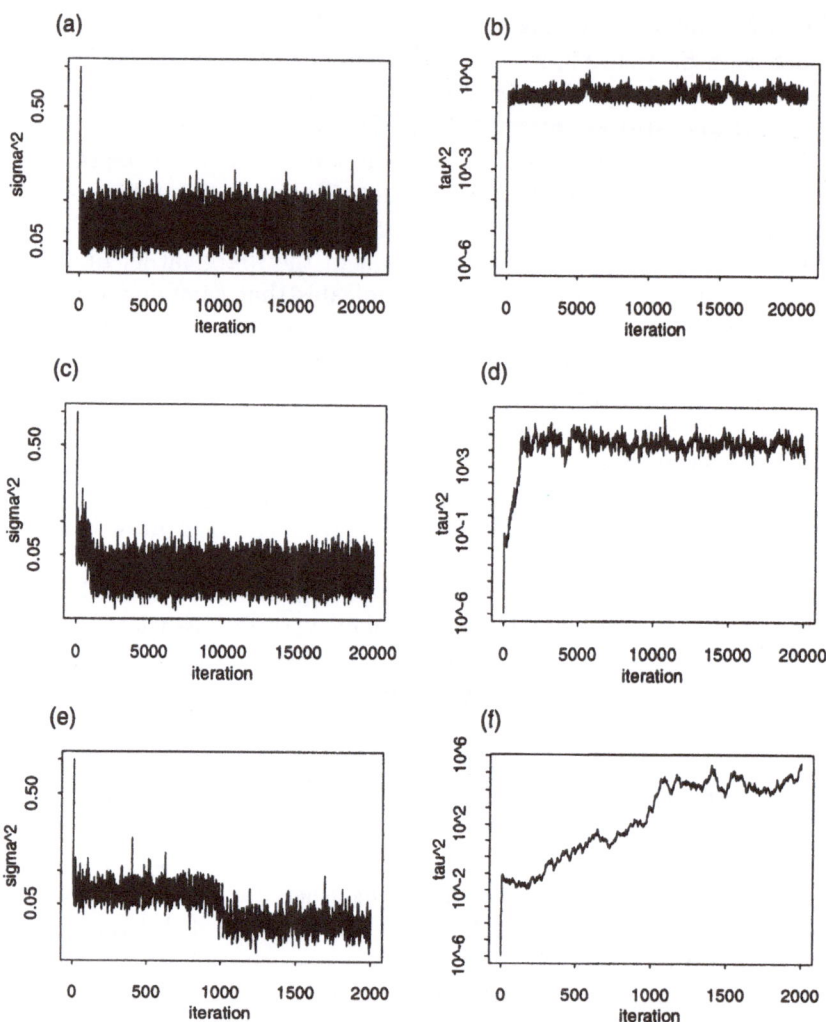

Figure 3: Part (a) plots the iterates of σ^2 and part (b) plots the iterates of τ^2 for sampling scheme 2.1. Parts (c) and (d) are the corresponding plots for sampling scheme 2.2. Parts (e) and (f) are the first $2,000$ iterates of parts (c) and (d) respectively.

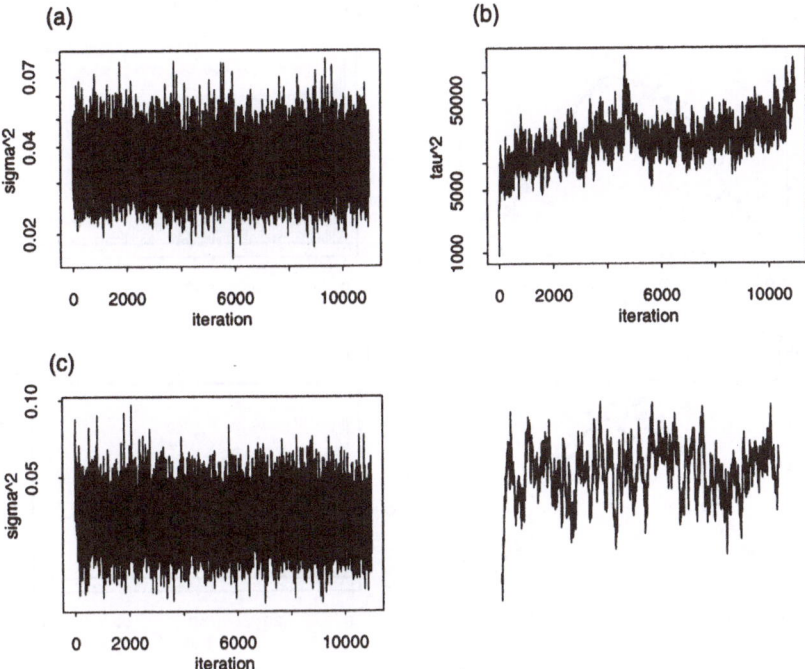

Figure 4: Part (a) shows the iterates of σ^2 and part (b) shows the iterates of τ^2 for sampling scheme 2.1. Parts (c) and (d) are the corresponding plots for sampling scheme 2.2.

expected when the Markov chains have converged. Also from Figure 5, the autocorrelations of $\log(\tau^2)$ are much higher for sampling scheme 2.2 than for sampling scheme 2.3. We can quantify this difference in autocorrelations as follows. We write the iterates in the sampling period as $X^{[j]}, (\sigma^2)^{[j]}, (\tau^2)^{[j]}, j = 1, \ldots, N$, where $N = 5000$. Once the Markov chain has converged the iterates form a stationary sequence, and thus we can estimate posterior means by the sample means in the sampling period. For example, we can estimate $E(\tau^2 | Y)$ by

$$\hat{\tau}^2 = N^{-1} \sum_{j=1}^{N} (\tau^2)^{[j]}.$$

As discussed in Carter and Kohn (1994a), we can use the sample autocovariance functions to estimate the variance of $\hat{\tau}^2$ as follows. Let $\gamma(k) = \operatorname{cov}\{(\tau^2)^{[j]}, (\tau^2)^{[j+k]}\}$, then

$$N \operatorname{var}(\hat{\tau}^2) \approx \sum_{|k| \leq C} (1 - |k|/N) \hat{\gamma}(k)$$

138

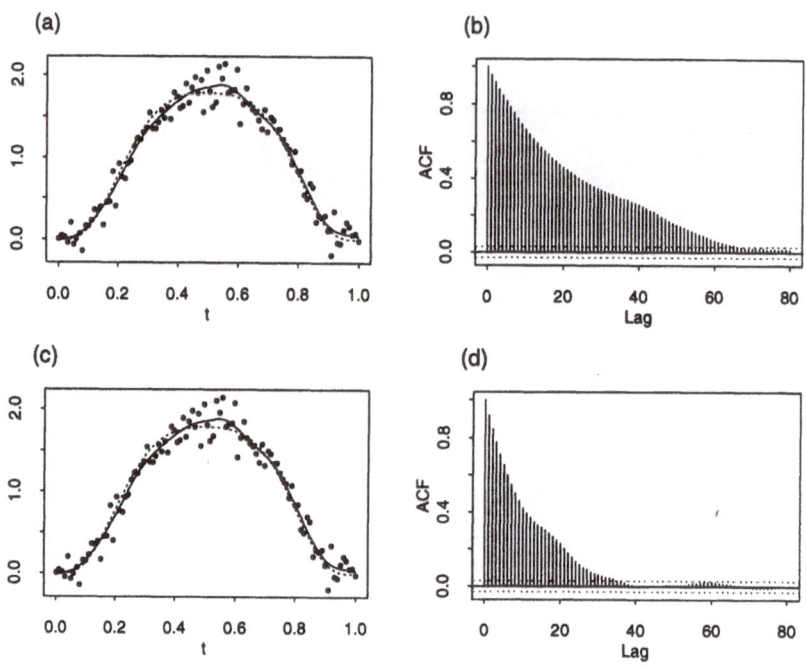

Figure 5: Part (a) shows the signal $f_1(t)$ (dashes) with the generated data (dots) and the Markov chain Monte Carlo estimate of $f_1(t)$ (solid). Part (b) shows the sample autocorrelation function for $\log(\tau^2)$. Parts (a) and (b) are for sampling scheme 2.2, and parts (c) and (d) are for sampling scheme 2.3.

where the constant C is chosen so that $\gamma(k) \approx 0$ for $|k| > C$. In this case we had $N = 5,000$, and we chose $C = 100$. Table 1 shows the results for the estimates of τ^2, $\log(\tau^2)$, $f(.1)$, $f(.25)$ and $f(.5)$. Column 1 gives the variable to be estimated, columns 2 and 3 give the sample variance and the variance of the mean for sampling scheme 2.2. Columns 4 and 5 are the corresponding estimates for sampling scheme 2.3. Column 6 is the ratio of columns 3 and 5, which is an estimate of the relative number of iterates to obtain a given precision. The numbers in column 6 range from approximately 1.5 to 3.5, so that the number of iterates for sampling scheme 2.2 would have to increase by a factor of approximately 3.5 to achieve the same accuracy as sampling scheme 2.3.

For this example, the extra computational time required for each iteration of sampling scheme 2.3 compared to sampling scheme 2.2 was negligible, and thus there is an overall gain in efficiency obtained from using sampling scheme 2.3.

variable	Sampling scheme 2.2		Sampling scheme 2.3		Ratio
	$\hat{\gamma}(0)$	$N \operatorname{var}(\cdot)$	$\hat{\gamma}(0)$	$N \operatorname{var}(\cdot)$	
τ^2	1.4×10^3	2.1×10^5	1.8×10^3	9.7×10^4	2.2
$\log(\tau^2)$	1.1	2.1×10^2	1.0	7.4×10^1	2.8
$f(.1)$	8.9×10^{-4}	6.5×10^{-3}	9.7×10^{-4}	1.9×10^{-3}	3.5
$f(.25)$	8.4×10^{-4}	2.1×10^{-3}	8.6×10^{-4}	1.4×10^{-3}	1.6
$f(.5)$	2.9×10^{-3}	7.7×10^{-3}	3.0×10^{-3}	3.9×10^{-3}	2.0

Table 2: Variance of the estimates of τ^2, $\log(\tau^2)$, $f(.1)$, $f(.25)$ and $f(.5)$.

3 Modelling outliers and discontinuities

In this section, we extend the Bayesian spline smoothing model discussed in Section 2 by using mixture of normal errors to model outliers and discontinuities. We then discuss Markov chain methods to generate from the posterior distribution of these models.

3.1 Description of models

We consider the signal plus noise observations given by (2.1), but relax the assumptions of Section 2.1 that the signal is a smooth function with Gaussian observation errors. Instead, we assume that the signal is a piecewise smooth function with observation errors that may contain outliers. To model these assumptions, we change the prior M.2.1 given in Section 2.1 in the following way. First, we only consider the signal at the discrete points $f(t_i), i = 1, \ldots, n$, instead of the continuous prior given by (2.2). We again use the linear state space model given by (2.4) and (2.5), but we modify the priors for both the observation errors and the state transition errors in the following way.

We consider two different priors corresponding to observation and state transition errors that are (a) finite mixtures of normals or (b) t distributed. The first prior, denoted by M.3.1, has $e(i) \sim N\{0, \sigma^2 K_1(i)\}$ and $u(i) \sim N\{0, \tau^2 K_2(i)U(i)\}$. Let $K(i) = \{K_1(i), K_2(i)\}', i = 1, \ldots, n$ and let $K = \{K(1), \ldots, K(n)\}$. We assume the $K(i)$ are independent and identically distributed discrete valued multivariate variables. Then, up to the scale factors σ^2 and τ^2 respectively, the $e(i)$ and $u(i)$ are finite mixtures of normals, and the $K(i)$ are indicator variables. To model outliers and discontinuities, we choose the prior $p\{K(i)\}$ so that both the observation errors $e(i)$ and the the state transition errors $u(i)$ have heavy tails, but we impose the restriction that an additive and an innovation outlier cannot occur simultaneously, as shown in Table 3.

The second prior, denoted by M.3.2, has $e(i) \sim N\{0, \sigma^2 \omega(i)^2\}$ and $u(i) \sim N\{0, \tau^2 \psi(i)^2 U(i)\}$ where the $\omega(i)^2$ and the $\psi(i)^2$ are independent and identically distributed inverse chi-squared variables with 3 degrees of freedom.

(j, k)	$(1, 1)$	$(10, 1)$	$(10^2, 1)$	$(1, 10)$	$(1, 10^2)$
$p\{K(i) = (j, k)'\}$	.95	.00625	.00625	.00625	.00625

(j, k)	$(1, 10^3)$	$(1, 10^4)$	$(1, 10^5)$	$(1, 10^6)$
$p\{K(i) = (j, k)'\}$	.00625	.00625	.00625	.00625

Table 3: Distribution of $K(i)$ for the prior M.3.1

We let $\Omega = \{\omega(1)^2, \ldots, \omega(n)^2\}'$ and $\Psi = \{\psi(1)^2, \ldots, \psi(n)^2\}'$. Then, up to the scale factors σ^2 and τ^2 respectively, the $e(i)$ and $u(i)$ are t distributed with 3 degrees of freedom, and hence have heavy tails. A t distributed prior to model heavy tailed errors is proposed by Meinhold and Singpurwalla (1989).

We also consider the following two priors that assume the signal is a piecewise smooth function with Gaussian observation errors. We use these priors in Section 3.3 to illustrate some differences between Markov chain Monte Carlo sampling schemes. Consider the linear state space model given by (2.4) and (2.5), but with state transition errors that are (a) finite mixtures of normals or (b) t distributed. The mixture of normal prior, denoted by M.3.3, has $u(i) \sim N\{0, \tau^2 K(i)U(i)\}$, where the $K(i)$ are independent and identically distributed discrete valued variables. We use the same notation for the discrete valued variables in M.3.1 and M.3.3 to simplify the discussion on Markov chain Monte Carlo sampling schemes in Section 3.3. To model innovation outliers we chose the prior $p\{K(i)\}$ so that the errors $u(i)$ have heavy tails, as shown in Table 4 below.

j	1	10	10^2	10^3	10^4	10^5	10^6
$p\{K(i) = j\}$	.95	.0083·	.0083·	.0083·	.0083·	.0083·	.0083·

Table 4: Distribution of $K(i)$ for the prior M.3.3

The t distributed prior, denoted by M.3.4, has $u(i) \sim N\{0, \tau^2\psi(i)^2U(i)\}$ where the $\psi(i)^2$ are independent and identically distributed inverse chi-squared variables with 3 degrees of freedom.

3.2 Markov chain Monte Carlo sampling schemes

We discuss several sampling schemes to generate from the models in Section 3.1; we discuss the efficiency of these sampling schemes in section 3.3.

For models M.3.1 and M.3.3, Carter and Kohn (1994a) show how to generate X, K, σ^2 and τ^2 from the following sampling scheme.

Sampling scheme 3.1 Generate from

$$(i)p(\tau^2|Y, G, K, \sigma^2) = p(\tau^2|G, K); (ii)p(X|Y, K, \sigma^2, \tau^2);$$
$$(iii)p(K|Y, X, \sigma^2, \tau^2); (iv)p(\sigma^2|Y, X, K, \tau^2) = p(\sigma^2|Y, G, K).$$

Sampling scheme 3.1 is a generalisation of sampling scheme 2.3; invariance with respect to $p(X, K, \sigma^2, \tau^2|Y)$ follows from the results in Carter and Kohn (1994b).

Carter and Kohn (1994b, c) give an $O(n)$ algorithm to generate the indicator variables $K(1), \ldots, K(n)$ one at a time from the conditional distributions $p\{K(t)|Y, K(s), s \neq t, \sigma^2, \tau^2\}$, which do not depend on the total state vector X. This suggests the following sampling scheme.

Sampling scheme 3.2 Generate from

$$(i)p(\tau^2|Y, G, K, \sigma^2) = p(\tau^2|G, K); (ii)p\{K(t)|Y, K(s), s \neq t, \sigma^2, \tau^2\} \; t = 1, \ldots, n;$$
$$(iii)P(X|Y, K, \sigma^2, \tau^2); (iv)p(\sigma^2|Y, X, K, \tau^2) = p(\sigma^2|Y, G, K).$$

Invariance with respect to $p(X, K, \sigma^2, \tau^2|Y)$ follows from the results in Carter and Kohn (1994b). We show in Section 3.3 that sampling scheme 3.2 can be substantially more efficient than sampling scheme 3.1, particularly for complex models.

Remark 3.1 Because sampling scheme 3.2 generates the indicator variables K without conditioning on the total state vector X, it is usually irreducible, even when sampling scheme 3.1 is reducible. This happens when the distribution of $X|Y, K, \sigma^2, \tau^2$ is singular with the singularity depending on K. See Carter and Kohn (1994b) for further discussion and examples.

We now consider priors M.3.1 and M.3.3. Carlin et al. (1992) show how to generate from $p(\Psi|Y, X, K, \sigma^2, \Omega, \tau^2)$ and $p(\Omega|Y, X, K, \sigma^2, \tau^2, \Psi)$. This suggests the following sampling scheme.

Sampling scheme 3.3 Generate from

$$(i)p(\tau^2|Y, G, K, \sigma^2, \Omega, \Psi) = p(\tau^2|G, K, \Psi); (ii)p(X|Y, K, \sigma^2, \Omega, \tau^2, \Psi);$$
$$(iii)p(\Psi|Y, X, K, \sigma^2, \Omega, \tau^2) = p(\Psi|X, K, \tau^2);$$
$$(iv)p(\sigma^2|Y, X, K, \Omega, \tau^2, \Psi) = p(\sigma^2|Y, G, K, \Omega);$$
$$(v)p(\Omega|Y, X, K, \sigma^2, \tau^2, \Psi) = p(\Omega|Y, G, K, \sigma^2).$$

For prior M.3.3 the generations involving Ω are omitted.

3.3 Comparison of sampling schemes

We first empirically compare sampling schemes 3.1 and 3.2 with prior M.3.3 and sampling scheme 3.3 with prior M.3.4. We use data generated from (2.1) with the function

$$f_2(t) = \begin{cases} 0 & \text{if } t \leq .5 \\ 1 & \text{if } t > .5 \end{cases}$$

where $t \in [0, 1]$. The observation error standard deviation was $\sigma = 0.15$, the sample size was $n = 100$, and the design was equally spaced over $[0, 1]$.

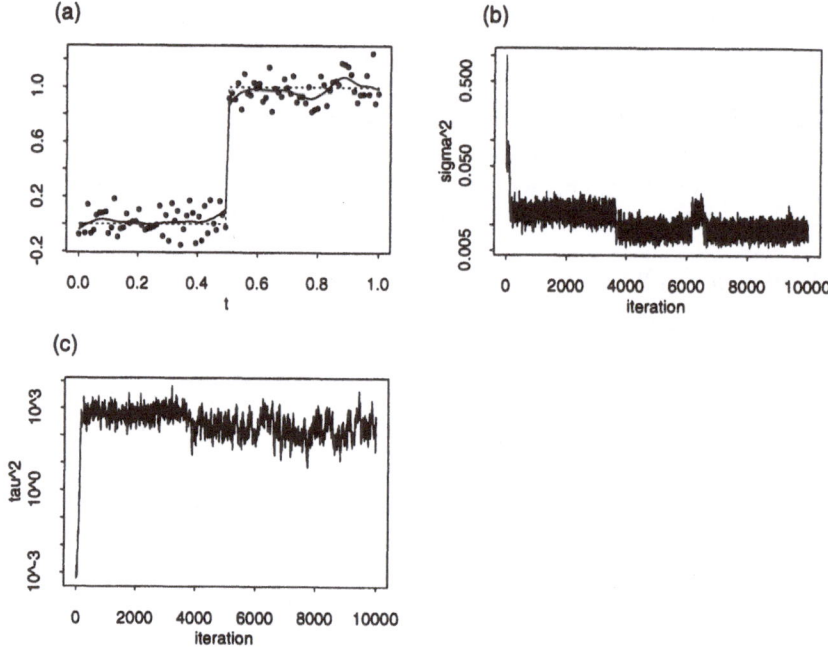

Figure 6: Part (a) shows the signal $f_2(t)$ (dashes) with the generated data (dots) and the Markov chain Monte Carlo estimate of $f_2(t)$ (solid). Part (b) shows the iterates of σ^2. Part (c) shows the iterates of τ^2. The sampling scheme is 3.1.

We ran the Markov chains for a variety of starting values. Figure 6 shows the results for a particular run of sampling scheme 3.1 with a warmup period of 5000 iterations followed by a sampling period of 5000 iterations. The starting values were $\sigma^2 = 1$, $K = (1, \dots, 1)'$ and $x(i) = E\{x(i)|Y, K, \sigma^2, \tau^2 = 1\}$, $i = 1, \dots, n$.

Figures 6(a)–(c) shows the function estimates for the sampling period, and the iterates of σ^2 and τ^2 on a log scale for both the warmup period and the sampling period. Figure 7 shows the corresponding results for a particular run of sampling scheme 3.2. The length of the warmup period and the sampling period are the same as in Figure 6, as are the starting values.

Figure 8 shows the corresponding results for a particular run of sampling scheme 3.3. The length of the warmup period and the sampling period are the same as in Figures 6 and 7. The starting values were $\sigma^2 = 1$, $\Psi = (1, \dots, 1)'$ and $x(i) = E\{x(i)|Y, \sigma^2, \tau^2 = 1, \Psi\}$, $i = 1, \dots, n$.

We first discuss the results for sampling scheme 3.2 shown in Figure 7 and then compare the results for sampling schemes 3.1 and 3.3 shown in Figures 6 and 8 respectively. From Figure 7 the Markov chain for sampling scheme 3.2 appears to converge after about 200 iterations. Similar results were obtained

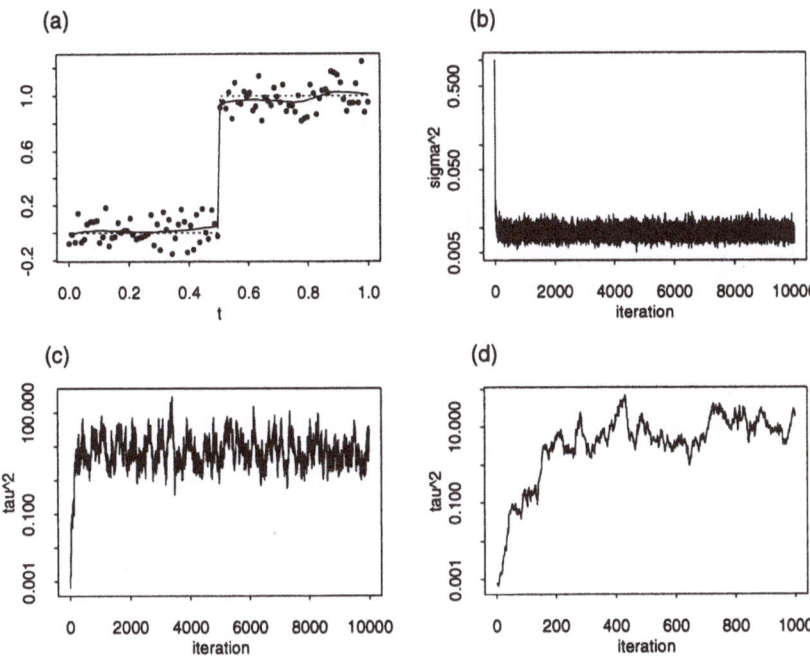

Figure 7: Part (a) shows the signal $f_2(t)$ (dashes) with the generated data (dots) and the Markov chain Monte Carlo estimate of $f_2(t)$ (solid). Part (b) shows the iterates of σ^2. Part (c) shows the iterates of τ^2. Part (d) shows the first 1000 iterates of τ^2. The sampling scheme is 3.2.

for other arbitrary starting values, suggesting that the results shown in Figure 7 represent the whole posterior distribution and not just a local mode. Comparing Figure 6 with Figure 7 shows that sampling scheme 3.1 has not converged, even after 10, 000 iterations. For instance, the iterates of $\log(\tau^2)$ for sampling scheme 3.1 (shown in Figure 6(c)) are non-stationary and differ in mean to the corresponding iterates for sampling scheme 3.2 (shown in Figure 7(c)). Furthermore, the iterates of $\log(\sigma^2)$ for sampling scheme 3.1 (shown in Figure 6(b)) come from a multi-modal distribution, whereas this is not the case for sampling scheme 3.2 (shown in Figure 7(b)). We found that for other arbitrary choices of starting values sampling scheme 3.1 did not converge within a reasonable number of iterations. We remark that we found it difficult to construct starting values that would make sampling scheme 3.1 converge quickly, even though we already had a good estimate of the whole posterior distribution from sampling scheme 3.2.

From Figure 8 the Markov chain for sampling scheme 3.3 appears to converge after about 3, 000 iterations. Similar results were obtained for other arbitrary starting values, suggesting that the results shown in Figure 8 represent the whole posterior distribution and not just a local mode. We

144

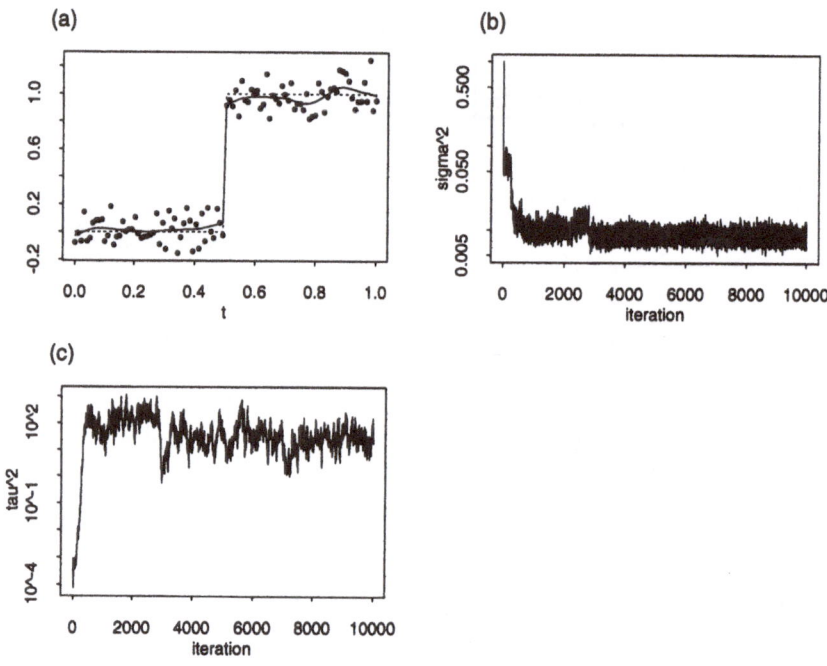

(a)

(b)

(c)

Figure 8: Part (a) shows the signal $f_2(t)$ (dashes) with the generated data (dots) and the Markov chain Monte Carlo estimate of $f_2(t)$ (solid). Part (b) shows the iterates of σ^2. Part (c) shows the iterates of τ^2. The sampling scheme is 3.3.

remark that there is no exact correspondence between the results for sampling scheme 3.3 and the results for sampling scheme 3.2 because the priors are different. We comment, however, that sampling scheme 3.2 converges more quickly than sampling scheme 3.3, although the performance of sampling scheme 3.3 was always reasonable.

We now empirically compare sampling schemes 3.1 and 3.2 with prior M.3.1 and sampling scheme 3.3 with prior M.3.2. We use data generated from (2.1) with the function $f_2(t)$ used above. The observation error standard deviation was $\sigma = 0.15$ and we have added three large outliers. The sample size was $n = 100$, and the design was equally spaced over $[0, 1]$.

We ran the Markov chains for a variety of starting values. Figure 9 shows the results for a particular run of sampling scheme 3.1 with a warmup period of $5,000$ iterations followed by a sampling period of $5,000$ iterations. The starting values were $\sigma^2 = 1$, $K = \{(1,1)' \ldots ,(1,1)'\}'$ and $x(i) = E\{x(i)|Y, K, \sigma^2, \tau^2 = 1\}$, $i = 1, \ldots, n$. Figure 9 shows the function estimates for the sampling period and the iterates of σ^2 and τ^2 on a log scale for both the warmup period and the sampling period. Figure 10 shows the corresponding results for a particular run of sampling scheme 3.2. The

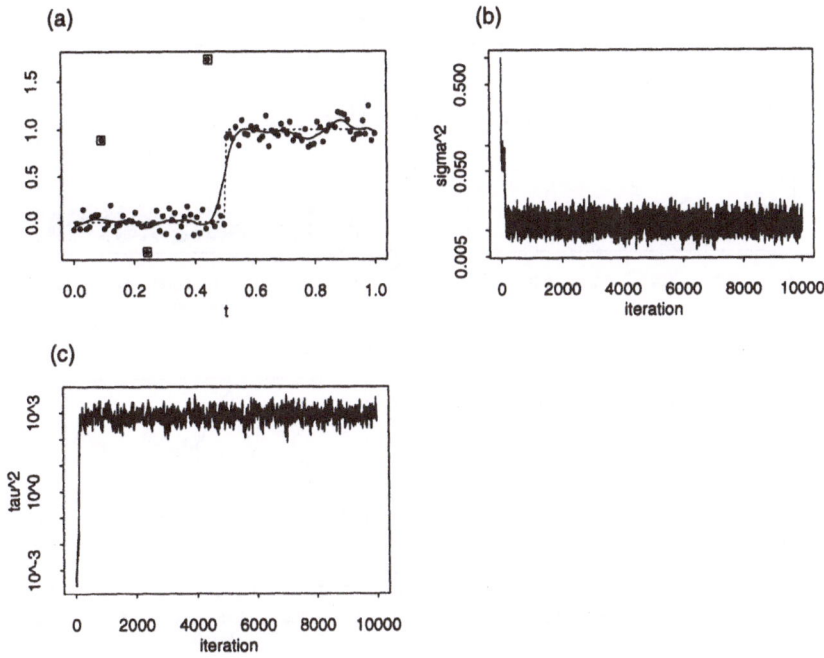

Figure 9: Part (a) shows the signal $f_2(t)$ (dashes) with the generated data (dots) and the Markov chain Monte Carlo estimate of $f_2(t)$ (solid). The outliers are indicated by squares. Part (b) shows the iterates of σ^2. Part (c) shows the iterates of τ^2. The sampling scheme is 3.1.

length of the warmup period and the sampling period are the same as in Figure 9, as are the starting values. Figure 11 shows the corresponding results for a particular run of sampling scheme 3.3. The length of the warmup period and the sampling period were the same as in Figures 9 and 10. The starting values were $\sigma^2 = 1$, $\Omega = (1, \ldots, 1)'$, $\Psi = (1, \ldots, 1)'$ and $x(i) = E\{x(i)|Y, \sigma^2, \Omega, \tau^2 = 1, \Psi\}, i = 1, \ldots, n$.

We first discuss the results for sampling scheme 3.2 shown in Figure 10 and then compare with the results for sampling schemes 3.1 and 3.3 shown in Figures 9 and 11 respectively. From Figure 10 the Markov chain for sampling scheme 3.2 appears to converge after about 200 iterations. Similar results were obtained for other arbitrary starting values, suggesting that the results shown in Figure 10 represent the whole posterior distribution and not just a local mode. Comparing Figure 9 with Figure 10 we see that the function estimates and the iterates for σ^2 and τ^2 are quite different, which shows that sampling scheme 3.1 has not converged, even after 10,000 iterations. We found that for other arbitrary choices of starting values sampling scheme 3.1 did not converge within a reasonable number of iterations.

Comparing Figure 10 with Figure 11 we see that the function estimates are

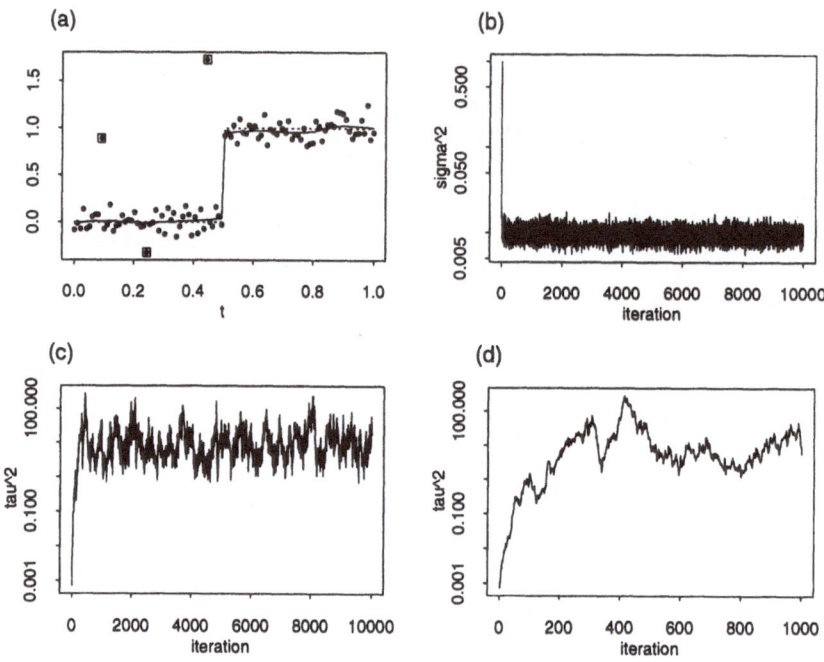

Figure 10: Part (a) shows the signal $f_2(t)$ (dashes) with the generated data (dots) and the Markov chain Monte Carlo estimate of $f_2(t)$ (solid). The outliers are indicated by squares. Part (b) shows the iterates of σ^2. Part (c) shows the iterates of τ^2. Part (d) shows the first 1000 iterates of τ^2. The sampling scheme is 3.2.

quite different. Since there is no exact correspondence between the results for sampling schemes 3.2 and 3.3 this does not conclusively show that sampling scheme 3.3 has not converged. We found, however, that other choices of starting values yielded different results, which shows that sampling scheme 3.3 often did not converge within a reasonable number of iterations.

4 Conclusion

For smooth functions, Bayesian spline smoothing seems to give very similar results to marginal likelihood spline smoothing. The advantage of the Bayesian approach is that it can be extended to state space models with normal mixture errors, which are robust against outliers and discontinuities in the function. There are several Markov chain Monte Carlo sampling schemes to analyse such Bayesian models with large differences in performance between them. The best performing sampling schemes, however, give Markov chains that converge rapidly to the posteror distribution. In particular, for

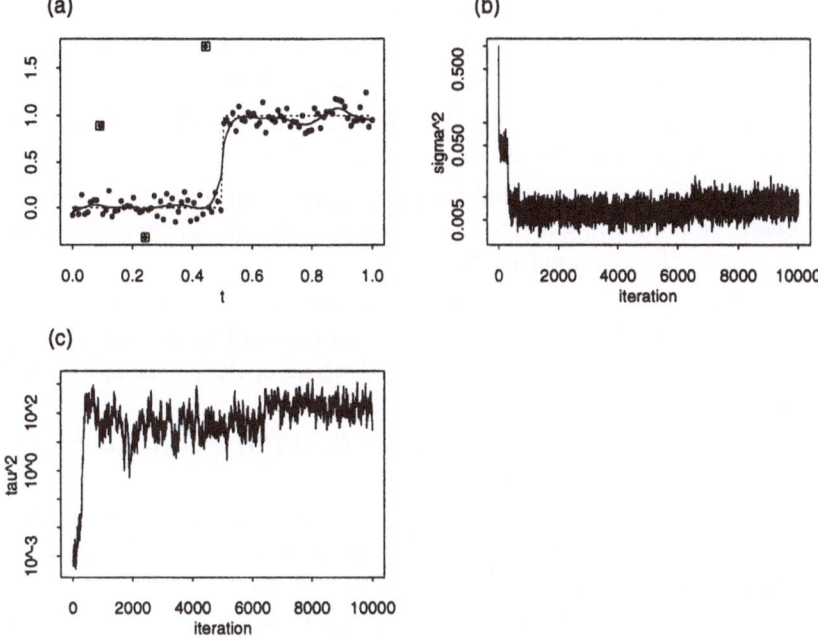

Figure 11: Part (a) shows the signal $f_2(t)$ (dashes) with the generated data (dots) and the Markov chain Monte Carlo estimate of $f_2(t)$ (solid). The outliers are indicated by squares. Part (b) shows the iterates of σ^2. Part (c) shows the iterates of τ^2. The sampling scheme is 3.3.

a Gaussian state space model with no outliers and level shifts, sampling scheme 2.3 should be used. In the presence of outliers and discontinuities in the regression function, sampling scheme 3.2 is the most reliable.

References

Carlin, B. P., Polson, N. G. and Stoffer, D. S. (1992). A Monte Carlo approach to nonnormal and nonlinear state space modelling. *Journal of the American Statistical Association*, **75**, 493–500.

Carter, C.K. (1993). On Markov chain Monte Carlo methods for state space modelling. Unpublished PhD thesis, University of New South Wales.

Carter, C. K. and Kohn, R. (1994a). On Gibbs sampling for state space models. *Biometrika*, **81**, 541–553.

Carter, C. K. and Kohn, R. (1994b). Markov chain Monte Carlo in conditionally Gaussian state space models. *Biometrika*, to appear.

Carter, C. K. and Kohn, R. (1994c). Bayesian methods for conditionally Gaussian state space models. Preprint.

De Jong, P. and Shephard, N. (1994). Efficient sampling from the smoothing density in time series models. *Biometrika*, to appear.

Frühwirth-Schnatter, S. (1994). Data augmentation and dynamic linear models. *Journal of Time Series Analysis,* 15, 183–202.

Hastie, T. J. and Tibshirani, R. J. (1990). *Generalized additive models.* New York: Chapman and Hall.

Kohn, R. and Ansley, C. F. (1987). A new algorithm for spline smoothing based on smoothing a stochastic process. *SIAM Journal of Scientific and Statistical Computing,* 8, 33–48.

Kohn, R., Ansley, C. F. and Tharm, D. (1991). The performance of cross-validation and maximum likelihood estimators of spline smoothing parameters. *Journal of the American Statistical Association,* 86, 1042–1050.

Lipsey, R. G., Sparks, G. R. and Steiner, P. O. (1976). *Economics* (2nd ed.) New York: Harper and Row.

Liu, J. S., Wong, W. H. and Kong, A. (1994). Covariance structure of the Gibbs sampler with applications to the comparisons of estimators and augmentation schemes. *Biometrika,* 81, 27–40.

Marion, J. B. (1970). *Classical dynamics of particles and systems.* (2nd ed.) New York: Academic Press.

Meinhold, R.J. and Singpurwalla, N.D. (1989). Robustification of Kalman filter models. *Journal of the American Statistical Association,* 84, 479–486.

McDonald, J. A. and Owen, A. B. (1986). Smoothing with split linear fits. *Technometrics,* 28, 195–208.

Müller, H. G. (1992). Change-points in nonparametric regression analysis. *Annals of Statistics,* 20, 737–761.

Shephard, N. (1994). Partial non-Gaussian state space. *Biometrika,* 81, 115–132.

Tierney, L. (1994). Markov chains for exploring posterior distributions. *Annals of Statistics,* to appear.

Wahba, G. (1978). Improper priors, spline smoothing and the problem of guarding against model errors in regression. *Journal of the Royal Statistical Society,* Ser. B, 40, 364–372.

Wahba, G. (1983). Bayesian "confidence intervals" for the cross-validated smoothing spline. *Journal of the Royal Statistical Society,* Ser. B, 45, 133–150.

Wecker, W. E. and Ansley, C. F. (1983). The signal extraction approach to nonlinear regression and spline smoothing. *Journal of the American Statistical Association,* 78, 81–89.

The Invariance of Statistical Analyses with Smoothing Splines with Respect to the Inner Product in the Reproducing Kernel Hilbert Space

Angelika van der Linde

Bremen Institute for Prevention Research and Social Medicine
Grünenstr. 120, D-28199 Bremen, Germany

Summary

Analyses with smoothing splines are conveniently formalized using a reproducing kernel Hilbert space. The inner product in such a space and the corresponding reproducing kernel – given a smoothness criterion – can be defined in infinitely many ways. The smoothing spline itself is known to be independent of the choice of the inner product. Furthermore it is proven that the usual statistical inferences with smoothing splines (under different distributional assumptions leading to signal extraction, Bayesian or minimax estimation) and particularly the smoothing error are invariant w.r.t. to the choice of the inner product. For a special inner product the smoothing spline is an orthogonal projection.

Keywords: Interpolation splines, duality, congruence, Bayesian estimation

1 Introduction

An abstract setting for spline interpolation and smoothing is the theory of reproducing kernel Hilbert spaces (RKHSs). It provides a common language for numerical analysts (Sard, 1967; Larkin, 1983; Meinguet, 1979, 1984) and statisticians (Kimeldorf and Wahba, 1970, 1971; Peele and Kimeldorf 1977; Weinert and Kailath, 1974; Wahba, 1984, 1990).

Interpolation and smoothing splines arise as solutions to optimization problems in a RKHS. Due to interest in computational advantages special RKHSs were suggested in early papers (Weinert, 1978; Weinert, Sidhu and Byrd, 1977). Special emphasis was given to the choice of an inner product according to which the interpolation spline turned out to be an orthogonal projection and to the construction of spaces in which smoothing can be reduced to interpolation.

From a statistical point of view a smoothing spline is an estimate of an unknown smooth function f based on "noisy" observations of values of f. For heuristics and statistical interpretations of smoothing splines Weinert's constructions are not helpful and different RKHSs provide more insight. Especially for Bayesian interpretations (Wahba 1978, 1983; Silverman 1985) RKHSs were preferred that allowed for a non-degenerate prior distribution on the rough part of f.

Of course all calculations depend on the chosen RKHS and usually the analysis is performed within a given RKHS. However, given the variety of representations one needs to make sure of their coincidence. Within a particular framework (splines based on linear differential operators allowing for a state space representation) Ansley and Kohn (1986) established a link between Wahba's and Weinert's work. In this paper the problem of invariance w.r.t. the choice of an inner product for a given set of functions is solved more generally.

For each of the three statistical characterizations of smoothing with splines – signal extraction, Bayesian or minimax estimation – invariance of the smoothing error and of criteria to determine the smoothing parameter is investigated.

Roughly speaking the idea is that invariance is due to the uniqueness of the smoothing spline as a function, thus being independent of the inner product, and that most of the terms relevant for statistical inference can be derived from dual or isometric transformations.

The paper is organized as follows. In section 2 the necessary formalism of RKHSs for interpolation and smoothing is provided. Abstract interpolation and smoothing errors are defined and shown to be invariant under the choice of the inner product. Also further results like an invariant decomposition of the smoothing error are derived. In section 3 statistical inferences about smoothing splines are linked to these results by applications of congruence mappings. In section 4 finally the smoothing spline as an orthogonal projection is discussed.

Proving the results of invariance the paper becomes a review of some statistical and computational features of smoothing with splines.

2 Splines in RKHSs

2.1 Function spaces

Let $(H, \langle \cdot, \cdot \rangle)$ be a complete inner product space consisting of real valued functions on a domain S. It possesses a reproducing kernel K iff evaluation functionals δ_t are bounded, the reproducing kernel describing their Riesz representers

$$\langle K(\cdot, t), h \rangle = h(t) \qquad \text{f.a. } t \in S, \ h \in H. \tag{2.1}$$

$(H, \langle \cdot, \rangle)$ with reproducing kernel K is called a RKHS and denoted by $H(K)$.

Smoothing splines are used to estimate an unknown function f subject to the requirement that f is smooth. Smoothness respectively roughness is defined by a semi-norm $J(\cdot)$ corresponding to a bilinear form $\langle \cdot, \cdot \rangle^\circ$ on a set of functions H with M-dimensional null space H_1 the elements of which are considered to be ultimately smooth. On an arbitrary but fixed complementary space H_2 (isomorphic to the quotient space) the bilinear form $\langle \cdot, \cdot \rangle^\circ$ defines an inner product $\langle \cdot, \cdot \rangle_2$. Thus $H = H_1 \oplus H_2$, and H_1, being a finite dimensional subspace, is easily provided with an inner product $\langle \cdot, \cdot \rangle_1$ and a corresponding reproducing kernel K_1. If also $(H_2, \langle \cdot, \cdot \rangle_2)$ can be shown to be a RKHS with kernel K_2 the total space H with $\langle \cdot, \cdot \rangle = \langle \cdot, \cdot \rangle_1 + \langle \cdot, \cdot \rangle_2$ may be addressed as RKHS $H(K)$ with $K = K_1 + K_2$. By construction H_1 and H_2 are orthogonal subspaces with corresponding orthogonal projections P_1 and P_2.

Examples for such a construction are given in (Wahba, 1990, ch. 1.2), (Meinguet, 1979).

The decomposition of H into $H_1 \oplus H_2$ is in no way unique. The many ways of providing H with the structure of a Hilbert space, given $\langle \cdot, \cdot \rangle^\circ$, are however equivalent in the sense that the same topology is induced on H.

To be more precise let

$$H_1 = \text{span } \{w_1, \ldots, w_M\}. \tag{2.2}$$

A particular choice of H_2 may be described by a *"decomposition design"* $\tilde{\Lambda} = \{\tilde{\lambda}_1, \ldots, \tilde{\lambda}_M\}$ of M linear functionals satisfying rg $((\tilde{\lambda}_i(w_j)))_{i,j=1,\ldots,M} = M$. Then

$$H_2 := \{h \in H | \tilde{\lambda}(h) = 0, \tilde{\lambda} \in \tilde{\Lambda}\} \tag{2.3}$$

$$\langle g, h \rangle := \sum_{j=1}^M \tilde{\lambda}_j(g)\tilde{\lambda}_j(h) + \langle g, h \rangle^\circ. \tag{2.4}$$

Suppose $H = H_1 \oplus H_2$ is a RKHS $H(K)$. In the sequel it will be assumed that $\{w_1, \ldots, w_M\}$ is *dual* to $\tilde{\Lambda}$ i.e. $\tilde{\lambda}_i(w_j) = \delta_{ij}$. Then $\{w_1, \ldots, w_M\}$ is an *orthonormal* basis of H_1.

Proposition 1. Let $\Lambda = \{\lambda_1, \ldots, \lambda_M\}$ denote another decomposition design of bounded linear functionals on $H(K)$. Then $\| \cdot \|$ and $\| \cdot \|_\Lambda$ defined analogously to (2.4) are equivalent.

Proof. Follows from Böhmer (1974, th. 4.12, p. 101). □

In particular $(H, \langle \cdot, \cdot \rangle_\Lambda)$ is also a RKHS $H(K^\Lambda)$ and

$$J(h) = \langle h, h \rangle^\circ = \| P_2 h \|^2 = \| P_2^\Lambda h \|_\Lambda^2. \tag{2.5}$$

2.2 Invariance in interpolation

Consider an *experimental design* $d = \{\ell_1, \ldots, \ell_n\}$ of bounded linear functionals on $H(K)$ with Riesz representers $L_1, \ldots, L_n$. Observations on $h \in H$ according to d are

$$h_d = (\ell_1(h), \ldots, \ell_n(h))' = (\langle L_1, h \rangle, \langle L_2, h \rangle, \ldots, \langle L_n, h \rangle)' \tag{2.6}$$

and any further aspect of interest is described by a bounded linear functional ℓ with representer L.

Set $d^+ = \{\ell, \ell_1, \ldots, \ell_n\}$.

Definition 1. The unique solution to the optimization problem in $H(K)$

(I) minimize $\| P_2 h \|^2$ subject to $h_d = x$, $x \in \mathbb{R}^n$

is called the *interpolation spline* with respect to x and denoted by $h_{I(x)}$.
(Böhmer 1974, p. 89/90). $\qquad\qquad\qquad\qquad\qquad\qquad\qquad\qquad\nabla$

The interpolation spline is an optimal approximant of h in the sense that values $\ell(h)$ are optimally approximated by $\ell(h_{I(x)})$. More formally we have

Definition 2. The *best approximating functional* in the sense of Sard for ℓ in span $\{\ell_1, \ldots, \ell_n\}$ is ℓ^* with Riesz prepresenter L^* if

(i) $\langle L - L^*, h \rangle = 0$ f.a. $h \in H_1$

(ii)
$$\begin{aligned} & \| P_2(L - L^*) \|^2 \\ = \ & \min\{\| P_2(L - \hat{L}) \|^2 \,|\hat{L} \in \text{ span } \{L_1, \ldots, L_n\}, L - \hat{L} \perp H_1\} \end{aligned}$$

(Böhmer, 1974, p. 197). $\qquad\qquad\qquad\qquad\qquad\qquad\qquad\qquad\qquad\nabla$

Proposition 2 (dual projection theorem).

For $h \in H(K)$
$$\langle L, h_{I(h_d)} \rangle = \langle L^*, h \rangle \tag{2.7}$$

Proof. Kimeldorf and Wahba (1971), Lemma 4.1. $\qquad\qquad\qquad\qquad\square$

We write
$$\ell^* = c^*(\ell)' \begin{pmatrix} \ell_1 \\ \vdots \\ \ell_n \end{pmatrix} \tag{2.8}$$

and thus
$$\langle L, h_{I(h_d)} \rangle = \langle L^*, h \rangle = c^*(\ell)' h_d. \tag{2.9}$$

Note that
$$|\langle L - L^*, h \rangle|^2 = |\langle P_2(L - L^*), P_2 h \rangle|^2 \leq \| P_2(L - L^*) \|^2 \| P_2 h \|^2 . \tag{2.10}$$

Definition 3. The *relative interpolation error* is
$$e_I(\ell) := \| P_2(L - L^*) \|^2 . \tag{2.11}$$

$\qquad\qquad\qquad\qquad\qquad\qquad\qquad\qquad\qquad\qquad\qquad\qquad\qquad\nabla$

We now consider invariance of the interpolation spline and the relative interpolation error with respect to a decomposition design Λ.

Theorem 1. The interpolation spline $h_{I(x)}$ does not depend on Λ.

Proof. The optimization problem in $H(K^\Lambda)$ reads the same as (I), and the interpolation function is unique as a function in H. $\qquad\qquad\qquad\square$

Thus also $c^*(\ell)$ is uniquely determined and independent of Λ. A *representation* of $c^*(\ell)$ in matrix calculus however depends on the choice of the decomposition design.

Introducing the matrices

$$
\left.
\begin{array}{l}
v(\ell) = (\langle P_1 L, w_1 \rangle, \ldots, \langle P_1 L, w_M \rangle)' \\
V_d = ((\langle P_1 L_i, w_j \rangle))_{i=1,\ldots,n, j=1,\ldots,M} \\
V_{d+} = \begin{pmatrix} v(\ell)' \\ V_d \end{pmatrix}
\end{array}
\right\}
\tag{2.12}
$$

$$
\left.
\begin{array}{l}
m(\ell) = (\langle P_2 L_i, P_2 L_1 \rangle, \ldots, \langle P_2 L, P_2 L_n \rangle)' \\
M_d = ((\langle P_2 L_i, P_2 L_j \rangle))_{i,j=1,\ldots,n} \\
M_{d+} = \begin{pmatrix} \langle P_2 L, P_2 L \rangle & m(\ell)' \\ m(\ell) & M_d \end{pmatrix}
\end{array}
\right\}
\tag{2.13}
$$

we have (Kimeldorf and Wahba, 1971)

$$
\begin{aligned}
c^*(\ell)' &= v(\ell)'(V_d' M_d^{-1} V_d)^{-1} V_d' M_d^{-1} \\
&\quad + m(\ell)' M_d^{-1}(I - V_d(V_d' M_d^{-1} V_d)^{-1} V_d' M_d^{-1})
\end{aligned}
\tag{2.14}
$$

if $P_2 L_1, \ldots, P_2 L_n$ are linearly independent implying $\tilde{\Lambda} \cap d = \emptyset$ because $P_2 L_i = 0$ for $\ell_i \in \tilde{\Lambda}$.

For another decomposition design Λ the Riesz representers in $H(K^\Lambda)$ in general do not coincide with $L, L_1, \ldots, L_n$. Also if $\Lambda \cap d \neq \emptyset$, M_d is singular and (2.14) does not hold.

In order to demonstrate invariance of the relative interpolation error two preliminary results are needed. Let N denote the set of bounded linear functionals on $H(K)$ annihilating H_1, i.e.

$$
\mu \in N \Rightarrow \mu(w_j) = 0 \qquad j = 1, \ldots, M.
$$

Lemma. For $\mu \in N$ the Riesz representer $m_2 \in H(K_2)$ differs from its Riesz representer $m_2^\Lambda \in H(K_2^\Lambda)$ by a function in H_1.

Proof. Let p_{m_2} denote the function in H_1 interpolating the values $\lambda_i(m_2)$, $\lambda_i \in \Lambda$, $i = 1, \ldots M$. Then $m_2 - p_{m_2} \in H(K_2^\lambda)$, and for $h \in H$

$$
\begin{aligned}
\langle m_2 - p_{m_2}, h \rangle_\Lambda &= \langle m_2 - p_{m_2}, h \rangle^\circ \\
&= \langle m_2, h \rangle^\circ = \langle m_2, h \rangle.
\end{aligned}
$$

$\square$

Lemma 2. For $\mu_i, \mu_j \in N$ with Riesz representers $m_{2i}, m_{2j} \in H(K_2)$, $m_{2i}^\Lambda, m_{2j}^\Lambda \in H(K_2^\Lambda)$

$$
\langle m_{2i}, m_{2j} \rangle = \langle m_{2i}^\Lambda, m_{2j}^\Lambda \rangle_\Lambda.
$$

$\square$

Theorem 2. The relative interpolation error is independent of Λ.

Proof.

$$
\| P_2(L - L^*) \|^2 = \| (1, -c^*(\ell)') \begin{pmatrix} L \\ L_1 \\ \vdots \\ L_n \end{pmatrix} \|^2.
\tag{2.15}
$$

Applying lemma 2 to $(1, -c^*(\ell)')$ $\begin{pmatrix} \ell \\ \ell_1 \\ \vdots \\ \ell_n \end{pmatrix} \in N$ yields the result. $\qquad\square$

Of special interest is a decomposition design $\Lambda \subset d$ such that the optimization problem (I) is equivalent to norm minimization in $H(K^\Lambda)$. In this case $h_{I(x)}$ is the orthogonal projection of any function h with $h_d = x$ onto span $\{L_1^\Lambda, \ldots, L_n^\Lambda\}$, L_i^Λ denoting the Riesz representer of L_i in $H(K^\Lambda)$. A suitable design is

$$\Lambda = d_M := \{\ell_1, \ldots, \ell_M\}.$$

2.3 Invariance in smoothing. Suppose values f_d of an unknown smooth function f are observed with "random noise", i.e. we have observations

$$y_d = f_d + e_d. \qquad (2.16)$$

It is assumed that the errors $e(\ell_i)$ in observing $\ell_i(f)$ are realizations of random variables $\varepsilon(\ell_i)$ with

$$E(\varepsilon_d) = 0, \qquad \text{cov } \varepsilon_d = \sigma^2 \, \Sigma_d, \quad \Sigma_d > 0. \qquad (2.17)$$

Definition 4. The unique solution to the optimization problem in $H(K)$

(S) $\qquad$ minimize $\quad (y_d - h_d)' \, \Sigma_d^{-1} (y_d - h_d) + \varrho \, \| P_2 h \|^2, \qquad \varrho \in \mathbb{R}^+$

is called the smoothing spline with respect to y_d and ϱ and is denoted by $h_{S(y_d)}^\varrho$. ϱ is called the smoothing parameter. (Böhmer, 1974, p. 89/90) $\qquad\nabla$

Smoothing can be reduced to interpolation constructing a special RKHS (cp. Kimeldorf and Wahba, 1970). We consider $\overline{H} = H(K) \times \mathbb{R}^n$ with a *weighted* inner product, $\eta, \tau^2 \in \mathbb{R}^+$.

$$\left\langle \begin{pmatrix} g \\ \gamma \end{pmatrix}, \begin{pmatrix} h \\ \delta \end{pmatrix} \right\rangle^- := \frac{1}{\eta} \langle P_1 g, P_1 h \rangle + \frac{1}{\tau^2} \langle P_2 g, P_2 h \rangle + \frac{1}{\sigma^2} \gamma' \, \Sigma_d^{-1} \delta \qquad (2.18)$$

and corresponding projections $\overline{P}_i, i = 1, 2, 3$.

$\overline{H}$ may be regarded as a set of functions on $S \times T_n, T_n = \{1, \ldots, n\}$. It is a RKHS with

$$\overline{K}(\cdot, t) = \begin{cases} \begin{pmatrix} \eta K_1(\cdot, t) + \tau^2 K_2(\cdot, t) \\ 0 \end{pmatrix} & t \in S \\ \begin{pmatrix} 0 \\ \sigma^2 \sigma_i \end{pmatrix} & t = i \in T_n \end{cases} \qquad (2.19)$$

where $\Sigma_d = (\sigma_1, \ldots, \sigma_n), \sigma_i \in \mathbb{R}^n$.

We denote the extension of ℓ to $H(\overline{K})$ using a bar, too, i.e.

$$\overline{\ell} \begin{pmatrix} h \\ \delta \end{pmatrix} = \ell(h) = \langle L, h \rangle$$

with Riesz representer

$$\overline{L} = \begin{pmatrix} \eta P_1 L \\ \tau^2 P_2 L \\ 0 \end{pmatrix}.$$

Let furthermore L_i^S represent ℓ_i^S with

$$\ell_i^S \begin{pmatrix} h \\ \delta \end{pmatrix} = \langle L_i, h \rangle + \delta_i,$$

$$L_i^S = \begin{pmatrix} \eta P_1 L_i \\ \tau^2 P_2 L_i \\ \sigma^2 \sigma_i \end{pmatrix}, \qquad i = 1, \ldots, n.$$

The optimization problem (S) may be rewritten

(S) minimize $e_d' \Sigma_d^{-1} e_d + \varrho \parallel P_2 h \parallel^2$ w.r.t. $h \in H(K)$.

For $\varrho = \sigma^2 / \tau^2$ it is equivalent to

$(\overline{S})$ minimize $\parallel (\overline{P}_2 + \overline{P}_3) \binom{h}{\delta} \parallel^2_{H(K)}$

     subject to $\langle L_i^S, \binom{h}{\delta} \rangle^- = y_i,$ $i = 1, \ldots n$

and its solution is $\begin{pmatrix} h^\ell_{S(y_d)} \\ \delta^* \end{pmatrix}$.

Applying proposition 2 to $H(\overline{K}) = H(\eta K_1) \oplus (H(\tau^2 K_2) \oplus \mathbb{R}^n)$ we find the best approximating functional

$$(\overline{L})^* = \overline{c}^*(\ell)' \begin{pmatrix} L_1^S \\ \vdots \\ L_n^S \end{pmatrix} \tag{2.20}$$

such that

$$\langle (\overline{L})^*, \begin{pmatrix} h \\ \delta \end{pmatrix} \rangle^- = \langle \overline{L}, \begin{pmatrix} h^\ell_{S(y_d)} \\ \delta^* \end{pmatrix} \rangle^- = \overline{c}^*(\ell)' y_d. \tag{2.21}$$

Define

$$L_S^* := \overline{c}^*(\ell)' \begin{pmatrix} L_1 \\ \vdots \\ L_n \end{pmatrix} \tag{2.22}$$

which is the "component of $(\overline{L})^*$ in $H(K)$".

Definition 5. The *smoothing error* is

$$e_S(\ell) := \parallel (\overline{P}_2 + \overline{P}_3)(\overline{L} - (\overline{L})^*) \parallel^2_{H(\overline{K})}. \tag{2.23}$$

∇

We then have

Theorem 3. (a) The smoothing spline $h^\ell_{S(y_d)}$ is invariant w.r.t. the decomposition design Λ. (b) The smoothing error is invariant w.r.t. the decomposition design Λ.

Proof. Analogous to that of theorem 1 and theorem 2. $\square$

In particular

$$(1, -\bar{c}^*(\ell)') \begin{pmatrix} \ell \\ \ell_1 \\ \vdots \\ \ell_n \end{pmatrix} \in N. \tag{2.24}$$

Again the *representation* of $\bar{c}^*(\ell)$ depends on the decomposition design. A formula is obtained replacing in (2.14) M_d by $(\tau^2 M_d + \sigma^2 \Sigma_d)$.

Setting further

$$\Sigma_{d+}^0 := \begin{pmatrix} 0 & 0 \\ 0 & \Sigma_d \end{pmatrix} \tag{2.25}$$

the smoothing error may be written

$$\begin{aligned} e_S(\ell) &= (1, -\bar{c}^*(\ell)')(\tau^2 M_{d+} + \sigma^2 \Sigma_{d+}^0)\left(\begin{smallmatrix} 1 \\ -\bar{c}^*(\ell) \end{smallmatrix}\right) \\ &= \tau^2(1, -\bar{c}^*(\ell)')M_{d+}\left(\begin{smallmatrix} 1 \\ -\bar{c}^*(\ell) \end{smallmatrix}\right) + \sigma^2 \bar{c}^*(\ell)' \Sigma_d \bar{c}^*(\ell). \end{aligned} \tag{2.26}$$

There is no inner product of the type (2.18) such that with a particular decomposition design like $\Lambda = d_M$ semi-norm minimization becomes equivalent to norm optimization, given the observations. A characterization of the smoothing spline not only as an interpolation spline but also as an orthogonal projection requires another inner product to be considered in section 4.

2.4 Further basic results The next two results reflect that the smoothing spline is an interpolation spline in estimated function values. Analogously to (2.21) consider $\langle (\overline{L}_i)^*, \left(\begin{smallmatrix} h \\ 0 \end{smallmatrix}\right)\rangle^- = \bar{c}^*(\ell_i)' y_d$, set

$$A(\varrho) := \begin{pmatrix} \bar{c}^*(\ell_1)' \\ \vdots \\ \bar{c}^*(\ell_n)' \end{pmatrix} \tag{2.27}$$

and note that $A(\varrho)$ does not depend on Λ.

Theorem 4.

$$\bar{c}^*(\ell)' = c^*(\ell)' A(\varrho). \tag{2.28}$$

Proof. Fixing the values $h_d = x$ in the smoothing problem (S), minimization is reduced to $\varrho \parallel P_2 h \parallel$ subject to $h_d = x$ and thus solved by an interpolation spline.

Therefore

$$\begin{aligned} &\bar{c}^*(\ell) y_d \\ = \; &\ell(h_{S(y_d)}^\varrho) = \ell(h_{I((h_{S(y_d)}^\varrho)_d)}) = \ell(h_{I(A(\varrho)y_d)}) \\ = \; &c^*(\ell)' A(\varrho) y_d \end{aligned}$$

for any $y_d \in \mathbb{R}^n$. $\qquad\square$

Correspondingly the smoothing error may be decomposed into the sum of an interpolation error and an estimation error.

Theorem 5. For every decomposition design Λ

$$e_S(\ell) = \tau^2 e_I(\ell) + \parallel (\overline{P}_2 + \overline{P}_3)(\overline{L}^* - (\overline{L})^*) \parallel^2_{H(\overline{K})} \tag{2.29}$$

Proof. $e_S(\ell), e_I(\ell)$ are independent of Λ. Also, $(\overline{P}_2 + \overline{P}_3)(\overline{L^*} - (\overline{L})^*)$ represents a linear functional on $\overline{H}$ that annihilates $\overline{H}_1$ and therefore its norm is independent of Λ. Due to invariance of $e_S(\ell)$ we can thus choose $\Lambda = d_M$ according to which the interpolation spline and with it the best approximating functional results from orthogonal projection in $H(K^{d_M})$. In order to avoid further indexing by d_M we indicate operating in $H(K^{d_M})$ only by indexing the inner product. We then have

$$\langle \overline{P}_2(\overline{L} - \overline{L^*}), (\overline{P}_2 + \overline{P}_3)(\overline{L^*} - (\overline{L})^*)\rangle_{H(\overline{K}^{d_M})}$$

$$= \tau^2 \langle P_2(L - L^*), P_2((c^*(\ell) - \overline{c}^*(\ell))' \begin{pmatrix} L_1 \\ \vdots \\ L_n \end{pmatrix})\rangle_{H(K^{d_M})}$$

$$= 0$$

because $L - L^* \perp_{K^{d_M}}$ span $\{L_1, \ldots, L_n\}$. $\qquad \square$

We invariantly decompose $e_S(\ell)$ further by

Corollary 5.1 . For every design Λ

$$e_S(\ell) = \tau^2 e_I(\ell) + \tau^2 \| P_2(L^* - L_S^*) \|^2 + \sigma^2 \overline{c}^*(\ell)' \Sigma_d \overline{c}^*(\ell). \qquad (2.30)$$

Proof.

$$\| (\overline{P}_2 + \overline{P}_3)(\overline{L^*} - (\overline{L}^*)) \|_{H(\overline{K})}^2$$

$$= \| P_2(\overline{L^*} - \overline{L})^*) \|_{H(\overline{K})}^2 + \| \overline{P}_3(\overline{L^*} - \overline{L})^*) \|_{H(\overline{K})}^2$$

$$= \tau^2 \| P_2(L^* - L_S^*) \|^2 + \sigma^2 \overline{c}^*(\ell)' \Sigma_d \Sigma_d^{-1} \Sigma_d \overline{c}^*(\ell)$$

and all three summands in (2.30) are invariant w.r.t. to Λ. $\qquad \square$

The next result relates the best approximating functional in the sense of Sard (definition 2) to the best linear approximation of $\overline{L}$ in $H(\overline{K})$.
We are going to variate η and thus use an additional index to denote dependence on η, if necessary. In particular we write $H(\overline{K}_\eta), \overline{L}(\eta), L_i^S(\eta)$ instead of $H(\overline{K}), \overline{L}, L_i^S$.

By definition the interpolation spline, the smoothing spline and the corresponding errors do not depend on η and therefore $\overline{c}^*(\ell), \overline{c}^*(\ell), A(\varrho)$ are the same for all η.

Let $\hat{\overline{L}}(\eta)$ be the orthogonal projection of $\overline{L}(\eta)$ onto span $\{L_1^S(\eta), \ldots, L_n^S(\eta)\}$ in $H(\overline{K}_\eta)$ and write

$$\hat{\overline{L}}(\eta) = \overline{c}_\eta(\ell)' \begin{pmatrix} L_1^S(\eta) \\ \vdots \\ L_n^S(\eta) \end{pmatrix}. \qquad (2.31)$$

We now need to be explicit about the inner product in H_1. The reproducing kernel $K_1(\cdot, t)$ corresponding to (2.4) is, using (2.12)

$$K_1(\cdot, t) = \sum_{j=1}^{M} w_j(t) w_j = v(\delta_t)' w, \quad w = (w_1, \ldots, w_M).$$

Then

$$P_1 L_i = v(\ell_i)'w,$$

$$\langle P_1 L_i, P_1 L_j \rangle = v(\ell_i)'v(\ell_j). \tag{2.32}$$

Theorem 6.

$$\lim_{\eta \to \infty} \bar{c}_\eta(\ell) = c^*(\ell) \tag{2.33}$$

$$\lim_{\eta \to \infty} \| \overline{L}(\eta) - \hat{L}(\eta) \|^2 = e_S(\ell). \tag{2.34}$$

Proof.

(i)

$$
\begin{aligned}
&\bar{c}_\eta(\ell) \\
=\ & (\langle \overline{L}(\eta), L_1^S(\eta) \rangle_{H(\overline{K}_\eta)}, \ldots, \langle \overline{L}(\eta), L_n^S(\eta) \rangle_{H(\overline{K}_\eta)})((\langle L_i^S(\eta) L_j^S(\eta) \rangle_{H(\overline{K}_\eta)}))^{-1} \\
=\ & (\eta v(\ell)'V_d' + \tau^2 m(\ell)')(\eta V_d V_d' + \tau^2 M_d + \sigma^2 \Sigma_d)^{-1}.
\end{aligned}
$$

The result then follows from the proof of theorem 1 in Wahba (1978).

(ii)

$$
\begin{aligned}
e_\eta(\ell) :=&\| \overline{L}(\eta) - \hat{L}(\eta) \|^2 \\
=\ & (1, -\bar{c}_\eta(\ell)')(\eta V_{d+} V_{d+}' + \tau^2 M_{d+} + \sigma^2 \Sigma_{d+}^0)\binom{1}{-\bar{c}_\eta(\ell)} \\
\leq\ & (1, -\bar{c}^*(\ell)')(\tau^2 M_{d+} + \sigma^2 \Sigma_{d+}^0)\binom{1}{-\bar{c}_\eta(\ell)} \\
=\ & e_S(\ell)
\end{aligned}
$$

because $\bar{c}_\eta(\ell)$ decribes the best linear combination of $L_1^S(\eta), \ldots, L_n^S(\eta)$. Thus the sequence $(e_n(\ell))_\eta$ is bounded. Without loss of generality we can choose a monotone sequence of positive numbers $\eta_k \to \infty$ yielding a *monotone* sequence $(e_{\eta_k}(\ell))_{k \in \mathbb{N}}$ because

$$
\begin{aligned}
e_{\eta_k}(\ell) \leq\ & (1, -\bar{c}_{\eta_{k+1}}(\ell)')(\eta_k V_{d+} V_{d+}' + \tau^2 M_{d+} + \sigma^2 \Sigma_{d+})\binom{1}{-\bar{c}_{\eta_{k+1}}(\ell)} \\
\leq\ & (1, -\bar{c}_{\eta_{k+1}}(\ell)'(\eta_{k+1} V_{d+} V_{d+}' + \tau^2 M_{d+} + \sigma^2 \Sigma_{d+})\binom{1}{-\bar{c}_{\eta_{k+1}}(\ell)} \\
=\ & e_{\eta_{k+1}}(\ell).
\end{aligned}
$$

Therefore the limit exists and

$$\lim_{\eta \to \infty} e_\eta(\ell) \leq e_S(\ell).$$

Further

$$e_\eta(\ell) \geq (1, -\bar{c}_\eta(\ell)')(\tau^2 M_{d+} + \sigma^2 \Sigma_{d+})\binom{1}{-\bar{c}_\eta(\ell)}$$

and in the limit using (i)

$$\lim_{\eta \to \infty} e_\eta(\ell) \geq e_S(\ell).$$

$\square$

The proof of theorem 6 works for any decomposition design Λ.

3   Statistical interpretations

We consider the ususal statistical settings for smoothing with splines, the signal extraction approach (e.g. Kimeldorf and Wahba, 1970), the Bayesian approach (Wahba, 1978, 1983) and the minimax approach (Li, 1982). All of these are related to the theory in RKHSs developed before by applying a congruence mapping.

Proposition 3. Let $\{U(t)|t \in S\} \subset L_2(\Omega, \mathcal{A}, P)$ be a zero mean stochastic process with covariance function K spanning the Hilbert subspace $L_2(\{U(t)|t \in S\})$. Then the congruence mapping

$$
\begin{array}{rcl}
\Phi : L_2(\{U(t)|t \in S\}) & \longrightarrow & H(K) \\
V & \longmapsto & h : S \ni s \mapsto \mathrm{cov}\,(V, U(s)) \in \mathbb{R}
\end{array}
\tag{3.1}
$$

particularly with

$$
U(t) \mapsto K(\cdot, t)
$$

is an isomorphism preserving inner products.

Proof. Parzen (1959, th. 5D, p. 302) $\square$

3.1 The Bayesian approach Considering

$$
\Phi_\eta^{-1} : H(\overline{K}_\eta) \to L_2(\{U(t)|t \in S\}).
\tag{3.2}
$$

we relate to $\overline{L}$ a random variable $X_\eta(\ell)$ with zero mean and

$$
\mathrm{var}\, X_\eta(\ell) = \eta v(\ell)' v(\ell) + \tau^2 \langle P_2 L, P_2 L \rangle
\tag{3.3}
$$

and to L_i^S a random variable $Y_\eta(\ell_i) = X_\eta(\ell_i) + \varepsilon(\ell_i)$ such that

$$
\begin{array}{l}
\mathrm{cov}\,(Y_\eta)_d = \eta V_d V_d' + \tau^2 M_d + \sigma^2 \Sigma_d \\
\mathrm{cov}\,(X_\eta(\ell), (Y_\eta)_d) = \eta v(\ell)' V_d' + \tau^2 m(\ell)'
\end{array}
$$

and

$$
E\begin{pmatrix} X_\eta(\ell) \\ (Y_\eta)_d \end{pmatrix} = 0
\tag{3.4}
$$

$$
\mathrm{Cov}\begin{pmatrix} X_\eta(\ell) \\ (Y_\eta)_d \end{pmatrix} = \eta V_{d+} V_{d+}' + \tau^2 M_{d+} + \sigma^2 \Sigma_{d+}^0
\tag{3.5}
$$

Referring to (2.16) and adding the assumption of *normal* distributions, (3.4) and (3.5) are the moments of the joint distribution of $\binom{\ell(f)}{Y_d}$ when the prior for $\langle L, f \rangle$ is described by $X_\eta(\ell)$ and for f_d by $(X_\eta)_d$.

Theorem 7

$$
\lim_{\eta \to \infty} E(\ell(f)|(Y_\eta)_d = y_d) = \bar{c}^*(\ell) y_d
\tag{3.6}
$$

$$
\lim_{\eta \to \infty} \mathrm{var}\,(\ell(f)|(Y_\eta)_d = y_d) = e_S(\ell)
\tag{3.7}
$$

$$
\begin{array}{l}
\lim_{\eta \to \infty} \mathrm{cov}\,(\ell(f), \tilde{\ell}(f)|(Y_\eta)_d = y_d) \\
= (1, -\bar{c}^*(\ell)')(\tau^2 M_{d+} + \sigma^2 \Sigma_{d+}^0)\begin{pmatrix} 1 \\ -\bar{c}^*(\tilde{\ell}) \end{pmatrix}.
\end{array}
\tag{3.8}
$$

Proof. Under the assumption of normality the conditional expectation is equal to the best (minimal variance) approximation, and the conditional variance equals $e_\eta(\ell)$. (3.6), (3.7) then follow from (2.33), (2.34). Furthermore

$$
\begin{aligned}
&\text{cov} \quad (\ell(f), \tilde{\ell}(f)|(Y_\eta)_d = y_d) \\
=\;&\text{cov} \quad (X_\eta(\ell) - \bar{c}_\eta(\ell)'(Y_\eta)_d, X_\eta(\tilde{\ell}) - \bar{c}_\eta(\tilde{\ell})'(Y_\eta)_d) \\
=\;&\text{cov} \quad (X_\eta(\ell) - (\bar{c}^*(\ell)' - \bar{c}^*(\ell)' + \bar{c}_\eta(\tilde{\ell})')(Y_\eta)_d, \\
&\qquad\quad (X_\eta(\tilde{\ell}) - (\bar{c}^*(\tilde{\ell}) - \bar{c}^*(\tilde{\ell})' + \bar{c}_\eta(\tilde{\ell})')(Y_\eta)_d) \\
=\;&\text{cov} \quad (X_\eta(\ell) - \bar{c}^*(\ell)'(Y_\eta)_d), X_\eta(\tilde{\ell}) - \bar{c}^*(\tilde{\ell})'(Y_\eta)_d) \\
&+\text{cov} \quad ((\bar{c}^*(\ell)' - \bar{c}_\eta(\ell)')(Y_\eta)_d, (\bar{c}^*(\tilde{\ell})' - \bar{c}_\eta(\tilde{\ell})')(Y_\eta)_d) \\
&+\text{cov} \quad (X_\eta(\ell) - \bar{c}^*(\ell)'(Y_\eta)_d, (\bar{c}^*(\tilde{\ell})' - \bar{c}_\eta(\tilde{\ell})')(Y_\eta)_d) \\
&+\text{cov} \quad ((\bar{c}^*(\ell)' - \bar{c}_\eta(\ell)')(Y_\eta)_d, X_\eta(\tilde{\ell}) - \bar{c}^*(\tilde{\ell})')Y_\eta)_d).
\end{aligned}
$$

The first term of the sum for any η is equal to

$$
(1, -\bar{c}^*(\ell)')(\tau^2 M_{d+} + \sigma^2 \Sigma^0_{d+}) \begin{pmatrix} 1 \\ -\bar{c}^*(\tilde{\ell}) \end{pmatrix}.
$$

because $(1, -\bar{c}^*(\ell)')V_{d+} = (1, -\bar{c}^*(\tilde{\ell})')V_{d+} = 0$. The third and fourth term for $\eta \to \infty$ tend to zero because of (2.33). By the Cauchy-Schwarz inequality the second term is bounded by the product of standard deviations. Thus it remains to show

$$
\lim_{\eta \to \infty} \text{var} \left((\bar{c}^*(\ell)' - \bar{c}_\eta(\ell)')(Y_\eta)_d \right)
$$
$$
\lim_{\eta \to \infty} \text{var} \left(((1 - \bar{c}_\eta(\ell)') - (1, -\bar{c}^*(\ell)')) \begin{pmatrix} X(\ell) \\ (Y_\eta)_d \end{pmatrix} \right) = 0.
$$

Repeating the argument this follows from (3.7). $\qquad\square$

Corollary 7.1 The limiting posterior distribution does not depend on the decomposition design.

Proof. The proof of theorem 7 is valid for any decomposition design. $\qquad\square$
According to the decomposition

$$
f_{d+} = (P_1 f)_{d+} + (P_2 f)_{d+}
$$

the limit $\eta \to \infty$ describes a partially improper prior on f_{d+}. Usually the prior distributions $(P_1 f)_{d+}$ and $(P_2 f)_{d+}$ are defined separately. Therefore non-degenerate priors on $(P_2 f)_{d+}$, particularly with non-singular M_d arising from $\Lambda \cap d = \emptyset$, are the only ones that are considered. No problem however occurs using a prior according to an arbitrary decomposition design.

In Bayesian inference also the predictive distribution, the marginal distribution of the observations given the prior, is often of interest for model checks or diagnostics. As the prior is partially improper one usually refers to the marginal distribution of $U = R(Y_\eta)_d$ with $RV_d = 0$ as an appropriate predictive distribution.

Theorem 8. The predictive distribution of

$$
U \sim N(0, \tau^2 R M_d R' + \sigma^2 R \Sigma_d R')
$$

is independent of Λ.

Proof. Lemma 2. □

3.2 The signal extraction approach The congruence mapping Φ_η^{-1} restricted to $(\overline{P}_2 + \overline{P}_3)(H(\overline{K}_\eta))$ links to $\overline{L}$ a random variable $Z(\ell)$ with

$$\text{var } Z(\ell) = \tau^2 \langle P_2 L, P_2 L \rangle$$

and to L_i^S a random variable $Y(\ell_i) = Z(\ell_i) + \varepsilon(\ell_i)$ such that

$$
\begin{aligned}
\text{cov } Y_d &= \tau^2 M_d + \sigma^2 \Sigma_d. \\
\text{cov } \begin{pmatrix} z(\ell) \\ Y_d \end{pmatrix} &= \tau^2 M_{d+} + \sigma^2 \Sigma_{d+}^0.
\end{aligned}
$$

Assigning means

$$E(Z(\ell)) = \ell(P_1 f) \tag{3.9}$$

it is well established (Kimeldorf and Wahba, 1970) that interpolation in $H(K_\eta)$ is congruent to unbiased prediction of $Z(\ell)$ given Z_d and smoothing in $H(\overline{K}_\eta)$ is congruent to unbiased smoothing of $Z(\ell)$ given Y_d for any η. Therefore invariance of the predictor, smoother and the corresponding errors w.r.t. Λ is directly obtained from theorems 1, 2 and 3. However, as far as the likelihoodfunction is concerned and possibly used for estimating the smoothing parameter, these is a dependence on Λ.

3.3 The minimax approach Restricting the congruence mapping to $\overline{P}_3(H(\overline{K}_\eta))$ only, one comes back to (2.16) with

$$E(Y_d) = f_d. \tag{3.10}$$

The smoothing spline solves (Li, 1982)

$$\sup_{\substack{h \in H(K_\eta) \\ J(h) \le \tau^2}} E(\ell(h) - \ell(h_{S(y_d)}))^2 = \min_{c \in \mathbb{R}^n} \sup_{\substack{h \in H(K_\eta) \\ J(h) \le \tau^2}} E(\ell(h) - c' y_d)^2 \tag{3.11}$$

and the right hand side of (3.11), the minimax error, can be shown to be equal to $e_S(\ell)$, invariantly w.r.t. Λ. Within this approach the distributional assumptions are independent of Λ and thus any likelihood argument is not affected. Cross-validatory methods for determining the smoothing parameter (e.g. Wahba, 1990, ch. 4 or Robinson and Moyeed, 1989) depend only on $A(\varrho)$ which is invariant w.r.t. Λ.

4 The smoothing spline as orthogonal projection

Having found the inner product (2.18) most convenient for statistical purposes we now turn to the smoothing spline as an orthogonal projection. A corresponding inner product on $\overline{H}$ was suggested by Weinert (1978) based on a decomposition of H according to $\tilde{\Lambda} = d_M$ which we fix in the sequel without further indexing. Set $d_M^S = \{L_1^S, \ldots, L_M^S\}$ and

$$\begin{aligned}
\binom{h}{\delta}_{d_M^S} &:= \left(\langle L_1^S, \binom{h}{\delta}\rangle^-, \ldots, \langle L_M^S, \binom{h}{\delta}\rangle^-\right)' \\
&= \left(\langle L_1, h\rangle + \delta_1, \ldots, \langle L_M, h\rangle + \delta_M\right)' \\
\Sigma_{d_M} &:= (I_M \ 0)\Sigma_d \binom{I_M}{0}.
\end{aligned}$$

Generalizing to an arbitrary positive definite Σ_d and weighing, Weinert's inner product then is

$$\langle \binom{g}{\gamma}, \binom{h}{\delta}\rangle_w^- := \frac{1}{\eta}\binom{g}{\gamma}_{d_M^S} \Sigma_{d_M}^{-1} \binom{h}{\delta}_{d_M^S} + \frac{1}{\tau^2}\langle P_2 g, P_2 h\rangle + \frac{1}{\sigma^2}\gamma'\Sigma_d^{-1}\delta. \quad (4.1)$$

The induced topology is equivalent to that of $\langle \cdot, \cdot \rangle^-$ and straightforward computations show that the corresponding reproducing kernel is

$$\overline{K}_w(\cdot, t) \ = \ \begin{cases} \begin{pmatrix} (\eta + \sigma^2)v(\delta_t)'\Sigma_{d_M}v(\cdot) \\ \tau^2 K_2(\cdot, t) \\ -\sigma^2\Sigma_d\binom{I_M}{0}v(\delta_t) \end{pmatrix} \in \begin{array}{c} H_1 \\ \times \\ H_2 \\ \times \\ \mathbb{R}^n \end{array} \quad t \in S \\[2em] \begin{pmatrix} -\sigma^2\sigma_i'\binom{I_M}{0}v(\cdot) \\ 0 \\ \sigma^2\sigma_i \end{pmatrix} \in \begin{array}{c} H_1 \\ \times \\ H_2 \\ \times \\ \mathbb{R}^n \end{array} \quad t = i \in T_n \end{cases}$$

$$(4.2)$$

ℓ and ℓ_i^S are represented by

$$L_w = \begin{pmatrix} (\eta + \sigma^2)v(\ell)'\Sigma_{d_M}v(\cdot) \\ \tau^2 P_2 L \\ -\sigma^2\Sigma_d\binom{I_m}{0}v(\ell) \end{pmatrix} \quad (4.3)$$

$$(L_i^S)_w = (L_i)_w + \overline{K}_w(\cdot, i) \quad (4.4)$$

where $(L_i)_w$ is formed analogously to L_w with ℓ replaced by ℓ_i on the right hand side of (4.3).

Then, given observations

$$y_i = \langle (L_i^S)_w, \binom{h}{\delta}\rangle_w^- \qquad i = 1, \ldots, n \quad (4.5)$$

the smoothing problem $(\overline{S})$ is equivalent to norm-minimization in $H(\overline{K}_w)$ subject to (4.5), and the smoothing spline $h_{S(y_d)}^\varrho, \varrho = \sigma^2/\tau^2$ results as the component in H from orthogonal projection in $H(\overline{K}_w)$.

Denote by L_w^* the best approximating functional to L_w in $H(\overline{K}_w)$ in the sense of Sard.

Theorem 9

$$e_S(\ell) =\parallel L_w - L_w^* \parallel^2_{H(\overline{K}_w)} \qquad (4.6)$$

Proof.

$$\langle L_w^*, \binom{h}{\delta}\rangle_{\overline{w}} = \ell(h^S_{S(y_d)}) = \bar{c}^*(\ell)' y_d = \langle \bar{c}^*(\ell) \begin{pmatrix} (L_1^S)_w \\ \vdots \\ (L_n^S)_w \end{pmatrix}, \binom{h}{\delta}\rangle_{\overline{w}}.$$

Thus $L_w - L_w^*$ represents $\ell_e := \ell - \bar{c}^*(\ell)' \begin{pmatrix} \ell_i^S \\ \vdots \\ \ell_i^S \end{pmatrix}$ in $H(\overline{K}_w)$ and annihilates H_1. Any such linear functional annihilating H_1 with Riesz representer $\begin{pmatrix} m_1 \\ m_2 \\ \mu \end{pmatrix}$ in $H(\overline{K})$ is represented in $H(\overline{K}_w)$ by

$$m_w = \begin{pmatrix} -\mu'\binom{I_m}{0} v(\cdot) \\ m_2 \\ \mu \end{pmatrix}$$

for

$$\langle m_w, \binom{h}{\delta}\rangle_{\overline{w}} = \frac{1}{\tau^2}\langle m_2, P_2 h\rangle + \frac{1}{\sigma^2}\mu' \Sigma_d^{-1}\delta = \langle m, \binom{h}{\delta}\rangle^-.$$

In particular

$$\parallel m \parallel^2_{H(\overline{K})} = \parallel (\overline{P}_2 + \overline{P}_3)m \parallel^2_{H(\overline{K})} = \parallel m_w \parallel^2_{H(\overline{K})}.$$

$\square$

In consequence advantages resulting from a representation as orthogonal projection may be exploited theoretically (simplifying proofs) or computationally. The main statistical interpretations remain valid, because they hold independently of the decomposition design.

References

[1] Ansley, C.F., Kohn, R. (1986). On the equivalence of two stochastic approaches to spline smoothing, J. Appl. Prob. 23A, 391-405

[2] Böhmer, K. (1974). Spline Funktionen, Teubner, Stuttgart

[3] Kimeldorf, G.S., Wahba, G. (1970). Spline functions and stoachstic processes, Sankya A 132, 173-180

[4] Kimeldorf, G.S., Wahba, G.(1971). Some results on Tschebycheffian spline functions, J. Math. Anal. Appl. 33, 82-95

[5] Larkin, F.M.(1983). The weak Gaussian distribution as a means of localization in Hilbert space, in: Gorenflo,R., Hoffmann, K.H., (Eds.), Applied Nonlinear Functionalanalysis, Verlag Peter Lang, Frankfurt a.M.

[6] Li, K.C.(1982). Minimaxity of the method of regularization on stochastic processes, Ann. Statist. 10, 937-942

[7] Meinguet, J.(1979) Multivariate interpolation at arbitrary points made simple, J. Appl. Math. Phy. 30, 292-304

[8] Meinguet, J.(1984). Surface spline interpolation. Basic theory and computational aspects, in: Singh, S.P.(1984), (Ed.), Approximation Theory and Spline Functions, Reidel Publ. Comp.,

[9] Parzen, E.(1959). Statistical inference on time series by Hilbert space methods I, Technical Report No 23, Statistics Department, Standford University, (reprinted in Parzen, E.(1967). Time Series Analysis, San Francisco)

[10] Peele, L., Kimeldorf, G.S.(1977) Prediction functions and mean estimation functions for a time series, Ann. Statist. 5, 709-721

[11] Robinson, T., Moyeed, R. (1989) Making robust the cross-validatory choice of smoothing parameter in spline smoothing regression, Comm. Statist.-Theor. Meth. 18, 523-539

[12] Sard, A. (1967). Optimal Approximation, J. Functional Analysis 1, 222-244

[13] Silverman, B., Some aspects of the spline smoothing approach to non-parametric regression curve fitting, J. Roy. Statist. Soc. B 47, 1-52 (with discussion)

[14] Wahba, G. (1978). Improper priors, spline smoothing and the problem of guarding against model errors in regressions, J. Roy. Statist. Soc. B 40, 364-372

[15] Wahba, G. (1984). Cross-validated spline methods for the estimation of multivariate functions from data on functionals, in David, H.A., (Ed.), Statistics. An Appraisal, Iowa State University Press

[16] Wahba, G. (1990). Spline models for observational data, Society for Industrial and Applied Mathematics, Philadelphia, P.A.

[17] Weinert, H.L. (1978). Statistical methods in optimal curve fitting, Comm. Statist. – Theor. Meth. 7, 417-435

[18] Weinert, H.L., Kailath, T. (1974). Stochastic interpretations and recursive algorithms for spline functions, Ann. Statist. 2, 787-794

[19] Weinert, H.L., Sidhu, G.S., Byrd, R.H. (1977). Stochastic error analysis of spline approximation, in Proc. IEEE conf. on Decision and Control, New York

A Note on Cross-Validation for Smoothing Splines

Gerhard P. Neubauer
Michael G. Schimek

Medical Biometrics Group, University of Graz Medical Schools,
A-8036 Graz, Austria

Summary

Two methods for choosing the smoothing parameter λ (generalized cross-validation, Wahba and Wold, 1975, and robustified generalized cross-validation, Robinson and Moyeed, 1989) are compared in a simulation study. It turns out that robustified cross-validation performs better. Computational problems of finding the cross-validation score are discussed. Findings from linearly transformed data lead to a reduction in costs. In addition we consider problems of variance estimation. Two competitors for error variance estimation turn out to perform equally well, and the estimate for the variance of the Bayesian prior appears to be useful for describing the complexity of the estimated function.

Keywords: Cross-validation, cubic smoothing splines, design variable, linear data transformation, non-parametric regression, robustified cross-validation, smoothing parameter, variance estimation.

1 Introduction

Cubic smoothing splines are widely used in nonparametric regression (Silverman, 1985; Wahba, 1990; Green and Silverman, 1994). They can be derived as the solution, $\hat{g}_\lambda$, to the penalized least squares problem

$$S(\lambda) = \sum_i (y_i - g(x_i))^2 + \lambda \int (g''(x))^2 dx, \tag{1}$$

where $\int (g''(x))^2$ serves as a roughness measure. The so-called smoothing parameter, λ $(0 < \lambda < \infty)$, controls the trade-off between fidelity to the data and the smoothness of the estimated curve. We have data interpolation for $\lambda = 0$ and the linear fit for $\lambda = \infty$. The choice of λ is therefore crucial for the shape of the estimated function. Generalized cross-validation (Craven and Wahba, 1979) is the standard automatic procedure for λ determination. It is known that cross-validation can cause under-smoothing. Therefore Robinson and Moyeed (1989) introduced a robustification of generalized cross-validation. They also illustrate the performance of their method by some simulations, where the values of error variance ($\sigma(\epsilon) = \{0.02, 0.04, 0.06, 0.08, 0.10\}$) are rather small. We study the performance of the generalized and the robustified generalized cross-validation in a simulation experiment under more realistic assumptions.

In addition we consider a computational problem of applying cross-validation methods, and problems concerning estimators for σ^2 and $\tau^2 = \sigma^2/\lambda$.

2 Basic concepts of cross-validation

The standard procedure for the choice of λ is cross-validation (Wahba and Wold, 1975). The principle is to leave the data points out one at a time and to select that value of λ under which the missing data points are best predicted by the remainder of the data. Let $\hat{g}_\lambda^{-i}$ be the smoothing spline calculated from all the data pairs except (x_i, y_i), for the value λ, then the cross-validation score is defined by

$$XVSC(\lambda) = n^{-1} \sum_{i=1}^{n} (y_i - \hat{g}_\lambda^{-i}(x_i))^2.$$

Minimizing $XVSC$ with respect to λ provides the smoothing parameter value. Although it is not necessary to evaluate $n+1$ splines for each value of λ the computational burden is heavy. To reduce the costs Craven and Wahba (1979) introduced a generalized score based on the relation

$$y_i - \hat{g}_\lambda^{-i}(x_i) = \frac{y_i - \hat{g}_\lambda(x_i)}{1 - a_{ii}^{(\lambda)}},$$

where $a_{ii}^{(\lambda)}$ is the ith diagonal element of the hat matrix A_λ satisfying $\hat{g} = A_\lambda y$ with $\hat{g} = (\hat{g}_\lambda(x_1), \hat{g}_\lambda(x_2), \ldots, \hat{g}_\lambda(x_n))^T$ and $y = (y_1, y_2, \ldots, y_n)^T$. Replacing the $a_{ii}^{(\lambda)}$ by their average value $\mu_1(\lambda) = n^{-1} trace(A_\lambda)$ yields the generalized cross-validation score

$$GXVSC(\lambda) = n^{-1} \sum_{i=1}^{n} \frac{(y_i - \hat{g}_\lambda(x_i))^2}{(1 - \mu_1(\lambda))^2}.$$

Craven and Wahba (1979) as well as others could demonstrate good statistical properties for $GXVSC$. But it is generally known that cross-validation can cause undersmoothing, data interpolation being the worst case.

Robinson and Moyeed (1989) have put forward a robustification of generalized cross-validation. The robust score is given by

$$RGXVSC(\lambda) = n^{-1} \frac{1 + \mu_2}{(1 - \mu_1)^2} RSS(\lambda),$$

where $\mu_2 = n^{-1} trace(A_\lambda^2)$ and $RSS(\lambda) = \sum_{i=1}^{n} (y_i - \hat{g}_\lambda(x_i))^2$.

3   Some problems of application

The minimum of the mentioned score functions cannot be found analytically. Instead λ is determined by an iterative grid search on a log-scale. To start one has to consider the real line. It is obvious that information about the position and the size of a starting interval could substantially reduce the computational costs.

Our two main questions are:

(1)   What is the gain of $RGXVSC$ compared to $GXVSC$ in terms of the properties of the estimated functions?

(2)   Is there any information in the data that helps to choose an appropriate starting interval for the search of the empirical minimum of the cross-validation score?

Dealing with question (2) we find that the scale of the data is important. Therefore a further question arises:

(2.1)   How is λ influenced by changes in the scale of the data.

Our simulated data are linearly transformed in order to reveal the influence of the scaling on the magnitude of λ.

For the estimation of the error variance in analogy to linear regression equivalent degrees of freedom (edf) are required. Most often $edf_1 = trace(I - A_\lambda)$ is used, e.g. in Wahba (1990). Based on a general approach for quadratic forms Buckley, Eagleson and Silverman (1988) propose to take $edf_2 = trace[(I - A_\lambda)^2]$ instead. We compare both estimators and denote the first by $\hat{\sigma}_1^2 (= RSS/edf_1)$ and the second by $\hat{\sigma}_2^2 (= RSS/edf_2)$.

It is known that $\lambda = \sigma^2/\tau^2$, with σ^2 the error variance, and τ^2 the variance of the Bayesian prior distribution (Wahba, 1990). According to Van der Linde (1995), τ^2 expresses the size of the class of possible regression functions. τ^2 can be interpreted as the smoothing parameter corrected for error variance, hence reflecting the complexity of the function. Estimating it through $\hat{\tau}^2 = \sigma^2/\hat{\lambda}$ (σ^2 known for simulations) should give some idea of the complexity of the estimated function. This leads to the following additional questions:

(3.1)   Do the estimators of error variance perform differently?

(3.2)   How does $\hat{\tau}^2$ reflect the complexity of the estimated function?

4 Outline of the simulation study

All the program routines were written in GAUSS and the computations performed on a 486 PC. For the simulations we created sets of independent design points for the variable X, uniformly distributed over the interval $[0, 1]$, errors $\epsilon_i \sim N(0, \sigma^2(\epsilon))$, and a response variable $Y = g(X) + \epsilon$ represented by functions $g(X)$ and an error term as specified. The considered values for $\sigma(\epsilon)$ were 0.1, 0.2, 0.3, and 0.4. The sample size is always $n = 75$ (a realistic number in practical data analysis).

We investigate a model M_1 based on a single error sample with different levels of error variance. Further we study a model M_2 with different error samples and error variances. As a special case of these models the deterministic model $Y = g(X)$ is considered as well. The data we analyse have signal-to-noise ratios $(snr = \sigma(g(X))/\sigma)$ between 0.963 and 6.082 and are therefore not dominated by the noise. We consider three functions, different in shape and complexity, $g_1(X) = 4|X - 0.5| - 1$, $g_2(X) = \sin(2\pi(1 - X)^2)$, and $g_3(X) = 21X^2(1 - X)^2 \sin((2\pi(X - 0.5))^3)$. The top row of *Figure 1* shows the three functions together with one set of data (M_1 with $\sigma(\epsilon) = 0.4$).

For all the above combinations generalized cross-validation and robustified generalized cross-validation for cubic smoothing splines with knots equal to the design points (Schimek et al., 1993) was applied. The minimum of a cross-validation function was obtained empirically by a successive search in interlocked intervals on a log-scale.

As major results of the simulations we report the estimated values $\hat{\lambda}$. To describe the quality of the curve estimation we use $SSD = \sum_i d_i^2$, the sum of squared distances $d_i = g(x_i) - \hat{g}(x_i)$, between the true and the estimated function. Furthermore we use plots of true and estimated functions to evaluate the outcomes.

5 Results

Presenting the results we follow the questions posed in section 3. We use the abbreviations $G = GXVSC$, $R = RGXVSC$, $g_j = g_j(X)$, and $\sigma^2 = \sigma^2(\epsilon)$.

5.1 G compared to R: results from the simulations

As an example for the majority of the results we plot the estimates from those data sets given in the first row of *Figure 1*. The rows two and three of *Figure 1* show the estimated functions (dotted lines) together with the true functions (solid lines). We see that under both methods G and R the true

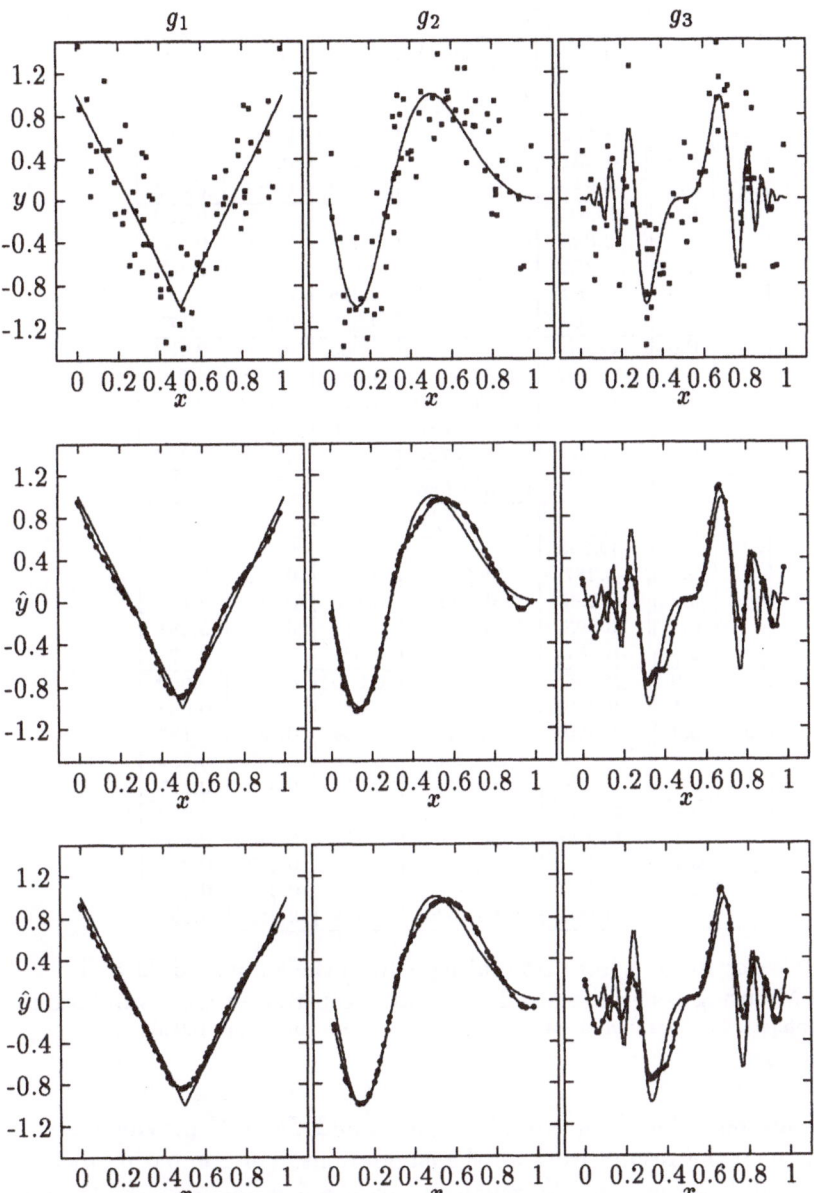

Figure 1: Data and estimates for model M_1 with $\sigma(\epsilon) = 0.4$.
Row 1: Data and true functions. Row 2: Estimates based on G. Row 3: Estimates based on R.
Dots: data. Solid lines: true functions. Dotted lines: estimated functions.

	σ	$\ln\hat\lambda$	$\hat\tau^2 = \sigma^2/\hat\lambda$	SSD	$\hat\sigma_1$	$\hat\sigma_2$
			Generalised cross-validation			
	0.0	-30.630	−	0.000	0.000	0.000
	0.1	-8.826	68.138	0.065	0.091	0.093
g_1	0.2	-7.862	103.875	0.146	0.183	0.185
	0.3	-7.331	137.343	0.248	0.275	0.279
	0.4	-6.940	165.361	0.372	0.368	0.372
	0.0	-28.912	−	0.000	0.000	0.000
	0.1	-9.835	186.844	0.061	0.090	0.092
g_2	0.2	-9.238	411.399	0.199	0.181	0.185
	0.3	-8.782	586.710	0.385	0.273	0.278
	0.4	-8.326	660.881	0.598	0.367	0.373
	0.0	-42.447	−	0.000	0.000	0.000
	0.1	*-29.501*	*6.491E10*	*0.621*	*0.015*	*0.017*
g_3	0.2	-12.964	17 086.196	1.129	0.198	0.207
	0.3	-12.163	17 252.880	2.075	0.286	0.296
	0.4	-11.572	16 973.736	3.169	0.374	0.386
			Robustified cross-validation			
	σ	$\ln\hat\lambda$	$\hat\tau^2 = \sigma^2/\hat\lambda$	SSD	$\hat\sigma_1$	$\hat\sigma_2$
	0.0	-30.598	−	0.000	0.000	0.000
	0.1	-8.267	38.971	0.061	0.092	0.093
g_1	0.2	-7.288	58.544	0.128	0.185	0.187
	0.3	-6.693	72.655	0.217	0.278	0.281
	0.4	-6.181	77.405	0.338	0.372	0.375
	0.0	-28.903	−	0.000	0.000	0.000
	0.1	-9.330	112.769	0.051	0.091	0.093
g_2	0.2	-8.500	196.621	0.164	0.184	0.187
	0.3	-7.938	252.290	0.330	0.278	0.282
	0.4	-7.523	296.165	0.551	0.372	0.377
	0.0	-42.434	−	0.000	0.000	0.000
	0.1	-14.133	13 475.995	0.411	0.111	0.118
g_3	0.2	-12.579	11 616.225	1.213	0.202	0.210
	0.3	-11.706	10 920.329	2.220	0.291	0.301
	0.4	-11.000	9 585.976	3.394	0.382	0.393

Table 1a: Results from estimating λ from the simulated data of Model M_1 using generalized and robustified generalized cross-validation
Emphasized entries denote cases of undersmoothing; note that 0.000 > zero.

functions are reproduced quite well. *Table 1a* and *Table 1b* provide detailed information on the results from all the simulations. For comparing method G to method R we use the discrepancy measure $\delta = \ln(\hat\lambda_R/\hat\lambda_G)$. We find very large values in two cases: M_1 with g_3 and $\sigma = 0.1$, and M_2 with g_1 and $\sigma = 0.2$. The columns one and two of *Figure 2* show the data, the true functions and the estimated functions for these two cases. It is obvious that $\hat\lambda_G$ undersmoothes the data. Checking all the plots a third case of undersmoothing is found where both G and R fail to give appropriate values for the smoothing parameter: M_2 with g_2 and $\sigma = 0.4$ (column three of *Figure 2*).

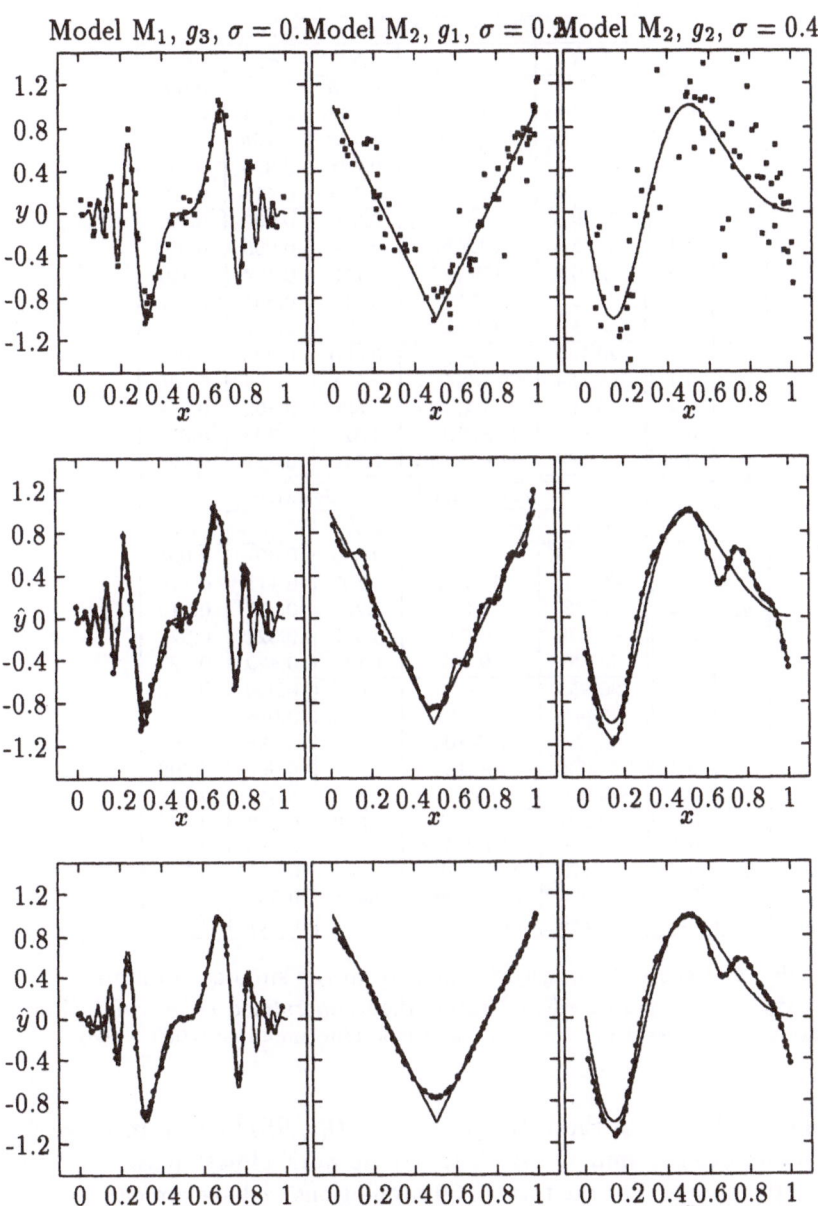

Model M_1, g_3, $\sigma = 0.2$ Model M_2, g_1, $\sigma = 0.2$ Model M_2, g_2, $\sigma = 0.4$

Figure 2: Three cases of undersmoothing.

Row 1: Data and true functions. Row 2: Estimates based on G. Row 3: Estimates based on R.

Dots: data. Solid lines: true functions. Dotted lines: estimated functions.

	σ	$\ln \hat{\lambda}$	$\hat{\tau}^2 = \sigma^2/\hat{\lambda}$	SSD	$\hat{\sigma}_1$	$\hat{\sigma}_2$
			Generalised cross-validation			
g_1	0.0	-18.557	–	0.000	0.001	0.001
	0.1	-7.830	25.161	0.092	0.114	0.115
	0.2	*-11.233*	*3025.770*	*0.733*	*0.190*	*0.196*
	0.3	-5.199	16.307	0.641	0.323	0.325
	0.4	-6.167	76.283	1.238	0.418	0.422
g_2	0.0	-21.266	–	0.000	0.000	0.000
	0.1	-8.727	61.720	0.119	0.098	0.099
	0.2	-8.562	209.189	0.279	0.194	0.197
	0.3	-7.327	137.005	0.719	0.304	0.308
	0.4	*-9.262*	*1684.992*	*2.345*	*0.376*	*0.383*
g_3	0.0	-20.130	–	0.000	0.003	0.004
	0.1	-15.154	38 134.391	0.275	0.084	0.030
	0.2	-13.564	31 106.347	1.509	0.203	0.214
	0.3	-12.694	29 322.118	2.327	0.339	0.354
	0.4	-11.786	21 025.526	7.340	0.411	0.426
			Robustified cross-validation			
g_1	0.0	-18.179	–	0.000	0.001	0.001
	0.1	-7.331	15.282	0.107	0.115	0.116
	0.2	-6.112	18.058	0.237	0.217	0.219
	0.3	-4.770	10.614	0.637	0.324	0.326
	0.4	-5.726	49.107	1.148	0.420	0.423
g_2	0.0	-20.959	–	0.000	0.000	0.000
	0.1	-8.441	46.340	0.136	0.098	0.100
	0.2	-8.196	145.104	0.282	0.195	0.198
	0.3	-7.072	106.148	0.722	0.305	0.308
	0.4	*-8.504*	*789.572*	*1.892*	*0.382*	*0.389*
g_3	0.0	-19.901	–	0.000	0.003	0.004
	0.1	-14.736	25 117.744	0.279	0.086	0.092
	0.2	-13.173	21 044.997	1.543	0.208	0.218
	0.3	-12.026	15 034.884	2.406	0.350	0.363
	0.4	-11.158	11 228.803	7.477	0.421	0.435

Table 1b: Results from estimating λ from the simulated data of
Model M_2 using generalized and robustified generalized cross-validation
Emphasized entries denote cases of undersmoothing; note that 0.000 > zero.

Concerning SSD we use the ratio $q_{SSD} = SSD_G/SSD_R$ as a measure for differences between G and R. Both G and R work almost perfect for the deterministic models. For the total of 30 simulations we have a mean $\bar{q}_{SSD} = 1.114$, i.e. on the average G produces estimates that are 11.4% worse than those from R. The ratios for the already mentioned situations where G failed were 1.510 and 3.093, i.e. estimates that are 51.0% and 209.3% worse when G is applied. In the case where both methods fail, we still have estimates that are 23.9% worse when G is used. Excluding all cases of undersmoothing we have a mean $\bar{q}_{SSD} = 1.023$ which we interpret as practically no difference between G and R for regular cases.

Finally, answering question (1), R is able to guard against undersmoothing in most cases.

5.2   An appropriate interval for starting the grid search

This section is structured according to theoretical considerations, simulation results and consequences for question (3.2).

5.2.1   Some theoretical considerations

The question is whether the data contain any relevant information that can be extracted before cross-validation is applied. Of special interest are aspects concerning the magnitude of λ. We know that λ controls the smoothness of the estimates, with data interpolation for $\lambda \to 0$ and with a straight line fit for $\lambda \to \infty$. Therefore the complexity of the function to be estimated is one aspect determining the magnitude of λ. If we think of a situation where data from the same true function are observed with different error variances, it is evident that a larger error variance (more noise) demands more smoothing and hence larger λ. So we identify the error variance as a second aspect concerning the magnitude of λ. Unfortunately we do not have any prior knowledge about the complexity of the true function or the error variance.

Based on the findings from subsection 5.1 there is another aspect relevant for question (2). We observe $\lambda < 0.001$ for our simulated data with $Y, X \in [0, 1]$, whereas for real data with $Y, X \in [a, b]$ we know that $\lambda \gg 1$ in most cases. Hence we assume that this remarkable difference in the magnitude of λ is due to the scale of X and/or Y.

5.2.2   Effects of design scale: results from simulations

To test the above assumption we transform the data M_2 with g_2 linearly, so that $(Y, X) \to (Y^t, X^t)$, determine $\hat{\lambda}^t$ and compare the results. Preliminary considerations show that only $X^t = cX$ $(c \neq 0)$ is of interest (NEUBAUER, 1994). We used $c = \{10, 5, 2\}$ for the transformation models T_1, T_2 and T_3. Generalised cross-validation is used to estimate λ^t. *Table 2* shows the corresponding results. Comparing $\hat{\lambda}^t$ with $\hat{\lambda}$ for the original data, we observe that $c \neq 1$ affects the values $\hat{\lambda}^t$. Moreover we find that $\hat{\lambda}^t \approx \hat{\lambda}^* = c^3 \hat{\lambda}$.

A similar result is presented in COLE and GREEN (1992). The authors propose

$$\lambda = \frac{n}{400} \frac{r(X)^3}{r(Y)^2}$$

as an initial choice for the smoothing parameter in an iterative procedure. They claim that the smoothing parameter depends on $r(X)$, the range of the design as well as on $r(Y)$, the range of the response. Neubauer (1994) shows that linear transformations of the response leave $\hat{\lambda}$ unchanged. Choosing initial $\hat{\lambda}$ by the mentioned procedure should therefore yield too small values, i.e., whenever $r(Y) > 1$.

$\hat{\lambda}$ reflects the amount of smoothing for a given set of data. If we are interested in comparing $\hat{\lambda}$ from different data sets, we should bear in mind that the magnitude of $\hat{\lambda}$ is also sensitive to the scale of the design. Otherwise such comparisons could be misleading. To eliminate the effects of the design scale $\hat{\lambda}_j$ should be estimated from the data (Y, X_j), then the coefficients c_j should be determined such that the ranges r_j are transformed to $c_j r_j = r^t = const$. The comparable estimates for λ can finally be obtained by the approximation $\hat{\lambda}_j^* = c_j^3 \hat{\lambda}_j$. Applying this approach to our data shows that $d = |\ln \hat{\lambda}^t - \ln \hat{\lambda}^*|$ is considerably small. The mean $\bar{\delta} = 0.431$ of the differences $\delta = \ln(\lambda_R/\lambda_G)$ (the two cases where only G fails are excluded) is by far larger than any of the $\bar{d}$ ($\bar{d}$ mean of the d values) in *Table 2*. Hence we can conclude that the above approximation works reasonably well.

σ	$\ln \hat{\lambda}$	T_1 ($c = 10$)		
		$\ln \hat{\lambda}^t$	$\ln \hat{\lambda}^*$	d
0	-28.903	-21.970	-21.995	0.025
0.1	-9.330	-2.401	-2.422	0.021
0.2	-8.500	-1.598	-1.592	0.006
0.3	-7.938	-1.080	-1.030	0.050
0.4	-7.523	-0.613	-0.615	0.002
	$\bar{d}$	–	–	0.021
σ	$\ln \hat{\lambda}$	T_2 ($c = 5$)		
		$\ln \hat{\lambda}^t$	$\ln \hat{\lambda}^*$	d
0	-28.903	-24.072	-24.074	0.002
0.1	-9.330	-4.509	-4.501	0.008
0.2	-8.500	-3.686	-3.671	0.015
0.3	-7.938	-3.136	-3.109	0.027
0.4	-7.523	-2.710	-2.694	0.014
	$\bar{d}$	–	–	0.013
σ	$\ln \hat{\lambda}$	T_3 ($c = 2$)		
		$\ln \hat{\lambda}^t$	$\ln \hat{\lambda}^*$	d
0	-28.903	-26.811	-26.823	0.012
0.1	-9.330	-7.247	-7.250	0.003
0.2	-8.500	-6.440	-6.420	0.020
0.3	-7.938	-5.847	-5.858	0.011
0.4	-7.523	-5.469	-5.443	0.026
	$\bar{d}$	–	–	0.014

Table 2: Results from the simulations
for the transformed data of model M_2 with g_2

5.2.3 An answer based on subsection 5.2.2

We know now that the magnitude of λ is sensitive to the scale, or the range r of the design variable X. An appropriate choice of the interval to start the search for the minimum of the cross-validation score can be made according to the following rule of thumb: The smaller the range r the smaller the resulting $\hat{\lambda}$.

Our findings suggest to expect $\hat{\lambda} < 1$ whenever $r \leq 10$ and $snr > 1$ (snr stands for "signal-to-noise ratio"). By extrapolating our findings we expect $\hat{\lambda} > 1$ whenever $r > 10$ and/or $snr < 1$. As the search for λ is performed on a log-scale we can now divide the real line at zero.

For empirical data snr is unknown and we make the above suggestion conservative by enlarging the intervals by for instance 10. To answer question (2) we propose the starting intervals $(-\infty, 10]$ for $r \leq 10$ and $[-10, \infty)$ for $r > 10$. In our implementation we use $-30 \approx -\infty$ and $30 \approx \infty$ and have therefore a reduction of around 33.3% in the size of the starting interval.

5.3 Variance estimation results

The mean $\bar{q}_{\sigma_k}$ of the ratios $q_{\sigma_k} = \hat{\sigma}_k/\sigma$ is used as a measure for the performance of the variance estimators. The means are $\bar{q}_{\sigma_1} = 0.931$ and $\bar{q}_{\sigma_2} = 0.980$. We find the estimators to perform almost equally, both slightly underestimating the true error variance.

For analysing the $\hat{\tau}^2$ we take the mean over the levels of error variances and the two models (excluding the four cases of undersmoothing). The results are given in *Table 3*. We see that with increasing functional complexity also $\hat{\tau}^2$ increases. Taking the ratios $\bar{\tau}_G^2/\bar{\tau}_R^2$ it shows, that G yields estimates about twice as large as those from R.

The ratios $\bar{\tau}_j^2/\bar{\tau}_1^2$ represent the complexity relative to the simplest function. Taking the square roots of these ratios we obtain values that approximately equal the number of extreme values (1,2,18) in each function, i.e. (1,1.95,15.91) for G, and (1,1.97,18.61) for R. At the moment we cannot say if this observation constitutes a stable pattern. It seems that τ^2 is reflecting the complexity of the estimated function in terms of extreme values.

In conclusion, the answers to the posed questions are: (3.1) There is no essential difference in the performance of the two estimators of error variance. (3.2) $\hat{\tau}^2$ is a measure for the complexity of the estimated function, that is probably related to the number of extreme values of the function.

	Means			Ratios $\bar{\tau}_j/\bar{\tau}_1$		
	$\bar{\tau}_1$	$\bar{\tau}_2$	$\bar{\tau}_3$	$\bar{\tau}_1/\bar{\tau}_1$	$\bar{\tau}_2/\bar{\tau}_1$	$\bar{\tau}_3/\bar{\tau}_1$
G:	84.64	321.96	21 413.45	1	3.80	253.00
R:	42.58	165.08	14 753.11	1	3.87	346.48

Ratios $\bar{\tau}_G/\bar{\tau}_R$	1.99	1.95	1.45

Table 3: Means and ratios of $\hat{\tau}^2$

6 Conclusions

The robustified cross-validation performs better than the generalized cross-validation, but may still yield values for λ that undersmooth the data. Therefore in data analysis always plots should be made to check the obtained estimation results.

The suggestions for appropriate starting intervals for the grid search are based on findings from simulations with linearly transformed data. We observe that $\hat{\lambda}$ depends on the design scale. We propose to reduce the real line to two overlapping intervals for search. The computational gain is around 33%.

The variance investigations show that there is essentially no difference between the two estimators of error variance, and support the usefulness of $\hat{r}^2$ in describing the complexity of the estimated function.

References

Buckley, M.J., Eagleson, G.K. and Silverman, B.W. (1988). The estimation of residual variance in nonparametric regression. *Biometrika, 75*, 189-199.

Cole, T.J. and Green, P.J. (1992). Smoothing reference centile curves: The LMS method and penalized likelihood. *Statistics in Medicine, 11*, 1305-1319.

Craven, P. and Wahba, G. (1979). Smoothing noisy data with spline functions. Estimating the correct degree of smoothing by the method of generalized cross-validation. *Numer. Math., 31*, 377-403.

Green, P.J. and Silverman, B.W. (1994). *Nonparametric Regression and Generalized Linear Models.* Chapman and Hall, London.

Neubauer, G.P. (1994). Effects of linear transformations on estimation in non-parametric spline regression. *Research Memorandum No.41*, Department of Statistics, University of Economics and Bussiness Administration, Vienna, Austria.

Robinson, T. and Moyeed, R. (1989). Making robust the cross-validatory choice of smoothing parameter in spline smoothing regression. *Commun. Statist. - Theory Meth., 18*, 523-539.

Schimek, M.G., Stettner, H., Haberl, J. and Neubauer, G.P. (1993). Approaches for fitting additive models with cubic splines and associated problems. *Bulletin de l'Institut International de Statistique, 49ème Session, Livraison 2*, 381-382.

Silverman, B.W. (1985). Some aspects of the spline smoothing approach to nonparametric regression curve fitting (with discussion). *J.Roy.Statist.Soc. B, 46*, 1-52.

Van der Linde, A. (1995). Computing the error for smoothing splines. *Computational Statistics, 10*, 143-154.

Wahba, G. (1990). *Spline models for observational data.* SIAM, Philadelphia, Pennsylvania.

Wahba, G. and Wold, S. (1975). A completely automatic French curve: Fitting spline functions by cross-validation. *Commun. Statist. 4*, 1-17.

Some Comments on Cross-Validation

Bernd Droge

Institute of Mathematics, Humboldt University,
Unter den Linden 6, D-10099 Berlin, Germany

Summary

A new variant of cross-validation, called full cross-validation, is proposed in order to overcome some disadvantages of the traditional cross-validation approach in general regression situations. Both criteria may be regarded as estimates of the mean squared error of prediction. Under some assumptions including normally distributed observations, the cross-validation criterion is shown to be outperformed by the full cross-validation criterion. Analogous modifications may be applied to the generalized cross-validation method, providing a similar improvement. This leads to the recommendation of replacing the traditional cross-validation techniques by the new ones for estimating the prediction quality of models or of regression function estimators.

Keywords: Parametric and nonparametric regression, selection of models and smoothing parameters, mean squared error of prediction, cross-validation, full cross-validation, generalized cross-validation.

1   Introduction

One of the main objectives in regression analysis is the prediction of future values of the response variable for some fixed values of the explanatory variables. This problem is closely related to that of estimating the unknown regression function (see e.g. Bunke and Droge, 1984b). The prediction is usually done on the basis of an estimate of the regression function. Frequently, the information available on the problem under consideration does not favour a specific estimate, so that we have to choose a good one from those being tentatively proposed.

For this selection process, many data-driven procedures have been discussed in the literature, and those based on the cross-validation criterion of Stone (1974) and the generalized cross-validation criterion of Craven and Wahba (1979) belong to the most popular of them. The use of both criteria may be motivated by being estimates of the mean squared error of prediction (MSEP), which provides an assessment of an estimator's prediction performance. For a more detailed discussion of this aspect and for a comparison of cross-validation and generalized cross-validation with other MSEP estimates in the linear, nonlinear and nonparametric regression problem, we refer to Bunke and Droge (1984a), Droge (1987) and Eubank (1988), respectively.

One advantage of cross-validation and generalized cross-validation over some other selection criteria such as Mallows' (1973) C_p is that they do not require estimation of the error variance. However, both criteria have also some disadvantages which have partly been elaborated, for example, in Bunke and Droge (1984a) for the linear regression case. For instance, cross-validation overestimates the MSEP in general, and increasing model misspecifications as well as increasing parameter dimensions are increasingly disadvantageous for cross-validation compared with other MSEP estimates such as C_p or some variant of the bootstrap. Additionally, cross-validation may fail in some nonlinear regression situations, as it will be illustrated in Section 3.

Our main goal in this paper is to propose a new variant of cross-validation, called full cross-validation, to overcome this failure. We address this issue in Section 3 after first describing the general framework of this paper in Section 2. In an analogous way, we introduce also a counterpart to the generalized cross-validation. The comparison of the different criteria as estimates of the MSEP is carried out in Sections 4 and 5 under the assumption of linear estimates and normally distributed observations. The comparisons show that the new variants of cross-validation outperform the traditional ones. Besides others, this is discussed in Section 6.

2   The Mean Squared Error of Prediction

We consider a regression situation where we have observations of a response variable Y at n predetermined values of a vector of, say k, explanatory variables $x = (x^1, \ldots, x^k)$. The resulting values of (x, Y), (x_i, y_i) for $i = 1, \ldots, n$, follow the model

$$y_i = f(x_i) + \varepsilon_i, \quad i = 1, \ldots, n, \tag{2.1}$$

where the ε_i are assumed to be zero mean, uncorrelated random errors having some common variance σ^2, and f is an unknown real-valued regression function that we wish to estimate.

The prediction problem could be formulated as predicting the values of uncorrelated random variables $y_1^*, \ldots, y_n^*$ with

$$\mathrm{E}(y_i^*) = f(x_i), \quad \mathrm{Var}(y_i^*) = \sigma^2,$$

using the vector of observations $y = (y_1, \ldots, y_n)^T$ and assuming that y and $y^* = (y_1^*, \ldots, y_n^*)^T$ are uncorrelated.

To estimate the regression function $f(.)$ we consider now a class of estimators $f_{n,\lambda}(.; y_1, \ldots, y_n)$ (or, in brief, $\hat{f}_\lambda(.)$), indexed by $\lambda \in \Lambda$. Λ represents some index set, and the values of its elements may be scalars, vectors or sets.

Then, y_i^* is usually predicted by $\hat{y}_i(\lambda) = \hat{f}_\lambda(x_i)(i = 1, \ldots, n)$. The performance of the predictor indexed by λ will be described by a mean squared error (MSEP),

$$MSEP(\lambda) = \frac{1}{n} \sum_{i=1}^{n} \mathrm{E}(y_i^* - \hat{y}_i(\lambda))^2 = \frac{1}{n} \mathrm{E}\|y^* - \hat{y}(\lambda)\|^2, \tag{2.2}$$

where $\hat{y}(\lambda) = (\hat{y}_1(\lambda), \ldots, \hat{y}_n(\lambda))^T$ and $\|.\|$ denotes the Euclidean norm of $\mathbb{R}^n$.

Clearly,

$$MSEP(\lambda) = \sigma^2 + MSE(\lambda),$$

where $MSE(\lambda) = \frac{1}{n} \sum_{i=1}^{n} \mathrm{E}(f(x_i) - \hat{y}_i(\lambda))^2$ is the mean squared error for estimating the regression function. Therefore, the prediction problem is closely related to the problem of estimating f.

Many of the existing regression function estimators in both parametric and nonparametric problems are linear in the observations (for a fixed index λ), that is, for each x there exists a n-vector $h(x, \lambda)$ with $\hat{f}_\lambda(x) = h(x, \lambda)^T y$. Thus we may write

$$\hat{y}(\lambda) = H(\lambda)y, \tag{2.3}$$

where the i-th row of the $n \times n$-matrix $H(\lambda) := ((h_{ij}(\lambda)))_{i,j=1,\ldots,n}$ is just $h(x_i, \lambda)^T$. The matrix $H(\lambda)$ will usually be symmetric (and, sometimes, non-negative definite). But nonlinear estimators may be investigated as well. The following examples illustrate the different nature of possible estimators.

Example 1 (*Selection of linear models and Fourier series regression estimation*). Suppose there are p known functions of the explanatory variables, say $g_1, \ldots, g_p$, associated with the response variable, and the aim is to approximate the regression function by an appropriate linear combination of some of these functions. Then Λ should be some subset of the power set of $\{1, \ldots, p\}$, and the problem is to select a "good" set $\lambda \in \Lambda$. For $p = k$ and $g_i(x) = x^i (i = 1, \ldots, p)$ one would obtain the variable selection problem in multiple linear regression. Using the least squares approach for fitting the models to the data gives, for each $\lambda \in \Lambda$, the following estimator of $f(x)$

$$\hat{f}_\lambda(x) = \sum_{i \in \lambda} \hat{\beta}_i^{(\lambda)} g_i(x),$$

where the coefficients $\hat{\beta}_i^{(\lambda)}$ are the minimizers of

$$\sum_{j=1}^{n} (y_j - \sum_{i \in \lambda} \beta_i g_i(x_j))^2 \qquad (2.4)$$

with respect to β_i ($i \in \lambda$). Then $H(\lambda)$ is just the projection onto the column space of the $n \times |\lambda|$-matrix $G(\lambda) = ((g_j(x_i)))_{i=1,\ldots,n}^{j \in \lambda}$. Notice that the problem of selecting an appropriate set of basis functions in Fourier series regression estimation is of the same structure, see e.g. Eubank (1988). ■

Example 2 (*Other projection-type estimators*, case $k = 1$). The following estimators lead also to projection matrices $H(\lambda)$.

The so-called *regressogram estimator* of f may be thought of as fitting piecewise constants on intervals of length λ^{-1} to the data, where the number λ of intervals is the smoothing parameter to be selected. Generalizing this approach, one could fit *piecewise polynomials* to the data by least squares, see e.g. Chen (1987). The role of λ is then played by the length of the intervals and the degree of the piecewise polynomials if it is not fixed in advance.

A *least squares spline estimator* of f is defined by estimating the coefficients of a spline of order m with knots at $\xi_1, \ldots, \xi_k$, say

$$s(x) = \sum_{j=1}^{m} \theta_j x^{j-1} + \sum_{j=1}^{k} \delta_j (x - \xi_j)_+^{m-1},$$

using least squares, for details we refer to Eubank (1988). Here, the set of knots $\lambda = \{\xi_1, \ldots, \xi_k\}$ may be regarded as the "smoothing parameter" to be selected, provided the order m of spline is fixed in advance. In general the order m is also a smoothing parameter, so that one could take $\lambda = \{\xi_1, \ldots, \xi_k, m\}$. ■

Example 3 (*Ridge regression*). Consider the parameter estimation problem in linear regression, that is, in Example 1 we include all functions g_i,

$i = 1, \ldots, p$, in the model. The ridge regression estimate is then $\hat{y}(\lambda) = H(\lambda)y$, where

$$H(\lambda) = G(G^T G + \lambda I)^{-1} G^T, \quad G := G(\{1, \ldots, p\})$$

and $\lambda \in \mathbb{R}^+$ is the so-called ridge parameter. It is well-known that the ridge estimator is the minimizer of the penalized least squares criterion

$$\sum_{j=1}^{n} (y_j - \sum_{i \in \lambda} \beta_i g_i(x_j))^2 + \lambda \sum_{j=1}^{p} \beta_j^2 \tag{2.5}$$

with respect to β_j $(j = 1, \ldots, p)$. Several generalizations of this estimator have been proposed, mainly by replacing λI by a matrix $K(\lambda)$ which may depend on some higher-dimensional parameter λ. ∎

Example 4 (*Smoothing splines*). Let $k = 1$ and $x_i \in [a, b]$ for $i = 1, \ldots, n$. If the regression function is assumed to be smooth in the sense that, for some positive integer m, it is in the Sobolev class $W_2^m[a, b]$ of all functions having $m - 1$ absolutely continuous derivatives and a square integrable m-th derivative, then a natural estimator of the regression function would be the minimizer of the penalized least squares criterion

$$n^{-1} \sum_{i=1}^{n} (y_i - f(x_i))^2 + \lambda \int_a^b f^{(m)}(t)^2 dt, \quad \lambda > 0 \tag{2.6}$$

over all functions $f \in W_2^m[a, b]$. The solution to this problem is known to be a linear estimator (polynomial spline of order $2m$ with knots at the x_i) that is usually called smoothing spline. An explicit form for $H(\lambda)$ can be found, for example, in Wahba (1978), whereas the properties of this matrix have been studied in Eubank (1984). ∎

Example 5 (*Nadaraya-Watson kernel estimation*). One of the most popular nonparametric regression curve estimates is the following kernel estimator introduced by Nadaraya (1964) and Watson (1964)

$$\hat{f}_\lambda(x) = \sum_{i=1}^{n} K(\lambda^{-1}(x - x_i))y_i \Big/ \sum_{i=1}^{n} K(\lambda^{-1}(x - x_i)),$$

where K is some kernel and $\lambda > 0$ the bandwidth to be selected. In general, the associated matrix $H(\lambda)$ will not be symmetric. ∎

Example 6 (*Selection of nonlinear regression models*). In order to approximate the unknown regression function we consider a class of nonlinear regression models

$$\{g_{\lambda\beta_\lambda} | \beta_\lambda \in B_\lambda\}, \lambda \in \Lambda = \{1, \ldots, p\},$$

where, for each $\lambda \in \Lambda$, the function $g_{\lambda\beta_\lambda}$ is known up to a finite-dimensional parameter β_λ belonging to some parameter space $\mathcal{B}_\lambda$. The problem is to select an appropriate model which can be identified by its index $\lambda \in \Lambda$. Usually the (nonlinear) least squares approach is used to fit the models to the data providing, for each $\lambda \in \Lambda$, the following estimator of $f(x)$:

$$\hat{f}_\lambda(x) = g_{\lambda\hat{\beta}_\lambda}(x),$$

where the parameter $\hat{\beta}_\lambda$ is the minimizer of

$$\sum_{j=1}^{n}(y_j - g_{\lambda\beta_\lambda}(x_j))^2 \tag{2.7}$$

with respect to β_λ. Then $\hat{f}_\lambda$ is of course no longer a linear estimator. ∎

3 Cross-Validation and Full Cross-Validation

In all examples introduced in the preceding section the index λ may be interpreted as a smoothing parameter, i.e. its choice will typically govern how smooth or wiggly the estimator will be. Ideally one would select an estimator by choosing that value of λ, which minimizes the MSEP defined in (2.2). Unfortunately, the MSEP is unknown. Therefore, in practice, one has to use an estimate of the MSEP, which is then minimized with respect to λ.

One of the most popular MSEP estimates in applications is the *cross-validation* (CV) criterion of Stone (1974) defined by

$$CV(\lambda) = \frac{1}{n}\sum_{i=1}^{n}(y_i - \hat{y}_{-i}(\lambda))^2, \tag{3.1}$$

where $\hat{y}_{-i}(\lambda)$ is the prediction of y_i^* leaving out the i-th data point, i.e.

$$\hat{y}_{-i}(\lambda) = f_{n-1,\lambda}(x_i; y_1, \ldots, y_{i-1}, y_{i+1}, \ldots, y_n).$$

Cross-validation is defined entirely with respect to samples of size $n - 1$, rather than n. Compared with various other criteria it has, however, the advantage that no estimate of the error variance is required.

For many linear estimators, cross-validation may be expressed in terms of the ordinary residuals $y_i - \hat{y}_i(\lambda)$. More precisely, we have the following

Lemma 1 *Let $\hat{f}_\lambda$ be a linear estimator of f fulfilling the compatibility condition*

$$\hat{y}_{-i}(\lambda) = f_{n,\lambda}(x_i; y_1, \ldots, y_{i-1}, \hat{y}_{-i}(\lambda), y_{i+1}, \ldots, y_n) \quad i = 1, \ldots, n. \tag{3.2}$$

Then, assuming $h_{ii}(\lambda) < 1$ for $i = 1, \ldots, n$, we have

$$CV(\lambda) = \frac{1}{n}\|y - \hat{y}(\lambda)\|_{C(\lambda)}^2, \tag{3.3}$$

where

$$C(\lambda) = \text{diag}[(1 - h_{11}(\lambda))^{-2}, \ldots, (1 - h_{nn}(\lambda))^{-2}].$$

Proof. Recalling (2.3) we obtain from (3.2) for an arbitrary $i \in \{1, \ldots, n\}$

$$\hat{y}_{-i}(\lambda) = \sum_{j \neq i} h_{ij}(\lambda) y_j + h_{ii}(\lambda) \hat{y}_{-i}(\lambda).$$

Consequently,

$$
\begin{aligned}
(1 - h_{ii}(\lambda)) \hat{y}_{-i}(\lambda) &= \sum_{j \neq i} h_{ij}(\lambda) y_j \\
&= \hat{y}_i - h_{ii}(\lambda) y_i,
\end{aligned}
$$

which provides the desired result:

$$y_i - \hat{y}_{-i}(\lambda) = (1 - h_{ii}(\lambda))^{-1}(y_i - \hat{y}_i). \qquad \blacksquare$$

Remark 1 The compatibility condition (3.2) is fulfilled for all examples of linear estimators considered in Section 2 (Examples 1–5). The corresponding expression (3.3) has been established previously for several estimators, see e.g. Stone (1974) for the model selection case and all examples with idempotent matrices $H(\lambda)$ (Examples 1, 2), Craven and Wahba (1979) for the spline smoothing problem (Example 4), and Eubank (1988, Exercise 2.5) for Example 3 (ridge regression). To verify condition (3.2) for the Nadaraya-Watson kernel estimator of Example 5, we start with noting that for $i \in \{1, \ldots, n\}$

$$\hat{y}_{-i}(\lambda) = \sum_{j \neq i} K(\lambda^{-1}(x_i - x_j)) y_j \Big/ \sum_{j \neq i} K(\lambda^{-1}(x_i - x_j)).$$

Then

$$
\begin{aligned}
f_{n,\lambda}&(x_i; y_1, \ldots, y_{i-1}, \hat{y}_{-i}(\lambda), y_{i+1}, \ldots, y_n) \\
&= \{\sum_{j \neq i} K(\lambda^{-1}(x_i - x_j)) y_j + K(0) \hat{y}_{-i}(\lambda)\} \Big/ \sum_{j=1}^{n} K(\lambda^{-1}(x_i - x_j)) \\
&= \{\hat{y}_{-i}(\lambda) \sum_{j \neq i} K(\lambda^{-1}(x_i - x_j)) + K(0) \hat{y}_{-i}(\lambda)\} \Big/ \sum_{j=1}^{n} K(\lambda^{-1}(x_i - x_j)) \\
&= \hat{y}_{-i}(\lambda). \qquad \blacksquare
\end{aligned}
$$

Cross-validation works well in many applications. However, there exist situations where it fails. This is illustrated by the following example.

Example 6 (*Selection of nonlinear regression models*, continued). Bunke et al. (1993, Example 4) applied the cross-validation approach to select among 15 sigmoidally shaped nonlinear regression models in order to describe the growth of bean root cells. One of the model candidates was

$$g_\beta(x) = \beta_1 \exp(\beta_2/(x + \beta_3)), \quad \beta = (\beta_1, \beta_2, \beta_3)^T. \tag{3.4}$$

A least squares fit of model (3.4) to the data (50 simulated observations) provides $\hat{\beta}_1 = 39.63, \hat{\beta}_2 = -0.51$ and $\hat{\beta}_3 = -0.01$, which is not too bad as illustrated in Figure 1.

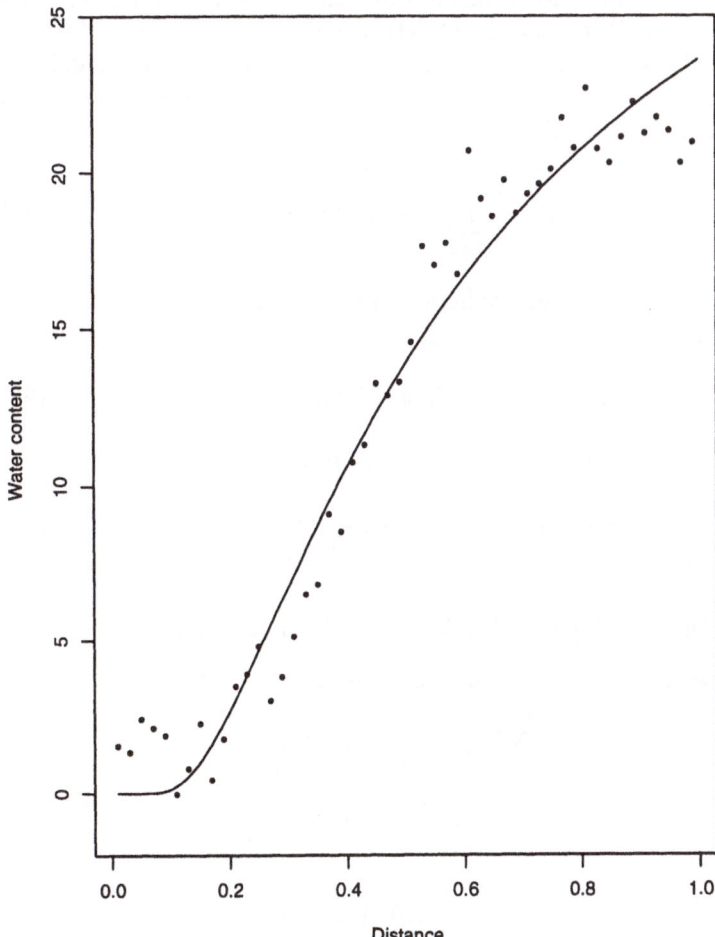

Figure 1: Dependence of water content on the distance from growing tip for bean root cells: The simulated data and the fitted model (3.4).

The associated value of the residual sum of squares,

$$RSS(\lambda) = n^{-1}\|y - \hat{y}(\lambda)\|^2, \tag{3.5}$$

is $RSS = 2.24$, whereas the cross-validation value is $CV = 7.72 \cdot 10^{21}$. The reason for this unacceptable value of the cross-validation criterion (3.1) is as follows: Model (3.4) has a pole at $x_0 = -\beta_3$, (that is, for $\beta_1 > 0$ and $\beta_2 < 0$, $g_\beta(x)$ tends to 0 as $x \downarrow x_0$ while it tends to ∞ as $x \uparrow x_0$,) and $x_1 = 0.01$ is somewhat greater than the estimate of β_3 leaving out the first observation, so that $g_{\hat{\beta}_{-1}}(x_1)$ becomes extremely large. ∎

To avoid the difficulties mentioned above and to adjust cross-validation for samples of size n, Bunke et al. (1993) proposed to substitute, for each $i \in \{1, \ldots, n\}$, the value y_i by $\hat{y}_i(\lambda)$ instead of leaving it out in defining the prediction at the i-th design point. The resulting criterion is called *full cross-validation* (FCV) criterion:

$$FCV(\lambda) = \frac{1}{n} \sum_{i=1}^{n} (y_i - \tilde{y}_i(\lambda))^2, \tag{3.6}$$

where $\tilde{y}_i(\lambda)$ is the least squares prediction of y_i^* with substituting y_i by $\hat{y}_i(\lambda)$ instead of deleting it, i.e.

$$\tilde{y}_i(\lambda) = f_{n,\lambda}(x_i; y_1, \ldots, y_{i-1}, \hat{y}_i(\lambda), y_{i+1}, \ldots, y_n). \tag{3.7}$$

For linear estimators $\hat{f}_\lambda$, (3.6) may be rewritten as

$$FCV(\lambda) = \frac{1}{n}\|y - \hat{y}(\lambda)\|^2_{D(\lambda)}, \tag{3.8}$$

where

$$D(\lambda) = \text{diag}[(1 + h_{11}(\lambda))^2, \ldots, (1 + h_{nn}(\lambda))^2].$$

To verify (3.8) we note that (2.3) and (3.7) imply

$$\begin{aligned}
\tilde{y}_i &= \sum_{j \neq i} h_{ij}(\lambda)y_j + h_{ii}(\lambda)\hat{y}_i \\
&= \hat{y}_i - h_{ii}(\lambda)y_i + h_{ii}(\lambda)\hat{y}_i,
\end{aligned}$$

and hence

$$y_i - \tilde{y}_i = (1 + h_{ii}(\lambda))(y_i - \hat{y}_i).$$

Remark 2 a) For linear estimators $\hat{f}_\lambda$ we obtain the following representation of both CV and FCV after some simple algebraic manipulation:

$$CV(\lambda) = RSS(\lambda) + \|y - H(\lambda)y\|^2_{\Gamma(\lambda)}, \tag{3.9}$$

and

$$FCV(\lambda) = RSS(\lambda) + \|y - H(\lambda)y\|^2_{\Delta(\lambda)}, \qquad (3.10)$$

where

$$\begin{aligned} \Gamma(\lambda) &= \text{diag}[h_{ii}(\lambda)(2 - h_{ii}(\lambda))(1 - h_{ii}(\lambda))^{-2}]^n_{i=1} \quad \text{and} \\ \Delta(\lambda) &= \text{diag}[h_{ii}(\lambda)(2 + h_{ii}(\lambda))]^n_{i=1}. \end{aligned}$$

b) It is easily seen that

$$C(\lambda) \geq D(\lambda) \geq I \quad \text{and} \quad \Gamma(\lambda) \geq \Delta(\lambda), \qquad (3.11)$$

and therefore we have for each realization y

$$CV(\lambda) \geq FCV(\lambda),$$

as it has been expected, since FCV uses in contrast to CV the observation y_i for predicting y_i^*. ∎

Craven and Wahba (1979) have proposed *generalized cross-validation* (GCV) as another useful method for selecting the parameter λ of linear estimates $\hat{f}_\lambda$. Assuming that $\text{tr}[H(\lambda)] < n$, the GCV criterion is defined by

$$GCV(\lambda) = \frac{1}{n}\|y - \hat{y}(\lambda)\|^2 / (1 - \frac{1}{n}\text{tr}H(\lambda))^2. \qquad (3.12)$$

Obviously, GCV weights the ordinary residuals $(y_i - \hat{y}_i)$ by the average of the weights used for the CV criterion under assumption (3.2) (cf. (3.3)). Applying this idea to FCV one obtains the following *generalized full cross-validation* (GFCV) criterion

$$GFCV(\lambda) = \frac{1}{n}\|y - \hat{y}(\lambda)\|^2(1 + \frac{1}{n}\text{tr}H(\lambda))^2. \qquad (3.13)$$

4   Comparisons of the Criteria

In this section the different cross-validation criteria introduced before are compared as estimates of the MSEP on the basis of their biases and variances, having consequences for the mean square errors. For simplicity we omit the index λ in the notation, since the comparisons are carried out for a fixed $\lambda \in \Lambda$. Furthermore, we confine ourselves to the case of linear estimates of the form (2.3). To make the variance formulae more accessible we assume to have normally distributed observations in (2.1), i.e.

$$\varepsilon_i \sim N(0, \sigma^2) i.i.d. \quad i = 1, \ldots, n. \qquad (4.1)$$

The results on cross-validation are derived under the compatibility condition (3.2) and $h_{ii} < 1$ for $i = 1, \ldots, n$.

As all MSEP estimates in the previous section were quadratic in the observations, their expectations and variances may be calculated in a straightforward manner with some algebra and using the well-known formulae

$$E\|z\|_A^2 = \|\mu\|_A^2 + \text{tr}[A\Sigma] \tag{4.2}$$

$$\text{Var}\|z\|_A^2 = 4\mu^T A\Sigma A\mu + 2\text{tr}[A\Sigma]^2, \tag{4.3}$$

which are valid under $z \sim N(\mu, \Sigma)$ and for a symmetric matrix A. This is done in the appendix, and the results are compiled in Tables 1 and 2. Concerning the bias formulae in Table 1 we note that an application of (4.2) yields the following expression for the MSEP

$$MSEP = \Omega + \sigma^2(1 + n^{-1}\text{tr}[H^T H]), \tag{4.4}$$

where, with $\mu = (f(x_1), \ldots, f(x_n))^T$,

$$\Omega = n^{-1}\|\mu - H\mu\|^2 \tag{4.5}$$

is the model bias term. Other notations used for brevity are

$$h = n^{-1}\text{tr}H, \quad t = n^{-1}\text{tr}[H^T H] \quad \text{and} \quad Q = I - H. \tag{4.6}$$

Table 1: Bias for different criteria as estimates of the MSEP.

Criterion	Bias(Criterion) = E(Criterion) − MSEP
RSS (3.5)	$-2\sigma^2 h$
CV (3.1)	$n^{-1}\|\mu - H\mu\|_\Gamma^2 + n^{-1}\sigma^2\text{tr}[\Gamma - 2\Gamma H + H^T\Gamma H - 2H]$
FCV (3.6)	$n^{-1}\|\mu - H\mu\|_\Delta^2 + n^{-1}\sigma^2\text{tr}[\Delta - 2\Delta H + H^T\Delta H - 2H]$
GCV (3.12)	$(1 - h)^{-2}h(2 - h)\Omega + \sigma^2(1 - h)^{-2}h(2t - h - ht)$
GFCV (3.13)	$h(2 + h)\Omega + \sigma^2 h(2t + ht - 3h - 2h^2)$

Table 2: Variance for different criteria as estimates of the MSEP.

Criterion	Var(Criterion)
RSS (3.5)	$4n^{-2}\sigma^2\mu^T[Q^T Q]^2\mu + 2n^{-2}\sigma^4\text{tr}[Q^T Q]^2$
CV (3.1)	$4n^{-2}\sigma^2\mu^T[Q^T CQ]^2\mu + 2n^{-2}\sigma^4\text{tr}[Q^T CQ]^2$
FCV (3.6)	$4n^{-2}\sigma^2\mu^T[Q^T DQ]^2\mu + 2n^{-2}\sigma^4\text{tr}[Q^T DQ]^2$
GCV (3.12)	$4n^{-2}\sigma^2(1 - h)^{-4}\mu^T[Q^T Q]^2\mu + 2n^{-2}\sigma^4(1 - h)^{-4}\text{tr}[Q^T Q]^2$
GFCV (3.13)	$4n^{-2}\sigma^2(1 + h)^4\mu^T[Q^T Q]^2\mu + 2n^{-2}\sigma^4(1 + h)^4\text{tr}[Q^T Q]^2$

For the bias comparisons we focus on the case of projection-type estimates, where the "hat matrix" H is symmetric and idempotent, that is,

$$H^T = H, \quad H^2 = H \quad \text{(and, consequently,} \quad h = t). \tag{4.7}$$

This is fulfilled for the Examples 1 and 2. Then there are essential simplifications in the bias formulae for the different criteria, which allow more detailed comparisons in the following theorem.

Theorem 1 *For linear estimators fulfilling (4.7) we have*

a) $\text{Bias}(CV) = n^{-1}\|\mu - H\mu\|_\Gamma^2 + n^{-1}\sigma^2 \sum_{i=1}^n h_{ii}^2(1 - h_{ii})^{-1} > 0,$

b) $\text{Bias}(FCV) = n^{-1}\|\mu - H\mu\|_\Delta^2 - n^{-1}\sigma^2 \sum_{i=1}^n h_{ii}^2(1 + h_{ii}),$

c) $\text{Bias}(GCV) = (1 - h)^{-2}h(2 - h)\Omega + \sigma^2 h^2(1 - h)^{-1} > 0,$

d) $\text{Bias}(GFCV) = h(2 + h)\Omega - \sigma^2 h^2(1 + h),$

e) $\text{Bias}(CV) > |\text{Bias}(FCV)|,$

f) $\text{Bias}(GCV) > |\text{Bias}(GFCV)|.$

Proof. The bias formulae follow immediately from Table 1 and (4.7). The positivity of the biases for CV and GCV is a consequence of $\Gamma > 0, h_{ii} < 1$ and $\Omega \geq 0, h < 1$, respectively. Finally, an application of the triangular inequality provides statements e) and f) by taking into account that

$$\Delta \leq \Gamma, \quad 1 + h_{ii} < (1 - h_{ii})^{-1} \quad (i = 1, \ldots, n)$$

and

$$2 + h < (2 - h)(1 - h)^{-2}, \quad 1 + h < (1 - h)^{-1},$$

respectively. ∎

Theorem 1 states that CV and GCV overestimate the MSEP, whereas FCV and GFCV may be positively or negatively biased in dependence of the deviation of the expected prediction $E\hat{y} = H\mu$ from the vector μ of the true values of the regression function (compared with some function of the error variance σ^2 and the "leverage values" h_{ii}). In any case, however, the absolute values of the biases of the traditional cross-validation variants (CV and GCV) are larger than those of the corresponding modified counterparts (FCV and GFCV). Additionally, with respect to the variances we show in the following theorem that GCV is outperformed by GFCV, and that FCV is better than CV at least in a minimax sense.

Theorem 2 *In the case of linear estimators we have*

a) *For any* $\kappa > 0,$ $\sup_{\|\mu\| \leq \kappa} \text{Var}(CV) \geq \sup_{\|\mu\| \leq \kappa} \text{Var}(FCV),$

b) $\text{Var}(GCV) > \text{Var}(GFCV).$

Proof. Statement b) is an immediate consequence of

$$\text{Var}(GCV) = (1 - h)^{-4}\text{Var}(RSS) \quad \text{and} \tag{4.8}$$
$$\text{Var}(GFCV) = (1 + h)^4\text{Var}(RSS), \tag{4.9}$$

since $(1 - h)^{-1} > (1 + h).$

Thus it remains to establish the relation a). To accomplish this we recall first that $C \geq D \geq 0$ and therefore,

$$A := Q^T C Q \geq B := Q^T D Q.$$

Now, let $\alpha_1 \geq \ldots \geq \alpha_n \geq 0$ and $\beta_1 \geq \ldots \geq \beta_n \geq 0$ be the eigenvalues of A and B, respectively. Then we have

$$\alpha_i \geq \beta_i \quad (i = 1, \ldots, n)$$

and the eigenvalues of A^2 and B^2 are just α_i^2 and $\beta_i^2 (i = 1, \ldots, n)$, respectively. This leads to

$$\text{tr}[Q^T C Q]^2 = \text{tr}[A^2] \geq \text{tr}[B^2] = \text{tr}[Q^T D Q]^2$$

and

$$\sup_{\|\mu\| \leq \kappa} \mu^T [Q^T C Q]^2 \mu = \sup_{\|\mu\| \leq \kappa} \mu^T A^2 \mu = \alpha_1^2 \kappa^2$$
$$\geq \beta_1^2 \kappa^2 = \sup_{\|\mu\| \leq \kappa} \mu^T B^2 \mu = \sup_{\|\mu\| \leq \kappa} \mu^T [Q^T D Q]^2 \mu,$$

which completes the proof. ∎

Remark 3 a) Clearly, statement b) of Theorem 2 remains true if one requires only the existence of fourth-order-moments of the error distribution instead of assumption (4.1).

b) (4.8) and (4.9) provide the following variance ratio

$$\frac{\text{Var}(GCV)}{\text{Var}(GFCV)} = (1 - h^2)^{-4},$$

which may be very large in particular for comparatively large values of h. For example, if one considers the problem of fitting a linear model with $p = 5$ parameters to a data set of size $n = 20$, then it follows $h = n^{-1} \text{tr} H = p/n = 0.25$ so that this variance ratio amounts approximately to 1.3. If the number of parameters involved in the model would be one half of the sample size (i.e., $p/n = 0.5$), then the variance of GCV would already be greater than the triple of the variance of GFCV.

c) In many cases, the variance of CV will be greater than that of FCV (not only in the minimax sense). This holds, for example, in the so-called balanced case where all leverage values h_{ii} are equal, that is,

$$h_{ii} = h = n^{-1} \text{tr} H \quad (i = 1, \ldots, n). \tag{4.10}$$

Then the results concerning the comparison between GCV and GFCV may be applied. Condition (4.10) is fulfilled, for instance, if in a linear regression model the underlying design is D- (or G-) optimal among all discrete designs. In case of the regressogram estimate (see Example 2) this condition would approximately hold for uniformly spaced design points. ∎

According to Eubank (1988, p. 30) the best motivation for GCV "is probably provided by the so called GCV Theorem first established by Craven and Wahba (1979)". This theorem states that for $h < 1$

$$\text{Bias}(GCV)/MSE \leq (2h + h^2/t)(1-h)^{-2} =: g_1$$

(see also Theorem 2.1 of Eubank, 1988). A similar result can be obtained for GFCV.

Theorem 3 *Assuming $h < 1$ we have*

$$|\text{Bias}(GFCV)|/MSE \leq \max\{h(2+h), (3+2h)h^2/t\} =: g_2.$$

Proof. Recalling $MSE = \Omega + \sigma^2 t$, the bias formula for GFCV in Table 1 provides

$$
\begin{aligned}
|\text{Bias}(GFCV)| &= |h(2+h)\Omega + \sigma^2 h(2t + ht - 3h - 2h^2)| \\
&= |h(2+h)MSE - \sigma^2 h^2(3+2h)| \\
&\leq \max\{h(2+h)MSE, \sigma^2(3+2h)h^2\}.
\end{aligned}
$$

Thus, the result follows from the fact that $\sigma^2/MSE \leq 1/t$. ∎

Remark 4 It is easily verified that a necessary and sufficient condition for $g_2 = (3+2h)h^2/t$ is given by $t \leq 2h - h/(2+h)$. This condition is fulfilled in the interesting case of a linearly admissible estimator, since then the matrix H is symmetric with all its eigenvalues in the closed interval $[0, 1]$ (see Theorem 3.1 of Rao, 1976) leading to $t \leq h$. Under this admissibility assumption it holds $g_1 \geq g_2$ if and only if $t \geq h - 2h^2 - h^3/2 + h^4$, for which $h \geq 0.327$ is a sufficient condition since $t \geq h^2$.

Furthermore, in view of Theorem 1 it is not surprising, that under the additional assumption (4.7) the upper bound for the relative bias of GFCV is smaller than that of GCV (i.e., $g_2 \leq g_1$). Both g_1 and g_2 can then be represented as $3h + O(h^2)$. In the special case of selecting linear regression models (Example 1), we would have $h \leq p/n$. ∎

5   Numerical Illustration

To get an impression of the differences between the considered cross-validation criteria, we compare their mean square errors (MSE's) as estimates of the MSEP in some special cases.

(a) Projection-type estimators (Examples 1 and 2)

At first, we examine the case of projection-type estimators in a balanced situation, that is, we assume (4.7) and (4.10). This leads to $CV = GCV$ and

$FCV = GFCV$, and Theorems 1 and 2 yield

$$\begin{aligned} \text{MSE}(CV) \ &:= \ \text{E}[CV - MSEP]^2 \\ &= \ [\text{Bias}(CV)]^2 + \text{Var}(CV) > \text{MSE}(FCV). \end{aligned}$$

For a sample size of $n = 20$ and for some values of the "model bias to noise ratio" $\rho := \Omega/\sigma^2$ and of the leverage value h, Table 3 contains the normalized MSE's of CV and FCV, which may be calculated according to

$$\begin{aligned} \text{M}(CV) \ &:= \ \text{MSE}(CV)/\sigma^4 \qquad\qquad\qquad\qquad\qquad (5.1) \\ &= \ h^2(2-h)^2(1-h)^{-4}\rho^2 + 2h^3(2-h)(1-h)^{-3}\rho \\ &\quad +4n^{-1}(1-h)^{-4}\rho + [h^4(1-h)n + 2]n^{-1}(1-h)^{-3} \\ \text{M}(FCV) \ &= \ h^2(2+h)^2\rho^2 - 2(1+h)n^{-1}[h^3(2+h)n - 2(1+h)^3]\rho \\ &\quad +(1+h)^2 n^{-1}[h^4 n + 2(1+h)^2(1-h)]. \end{aligned}$$

Table 3: Normalized MSE of CV and FCV as estimates of the MSEP for $n = 20$.

h	$\rho = 0$		$\rho = 0.1$		$\rho = 0.3$		$\rho = 0.5$		$\rho = 1$	
	CV	FCV	CV	FCV	CV	FCV	CV	FCV	CV	FCV
.02	.106	.106	.128	.128	.171	.171	.215	.215	.325	.324
.05	.117	.115	.141	.140	.191	.189	.243	.239	.374	.368
.10	.137	.132	.169	.161	.235	.222	.306	.287	.502	.464
.15	.163	.149	.205	.184	.298	.259	.402	.342	.714	.586
.20	.198	.168	.255	.207	.390	.297	.549	.403	.059	.734
.30	.308	.214	.429	.259	.736	.379	1.13	.538	2.49	1.10
.40	.534	.281	.815	.324	1.57	.465	2.57	.680	6.19	1.54
.50	1.05	.394	1.76	.417	3.72	.557	6.40	.822	16.2	2.03
.70	6.37	.944	13.2	.832	32.9	.822	60.8	1.10	166.3	3.04
.90	166	2.50	624	2.02	2129	1.48	4418	1.49	13570	3.88

Figure 2 shows the relative MSE-differences between CV and FCV, defined by

$$\text{RMD}(CV) := [\text{MSE}(CV) - \text{MSE}(FCV)]/\text{MSE}(FCV), \qquad (5.2)$$

over the range of $0 \leq h \leq 0.4, 0 \leq \rho \leq 2$ and for $n = 20$. Notice that larger values of h correspond typically to better data fits by the associated estimator and, hence, to smaller values of ρ. Furthermore, for fixed values of h and ρ, RMD(CV) as well as the MSE's of CV and FCV decrease monotonously in n.

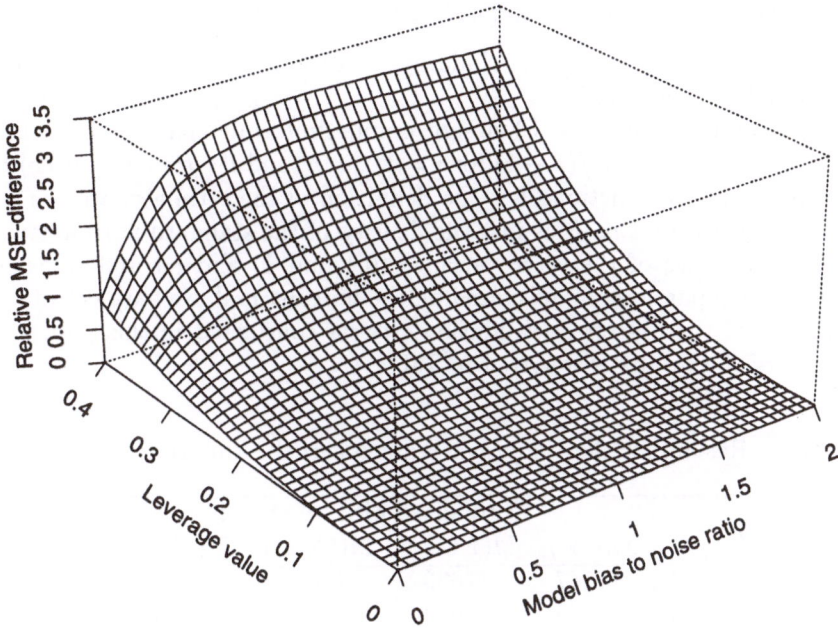

Figure 2: Relative MSE-difference between CV and FCV as function of the leverage value h and the model bias to noise ratio ρ for $n = 20$.

(b) Nadaraya-Watson kernel estimators (Example 5)

Now we consider the case of the Nadaraya-Watson kernel estimator, which is not covered by Theorem 1.

We assume to have $n = 51$ design points uniformly spaced on the unit interval, i.e. $x_i = (i - 1)/(n - 1)$ for $i = 1, \ldots, n$, and let $f(x) = sin(\pi x)$ be the true regression function. For the definition of the estimator we use the Gaussian kernel, that is, K is just the density of the standard normal law.

For two values of the error variance (0.1 and 1) and for 100 values of the smoothing parameter equispaced on the unit interval, we have calculated the MSE's of the four cross-validation criteria considered in this paper. A selection of the results is reported in Table 4.

Table 4: Results for the Nadaraya-Watson kernel estimator for $n = 51$.

σ^2	λ	Normalized MSE of criteria			
		M(CV)	M(FCV)	M(GCV)	M($GFCV$)
.1	.02	.12147	.17689	.12060	.17641
	.05	.06164	.05634	.06118	.05653
	.07	.05501	.05149	.05450	.05151
	.09	.05229	.04973	.05166	.04954
	.11	.05138	.04936	.05058	.04893
	.15	.05294	.05146	.05168	.05046
	.20	.06057	.05940	.05874	.05772
	.30	.08921	.08833	.08692	.08604
	.50	.12225	.12172	.12101	.12040
1	.02	.12103	.17806	.12045	.17667
	.05	.06103	.05656	.06085	.05658
	.08	.05200	.04978	.05188	.04980
	.11	.04857	.04726	.04847	.04726
	.14	.04690	.04602	.04680	.04599
	.17	.04608	.04543	.04598	.04538
	.20	.04584	.04533	.04572	.04526
	.30	.04726	.04697	.04713	.04686
	.50	.04973	.04958	.04965	.04950

For $\sigma^2 = 0.1$, Figure 3 displays the normalized MSE (5.1) of CV as well as the relative MSE-differences of the criteria compared with CV, which may be defined in analogy to (5.2).

We see that all criteria have their best properties as estimates of the MSEP for values of the smoothing parameter which are a little bit greater than the minimizers of the MSEP, which have been found to be $\lambda = 0.07$ and $\lambda = 0.14$ for $\sigma^2 = 0.1$ and $\sigma^2 = 1$, respectively.

In this example, the MSE's of the considered criteria differ only slightly

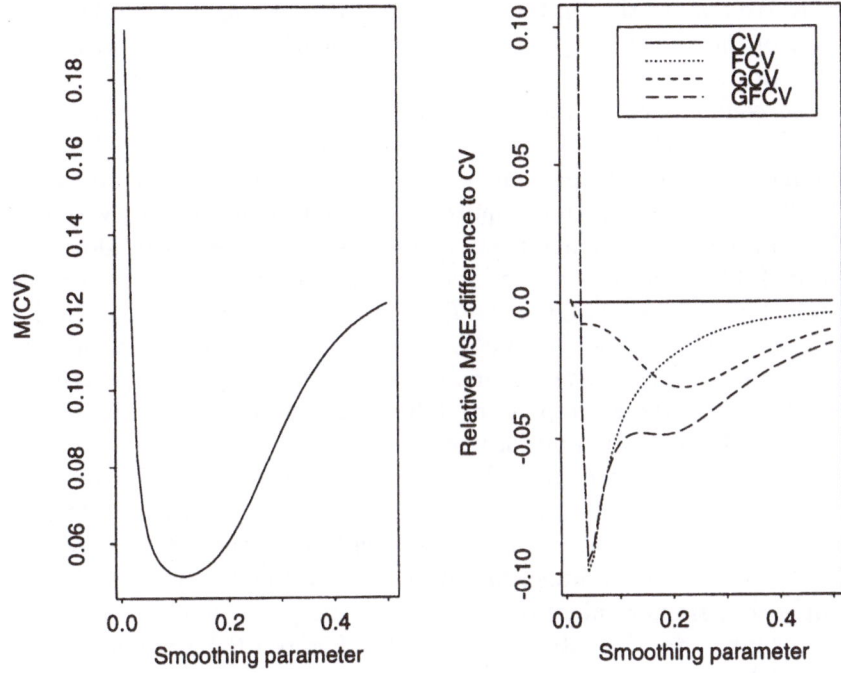

Figure 3: Normalized MSE of CV and relative MSE-differences of the different criteria compared with CV as function of the smoothing parameter λ in case of the Nadaraya-Watson kernel estimator when $\sigma^2 = 0.1$.

except for the case of nearly no smoothing, i.e. of very small smoothing parameters, where FCV and GFCV behave worse than CV and GCV. However, the region when the traditional cross-validation approach is superior to the new one is rather small. More precisely, the MSE begins to favour FCV (and GFCV) at about $\lambda = 0.03$. Finally, it turns out that CV and FCV are outperformed by their "generalized" variants.

6   Discussion

1. In the case of linear estimators fulfilling the compatibility condition (3.2), the weighting of the ordinary residuals in the full cross-validation criterion (3.8) may be seen as first-order-approximation to the corresponding weighting in the cross-validation criterion (3.3), since $1/(1-h_{ii}) = 1+h_{ii}+O(h_{ii}^2)$. The same holds for the comparison between GCV and GFCV. Furthermore, there are no problems with high leverage values when expressing FCV in terms of the ordinary residuals. This is in contrast to CV, where (3.3) is not defined

if $h_{ii} = 1$. We remark that in the linear regression case $h_{ii} = 1$ implies that the vector of the regression parameters is not identifiable when leaving out the i-th data point.

2. The comparisons in the preceding sections show that the proposed criteria FCV and GFCV outperform the traditional counterparts CV and GCV in many cases by having a smaller mean square error in estimating the MSEP. All these criteria do not require the estimation of the error variance, but CV may fail in some nonlinear regression situations. This does not necessarily lead to the recommendation of selecting models or smoothing parameters by minimizing FCV or GFCV (for some simulation results in this direction we refer e.g. to Droge, 1994). Nevertheless, after finally reaching at a hopefully appropriate model or smoothing parameter by some procedure, a good estimate of the corresponding MSEP such as FCV is needed as an assessment of the prediction performance.

3. The application of model selection procedures based on CV or FCV to several nonlinear regression problems in Bunke et al. (1993) suggests that both procedures provide almost the same results (i.e., the same ranking of models and, for parameter dimensions which are not too large compared with the sample size, similar values of the criteria) except for models where CV fails. For example, for the data set presented in Figure 1 the Morgan-Mercer-Flodin model

$$g_\beta(x) = (\beta_1 x^{\beta_2} + \beta_3\beta_4)/(\beta_4 + x^{\beta_2})$$

was the first choice among 15 sigmoidally shaped models with $FCV = 0.90$ and $CV = 0.91$, whereas model (3.4) ranked 13th and 15th according to FCV and CV, respectively, with $FCV = 2.30$ and $CV = 7.72 \cdot 10^{21}$.

4. The asymptotic properties of model selection procedures based on different criteria have been investigated by several authors, see e.g. Nishii (1984) and Müller (1993). Nishii (1984) showed, for example, that under the assumption of normally distributed errors and the existence of an adequate (true) linear model the minimum-CV-procedure is consistent in the sense that the probability of selecting a model not including the adequate one tends to zero as the sample size approaches infinity. This result has been generalized by Müller (1993) to the case of errors as in (2.1) and inadequate linear models defining a pseudo-true (instead of an adequate) model in an appropriate way and assuming some additional conditions on the design and the unknown regression function. Now it is easy to check that under the same assumptions the results remain true for the minimum-FCV-procedure.

Appendix: Proof of the Bias and Variance Formulae in Section 4

A1. Calculation of Biases in Table 1

At several steps we make use of formula (4.2) to calculate the expectation of quadratic forms.

1. We start with calculating the expectation of the RSS (3.5) using the notation (4.5), (4.6),

$$
\begin{aligned}
ERSS &= n^{-1}E\|\mu - H\mu\|^2 + n^{-1}\sigma^2\text{tr}(I - H)^T(I - H) \\
&= \Omega + \sigma^2(1 - 2h + t),
\end{aligned}
\tag{A1}
$$

which provides together with (4.4) the bias of RSS:

$$
\begin{aligned}
\text{Bias}(RSS) &= ERSS - MSEP \\
&= -2n^{-1}\sigma^2\text{tr}\,H = -2\sigma^2 h.
\end{aligned}
\tag{A2}
$$

2. Recalling (3.9) and (A2), and noting that $\text{tr}\,\Gamma H = \text{tr}\,H^T\Gamma$ we obtain

$$
\begin{aligned}
\text{Bias}(CV) &= \text{Bias}(RSS) + n^{-1}E\|y - Hy\|_\Gamma^2 \\
&= -2n^{-1}\sigma^2\text{tr}\,H + n^{-1}\|\mu - H\mu\|_\Gamma^2 + n^{-1}\sigma^2\text{tr}[Q^T\Gamma Q] \\
&= n^{-1}\|\mu - H\mu\|_\Gamma^2 + n^{-1}\sigma^2\text{tr}[\Gamma - 2\Gamma H + H^T\Gamma H - 2H].
\end{aligned}
$$

3. A comparison of (3.9) and (3.10) shows that the bias formula for FCV is the same as for CV, with the only exception that Γ has to be replaced by Δ.

4. Observing

$$
(1 - h)^{-2} = 1 + (1 - h)^{-2}h(2 - h),
$$

we derive from (3.12)

$$
\text{Bias}(GCV) = \text{Bias}(RSS) + (1 - h)^{-2}h(2 - h)ERSS.
$$

Some algebraic manipulation gives

$$
(1 - h)^{-2}h(2 - h)(1 - 2h + t) - 2h = (1 - h)^{-2}h(2t - h - ht),
$$

so that the formula for Bias(CV) in Table 1 follows from (A1) and (A2).

5. From (3.13), (4.6), (A1) and (A2) we derive

$$
\begin{aligned}
\text{Bias}(GFCV) &= \text{Bias}(RSS) + (2h + h^2)ERSS \\
&= -2\sigma^2 h + h(2 + h)\Omega + \sigma^2(2h + h^2)(1 - 2h + t) \\
&= h(2 + h)\Omega + \sigma^2 h(2t + ht - 3h - 2h^2).
\end{aligned}
$$

A2. Calculation of Variances in Table 2

For calculating the variances we apply (4.3) which holds under (4.1).
Recalling $y - Hy \sim N(Q\mu, \sigma^2 QQ^T)$ we obtain from (3.5)

$$
\begin{aligned}
\text{Var}RSS &= n^{-2}\text{Var}\|y - Hy\|^2 \\
&= 4n^{-2}\sigma^2 \mu^T [Q^T Q]^2 \mu + 2n^{-2}\sigma^4 \text{tr}[Q^T Q]^2.
\end{aligned}
$$

(3.3) gives

$$
\begin{aligned}
\text{Var}(CV) &= n^{-2}\text{Var}\|y - Hy\|_C^2 \\
&= 4n^{-2}\sigma^2 \mu^T [Q^T CQ]^2 \mu + 2n^{-2}\sigma^4 \text{tr}[Q^T CQ]^2.
\end{aligned}
$$

The variance formula for FCV follows analogously using (3.8). Finally we note that the variances of GCV and GFCV may be obtained by multiplying $\text{Var}RSS$ by $(1-h)^{-4}$ and $(1+h)^4$, respectively (see (3.12), (3.13) and (4.6)).

Acknowledgement

I am grateful to the referee for his comments on the first draft of the manuscript. This work was supported by the Deutsche Forschungsgemeinschaft, Sonderforschungsbereich 373 "Quantifikation und Simulation ökonomischer Prozesse", Humboldt-Universität zu Berlin, Berlin, Germany.

References

BUNKE, O. and DROGE, B. (1984a). Bootstrap and cross-validation estimates of the prediction error for linear regression models. *Ann. Statist.* **12**, 1400-1424.

BUNKE, O. and DROGE, B. (1984b). Estimators of the mean squared error of prediction in linear regression. *Technometrics* **26**, 145-155.

BUNKE, O., DROGE, B. and POLZEHL, J. (1993). Model selection and variable transformations in nonlinear regression. CORE Discussion Paper No. 9327, C.O.R.E., UCL, Belgium.

CHEN, K.-W. (1987). Asymptotically optimal selection of a piecewise polynomial estimator of a regression function. *J. Multiv. Anal.* **22**, 230-244.

CRAVEN, P. and WAHBA, G. (1979). Smoothing noisy data with spline functions: estimating the correct degree of smoothing by the method of generalized cross-validation. *Numer. Math.* **31** 377-403.

DROGE, B. (1987). A note on estimating the MSEP in nonlinear regression. *Statistics* **18**, 499-520.

DROGE, B. (1994). Some simulation results on cross-validation and competitors for model choice. Discussion Paper No. 30, Sonderforschungsbereich 373, Humboldt-Universität, Berlin.

EUBANK, R.L. (1984). The hat matrix for smoothing splines. *Statist. and Prob. Letters* **2**, 9-14.

EUBANK, R.L. (1988). *Spline Smoothing and Nonparametric Regression.* Marcel Dekker, New York.

MALLOWS, C.L. (1973). Some comments on C_p. *Technometrics* **15**, 661-675.

MÜLLER, M. (1993). Asymptotische Eigenschaften von Modellwahlverfahren in der Regressionsanalyse. Doctoral Thesis, Department of Mathematics, Humboldt University, Berlin (in German).

NADARAYA, E.A. (1964). On estimating regression. *Theor. Probab. Appl.* **9**, 141-142.

NISHII, R. (1984). Asymptotic properties of criteria for selection of variables in multiple regression. *Ann. Statist.* **12**, 758-765.

RAO, R.C. (1976). Estimation of parameters in a linear model. *Ann. Statist.* **4**, 1023-1037.

STONE, M. (1974). Cross-validatory choice and assessment of statistical predictions. *J. Roy. Statist. Soc. B* **36**, 111-147.

WAHBA, G. (1978). Improper priors, spline smoothing and the problem of guarding against model errors in regression. *J. Roy. Statist. Soc. B* **40**, 364-372.

WATSON, G.S. (1964). Smooth regression analysis. *Sankhya A* **26**, 359-372.

Extreme Percentile Regression

Ori Rosen
Ayala Cohen

Faculty of Industrial Engineering and Management
Technion - Israel Institute of Technology
Haifa 32000, Israel

Summary

In this study we propose a method for fitting extreme percentile regression
based on the r extremes corresponding to each value of a scalar covariate.
The estimation is performed by maximum penalized likelihood, exploiting
results from extreme value theory. Confidence bands for the true conditional
percentile are constructed using a bootstrap algorithm.

Keywords: extreme value theory, maximum penalized likelihood, bootstrap

1 Introduction

The aim of building a regression model is usually to estimate the conditional
mean of a dependent variable, given covariate values. Sometimes however,
it is also important to learn about conditional percentiles. In the case of
growth curves, for instance, there is an interest in modelling the dependence
of percentiles of a certain variable, such as weight, on a covariate such as age.
In recent years there has been growing interest in the subject of percentile
regression. Koenker and Basset (1978) generalize a minimization problem,
which yields the ordinary sample quantiles, to a linear model. Thus a new
class of statistics, called regression quantiles, is generated. They suggest estimators which have comparable efficiency to least squares for Gaussian linear
models, while substantially out-performing the least squares estimator over a
wide class of non-Gaussian error distributions. Hendrix and Koenker (1992)
suggest methods for estimating nonparametric models for conditional quantiles based on the regression quantile methods of Koenker and Basset (1978).
Spline parametrizations of the conditional quantile functions are used. The
methods are illustrated by estimating hierarchical models for household elec-

tricity demand. Koenker et al. (1994) explore a class of quantile smoothing splines and discuss their computation by linear programming techniques. Efron (1991) considers the problem of estimating regression percentiles by asymmetric least squares. This is a variant of ordinary least squares, in which the squared error loss function is given different weight depending on whether the residual is positive or negative. Cole (1988) describes a method for fitting smooth centile curves to reference data, based on transforming the dependent variable into a normal variate by the power transformation family of Box and Cox. The data are defined by ranges of values of the independent variable. In each range, the transformation parameters are estimated and then smoothed against the independent variable. The smooth parameter curves are used in turn to define a smooth curve for the 100αth centile, which is in effect, a conditional percentile. Based on Cole's method, Cole and Green (1992) fit smooth parameter curves as cubic splines by penalized likelihood. These curves are then utilized for building the centile curves. Healy et al. (1988) estimate centile curves by fitting polynomials. Their method ensures smooth curves as well as smooth behaviour of the intervals between centiles at a fixed covariate value. Goldstein and Pan (1992) present a nonparametric procedure for the joint smoothing of a series of percentile curves. It allows separate curves to be fitted over contiguous age ranges, constraining them to join smoothly. In addition, covariates can be incorporated.

Our interest is in estimating extreme conditional percentiles, such as the 0.99 percentile. This subject does not seem to have attracted much attention in the literature, although it may play a crucial role in some applications. For instance, it may be of interest to investigate children's excellence in some sport such as running. Only a fixed number of children, who are the top runners from each age stratum, will be included in the sample. The statistical analysis will result in estimating large quantiles as a function of runner age.

The data on which we base the estimation are the r largest values of a dependent variable for each of N distinct values of a scalar covariate. In section 2 we describe the construction of a likelihood function for these data based on extreme-value theory. In section 3, the parameters are estimated as smooth functions of the covariate by maximum penalized likelihood. These functions are then used for estimating conditional extreme percentiles. Section 4 includes a bootstrap algorithm for constructing confidence bands for the true conditional extreme percentile. In section 5 we illustrate our method on sea-level data. Section 6 concludes with a discussion.

2   Extreme-value theory

Let $Y_1, \ldots, Y_n$ be independent identically distributed random variables with a continuous distribution function F and let $M_n = \max(Y_1, \ldots, Y_n)$. Suppose we can find normalizing constants $a_n > 0$ and b_n, such that

$$P\left\{\frac{M_n - b_n}{a_n} \leq x\right\} \to G(x)$$

as $n \to \infty$, where G is a nondegenerate distribution function. Then G is necessarily one of three possible distribution functions (see for instance Arnold et al., 1992, p. 210 or Castillo, 1988, p. 100). We say that F belongs to the domain of G and denote it by $F \in \mathcal{D}(G)$. In the present work we restrict ourselves to the case $G = \Lambda$, where

$$\Lambda(x) = \exp(-\exp(-(x - \mu)/\sigma)), \quad -\infty < x < \infty \tag{1}$$

is the Gumbel distribution function. This is not an unreasonable assumption, since many commonly used distributions belong to this domain. Among them are the normal, lognormal, logistic, gamma, exponential and Weibull. In principle, one could have used instead of (1) the generalized extreme value distribution, with distribution function

$$H(x; \mu, \sigma, \gamma) = \exp\{-[1 - \gamma(x - \mu)/\sigma]^{1/\gamma}\}. \tag{2}$$

The index γ is a real parameter and x is such that $1 - \gamma(x - \mu)/\sigma > 0$. Equation (2) reduces to (1), as a special case, when $\gamma = 0$.

Let $Y_{(1)}, \ldots, Y_{(r)}$ be the r largest observations, such that $Y_{(1)} \geq \ldots \geq Y_{(r)}$. It is known (Weissman 1978) that the joint distribution of

$$\left(\frac{Y_{(1)} - b_n}{a_n}, \ldots, \frac{Y_{(r)} - b_n}{a_n}\right)$$

has for large n a limit distribution, which in the case that $F \in \mathcal{D}(\Lambda)$, has the density

$$f(x_1, \ldots, x_r; \mu, \sigma) = \sigma^{-r} \exp\left\{-\exp\left(-\frac{x_r - \mu}{\sigma}\right) - \sum_{j=1}^{r}\left(\frac{x_j - \mu}{\sigma}\right)\right\} \tag{3}$$

$$x_1 \geq \ldots \geq x_r.$$

Treating (3) as an approximate likelihood, Weissman estimates the parameters μ and σ by maximum likelihood, using the r largest observations from a sample of size n. He then estimates a large percentile of F by

$$\hat{\eta}_{1-c/n} = \hat{\sigma}(-\log c) + \hat{\mu}. \tag{4}$$

For example, if $n = 1000$ and $c = 10$, then

$$\hat{\eta}_{1-10/1000} = \hat{\eta}_{0.99} = \hat{\sigma}(-\log 10) + \hat{\mu}$$

is the maximum likelihood estimator of the 0.99 quantile of F. Boos (1984) gives practical recommendations on how to use Weissman's estimator (4) effectively. These include the choice of the ratio r/n as a function of n and the parent distribution F. For instance, in the case of parent distributions with approximately exponential tails, r/n should be 0.2 for $50 \leq n \leq 500$ and $p \geq 0.95$.

The data on which we base the estimation are the r largest observations $Y_{i1}, \ldots, Y_{ir}$ $(Y_{i1} \geq \ldots \geq Y_{ir})$ for each of N $(1 \leq i \leq N)$ distinct values of a scalar covariate t. The asymptotic joint density of these random variables is given by (3), expressed in terms of $\mu(t_i)$ and $\sigma(t_i)$. If we assume that the data for separate values of the covariate are independent, then the product of the N joint densities is the approximate likelihood based on all the observations. Let t_i $(1 \leq i \leq N)$ be the distinct ordered covariate values and assume that they lie within an interval $[a, b]$. The approximate log-likelihood can be written as

$$l(\boldsymbol{\mu}, \boldsymbol{\sigma}) = \sum_{i=1}^{N} \left\{ -r \log \sigma(t_i) - \exp\left(-\frac{y_{ir} - \mu(t_i)}{\sigma(t_i)}\right) - \sum_{j=1}^{r} \left(\frac{y_{ij} - \mu(t_i)}{\sigma(t_i)}\right) \right\}$$
(5)

where $\boldsymbol{\mu}$ and $\boldsymbol{\sigma}$ denote the vectors of values of the corresponding curves at the covariate values t_i. We assume that $\mu(t)$ and $\sigma(t)$ are smooth functions of the covariate t. In the next section we estimate them by maximum penalized likelihood.

Smith (1986) assumes certain parametric forms for $\mu(t)$ and $\sigma(t)$, e.g.,

$$\begin{aligned} \mu(t) &= \alpha + \beta t/N \\ \sigma(t) &= \sigma \end{aligned}$$
(6)

and estimates α, β and σ by maximum likelihood using sea-level data. In section 5 we illustrate our analysis on these data.

Smith (1989) reviews several methods of analyzing extreme values based on the extreme value limit distributions or related families. He also proposes some extentions based on the point-process view of high-level exceedances. The techniques are illustrated with an analysis of ozone data. Smith (1993) applies and extends methods whose aim is to find suitable probability distributions for exceedance times and excess values, which can then be estimated by maximum likelihood. In these works Smith does not estimate extreme percentiles, neither does he use maximum penalized likelihood estimation.

3 Maximum penalized likelihood

An attempt to maximize the log-likelihood (5) over all smooth functions $\mu(t)$ and $\sigma(t)$ is useless. It is always possible to choose $\mu(t)$ and $\sigma(t)$ such that they interpolate the data in the sense that the fitted values are identical to the observed responses. Following Cole and Green (1992) and Green and

Silverman (1994), rather than maximize the log-likelihood alone, $\mu(t)$ and $\sigma(t)$ are chosen to maximize the penalized log-likelihood

$$\Pi = l - \frac{1}{2}\alpha_\mu \int (\mu''(t))^2 dt - \frac{1}{2}\alpha_\sigma \int (\sigma''(t))^2 dt \ .$$

The functions $\mu(t)$ and $\sigma(t)$ are in the space of functions, which are differentiable on $[a, b]$ and have absolutely continuous derivatives. The parameters α_μ and α_σ are smoothing parameters. The penalized likelihood approach allows one to balance fidelity to the data (high values of the log-likelihood) with smoothness of the fitted curves (low values of the roughness penalties). It is common to perform the maximization by Fisher scoring. Given μ and σ at the current iteration, updated estimates are obtained by solving the following system of equations:

$$\begin{pmatrix} W_\mu + \alpha_\mu K & W_{\mu\sigma} \\ W_{\sigma\mu} & W_\sigma + \alpha_\sigma K \end{pmatrix} \begin{pmatrix} \mu^{new} - \mu \\ \sigma^{new} - \sigma \end{pmatrix} = \begin{pmatrix} u_\mu - \alpha_\mu K\mu \\ u_\sigma - \alpha_\sigma K\sigma \end{pmatrix}. \quad (7)$$

The u's are the first derivatives of the log-likelihood. The diagonal matrices W are the expectations of the negative second derivatives of the log-likelihood with respect to the variables specified in the subscripts. Explicit expressions for the u's and W's can be deduced from Smith (1986). For completeness, we include them in the appendix. The matrix K is an $N \times N$ square matrix dependent on the t_i's only (see Green and Silverman, 1994, pp. 12-13). The system (7) consists of $2N$ equations, therefore it may not be practical to solve this system directly. Instead, we use the backfitting algorithm, an iterative method, which converges under certain conditions (see Green and Silverman, 1994, pp. 67-69). The two block rows of (7) can be rewritten in the form of updating equations for μ and σ. The resulting μ and σ are obtained by applying weighted cubic smoothing splines:

$$\mu^{new} = S_1(y_1 - W_\mu^{-1} W_{\mu\sigma} \sigma^{new}) \quad (8)$$

$$\sigma^{new} = S_2(y_2 - W_\sigma^{-1} W_{\sigma\mu} \mu^{new}), \quad (9)$$

where S_1 and S_2 are the smoother matrices

$$S_1 = (W_\mu + \alpha_\mu K)^{-1} W_\mu \quad (10)$$

$$S_2 = (W_\sigma + \alpha_\sigma K)^{-1} W_\sigma \quad (11)$$

and

$$y_1 = \mu + W_\mu^{-1} u_\mu + W_\mu^{-1} W_{\mu\sigma} \sigma \quad (12)$$

$$y_2 = \sigma + W_\sigma^{-1} u_\sigma + W_\sigma^{-1} W_{\sigma\mu} \mu \ . \quad (13)$$

We first obtain initial solutions by finding maximum likelihood estimates for $\mu(t_i)$ and $\sigma(t_i)$, based on $Y_{i1}, \ldots, Y_{ir}$, $i = 1, \ldots, N$, according to Weissman's method. The solution then involves two loops:

1. **Inner loop (backfitting)** - The vectors $\boldsymbol{y}_1$ and $\boldsymbol{y}_2$ are calculated at the initial estimates (equations (12) and (13)). A new vector $\boldsymbol{\mu}^{new}$ is obtained by (8), applying a weighted cubic smoothing spline to the pairs $(t_i, (\boldsymbol{y}_1 - W_\mu^{-1} W_{\mu\sigma} \boldsymbol{\sigma}^{new})_i)$ with weights $(W_\mu)_{ii}$, $i = 1, \ldots, N$. A new vector $\boldsymbol{\sigma}^{new}$ is found in turn by substituting $\boldsymbol{\mu}^{new}$ in (9). This cycling between (8) and (9) continues until

$$\sum_{i=1}^N |\sigma^{new}(t_i) - \sigma(t_i)| < 0.01 .$$

2. **Outer loop** - As the inner loop converges, $\boldsymbol{y}_1$ and $\boldsymbol{y}_2$ are recalculated at the values of $\boldsymbol{\mu}$ and $\boldsymbol{\sigma}$ from the final iteration of the inner loop. The penalized log-likelihood is calculated at these estimates as well. New vectors $\boldsymbol{\mu}$ and $\boldsymbol{\sigma}$ are obtained from the inner loop. The whole process continues until the absolute difference between the current penalized log-likelihood and the old one is less than 0.001.

The amount of smoothing is governed by the smoothing parameters α_μ and α_σ. Practically, we specify for each updating equation its degrees of freedom. In general, if S is a weighted cubic smoothing spline matrix (given by (10) and (11) for our two updating equations), then a common definition for its degrees of freedom is

$$df = trace(S)$$

(see e.g., Hastie and Tibshirani, 1990). The degrees of freedom decrease as the smoothing parameter increases. It should be noted that since the matrices W_μ and W_σ depend on both μ and σ (see appendix), specifying df for one updating equation will affect the values of both α_μ and α_σ.

Using the fitted curves $\hat{\mu}(t)$ and $\hat{\sigma}(t)$, we estimate large percentiles by

$$\hat{\eta}_{1-c/n}(t) = \hat{\sigma}(t)(-\log c) + \hat{\mu}(t) .$$

In section 4 we construct confidence bands for the true conditional percentile, $\eta(t)$, by a bootstrap algorithm.

4 Bootstrap confidence bands

In this section we present our bootstrap algorithm for constructing confidence bands for the true conditional percentile. To this end, we first quote a result from extreme-value theory (see David, 1981, p. 266):

Let $\lambda(x) = -\log G(x)$, where $G(x)$ is the limit distribution function, mentioned in section 2, and Z_i, $i = 1, \ldots, r$, independent identically distributed exponentials with unit means. Then under the same assumptions as those specified in section 2, the asymptotic joint distribution of the r upper extremes, whose density is given by (3), is identical to the joint distribution of

$$\lambda^{-1}(Z_1), \ \lambda^{-1}(Z_1 + Z_2), \dots, \ \lambda^{-1}(Z_1 + \dots + Z_r) \ . \tag{14}$$

The function λ^{-1} is the inverse function of λ. For distribution functions F, belonging to the Gumbel domain ($F \in \mathcal{D}(\Lambda)$), the function λ^{-1} is given by

$$\lambda^{-1}(x) = \mu - \sigma \log x \ . \tag{15}$$

Our parametric bootstrap algorithm is as follows. Repeat the following two stages B times ($1 \le b \le B$):

1. For each design point, generate a sample of size r from the asymptotic distribution, whose density is given by (3). Thus the structure of the b'th bootstrap sample is

$$
\begin{array}{cccc}
X_{11}^{*b} & X_{12}^{*b} & \dots & X_{1r}^{*b} \\[4pt]
X_{21}^{*b} & X_{22}^{*b} & \dots & X_{2r}^{*b} \\[4pt]
\vdots & \vdots & \vdots & \vdots \\[4pt]
X_{N1}^{*b} & X_{N2}^{*b} & \dots & X_{Nr}^{*b} \ ,
\end{array}
$$

   where the i'th row is comprised of the r largest observations from the asymptotic distribution, generated according to equations (14) and (15), with parameters $\hat{\mu}(t_i)$ and $\hat{\sigma}(t_i)$. These parameters are the components of the curves $\hat{\mu}$ and $\hat{\sigma}$, estimated by maximum penalized likelihood from the original sample, at the covariate value t_i.

2. Find penalized maximum likelihood estimates $\hat{\mu}^{*b}$ and $\hat{\sigma}^{*b}$, based on the sample generated in stage 1, and compute an extreme percentile $\hat{\eta}^{*b}$.

3. Using $\hat{\eta}^{*b}$, $b = 1, \dots, B$, construct confidence bands for $\eta(t)$, in the following way (McDonald 1982, see Härdle 1991):

   (a) Define
   $$T = \sup_t |\hat{\eta}(t) - \eta(t)|$$
   and its bootstrap analog
   $$T^* = \sup_t |\hat{\eta}^*(t) - \hat{\eta}(t)| \ .$$

   (b) Let $\hat{F}_T^*$ be the empirical distribution function of T^*. Based on T^{*b}, $1 \le b \le B$, find $T_{\alpha/2}^*$, such that $\hat{F}_T^*(T_{\alpha/2}^*) = \alpha/2$ and similarly $T_{1-\alpha/2}^*$.

   (c) Lower and upper confidence bands are formed by
   $$
   \begin{aligned}
   L(t) &= \hat{\eta}(t) - T_{\alpha/2}^* \\
   U(t) &= \hat{\eta}(t) + T_{1-\alpha/2}^* \ .
   \end{aligned}
   $$

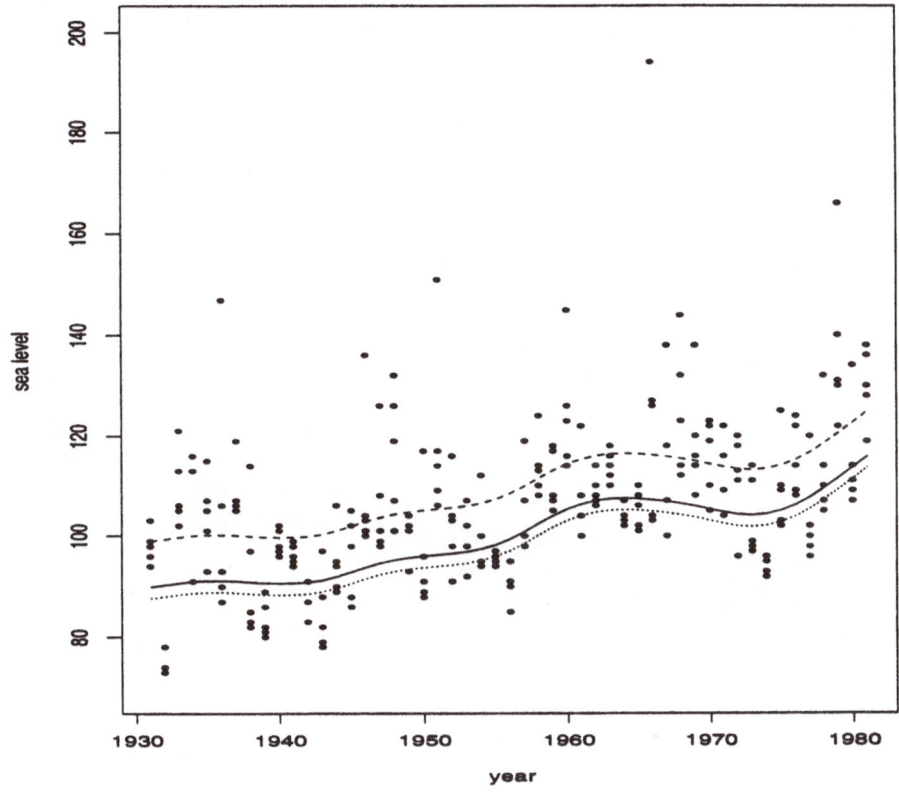

Figure 1: Venice data overlaid with 0.99 percentile (solid
line) and 0.95 confidence bands

5 Illustration

In this section we illustrate our method on the Venice sea-level data, given in
Smith (1986). This is a part of a larger set of data pertaining to the period
1872-1981. In a preliminary analysis, Smith shows that up to 1930, the sea
level is roughly constant with irregular fluctuations, therefore the given data
refer only to the years 1931-1981. For each year, the ten largest sea levels
are given. Smith (1986) also confirms by means of probability plots, the
assumptions underlying equation (3). In the case of the linear trend model
(equation (6)), the fit is particularly good for $r = 5$.

The data we analyze are the 5 largest sea levels for each year. Figure 1
shows these data overlaid with the maximum penalized likelihood estimate
of the 0.99 conditional percentile (solid line) along with 0.95 level bootstrap

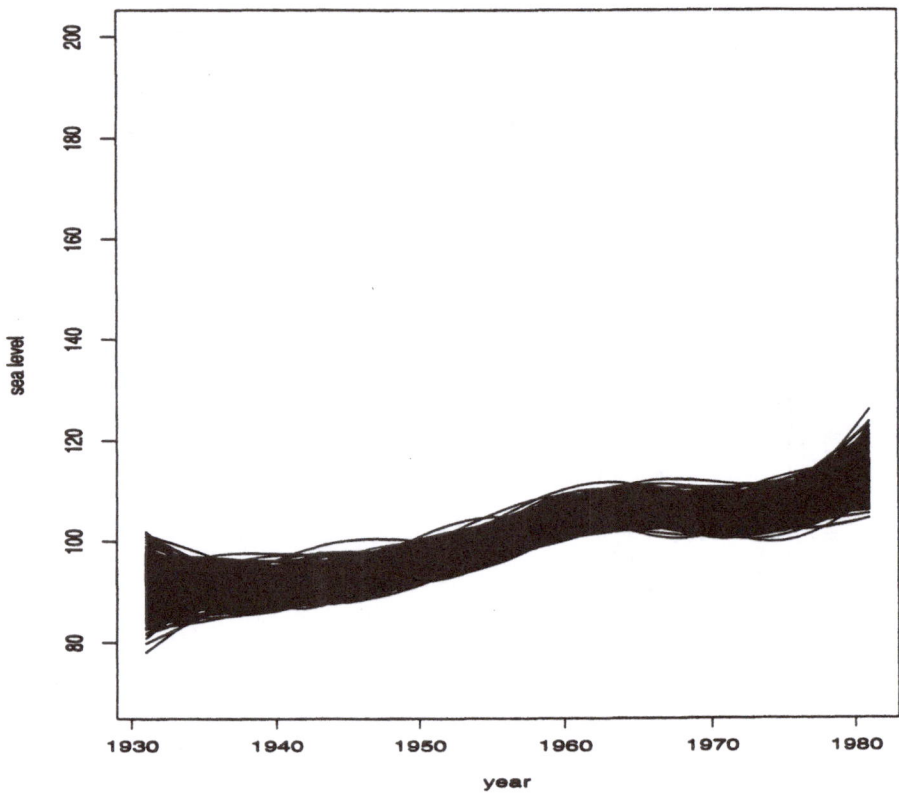

Figure 2 (a): 1000 bootstrap percentile curves

confidence bands.

The 0.99 conditional percentile was estimated using 7 *df* for μ and 5 *df* for σ. This choice is merely intended to demonstrate the method and is by no means optimal in any respect. It is possible to identify three parts in the percentile curve estimate. The first is roughly constant up to about 1942, whereas the second part (1942-1963) reveals an upward trend with slight periodicity. The third part shows mainly periodicity with an upward trend beginning in the mid 1970s.

Smith (1986) gives a plot displaying the annual maxima (i.e., the maximum sea level for each year) overlaid with so-called median predicted values, given in our notation by

$$\mu(t) - \sigma(t) \log(\log 2) \ .$$

This function describes the sea level which will be exceeded with probability

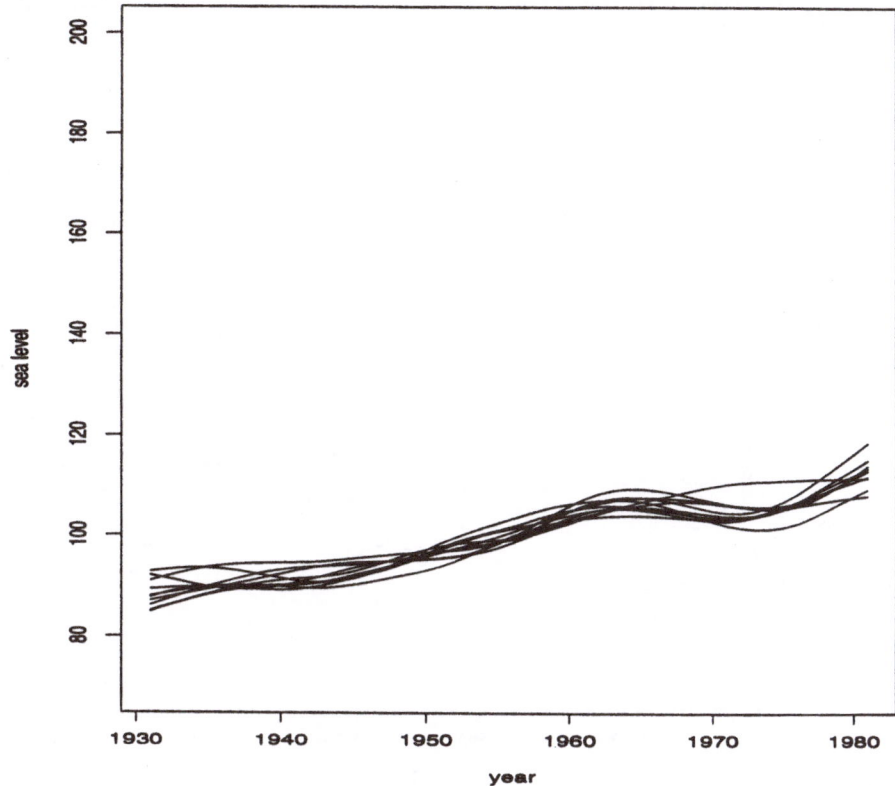

Figure 2 (b): 10 bootstrap percentile curves

0.5 in year t. The overlaying curves are median predicted values, corresponding to the linear trend model (equation (6)) and to three other parametric
models (quadratic and two sinusoidal with two distinct periods). Although
these curves differ in meaning from the percentile curves, they share with
them a similar functional structure. Therefore it is interesting to compare
Smith's parametric fit and the semi-parametric approach we use. It seems
that his linear and quadratic models succeed in detecting the trend, but fail
to capture the seasonality. On the other hand, the two sinusoids tend to
overemphasize the seasonal component.

In figure 2 (a), 1000 bootstrap curves are shown. These are the 0.99
conditional percentiles generated according to the algorithm described in the
previous section. They are used for constructing the confidence bands for the
true conditional percentile. As an additional exposition, figure 2 (b) shows
10 of the 1000 curves.

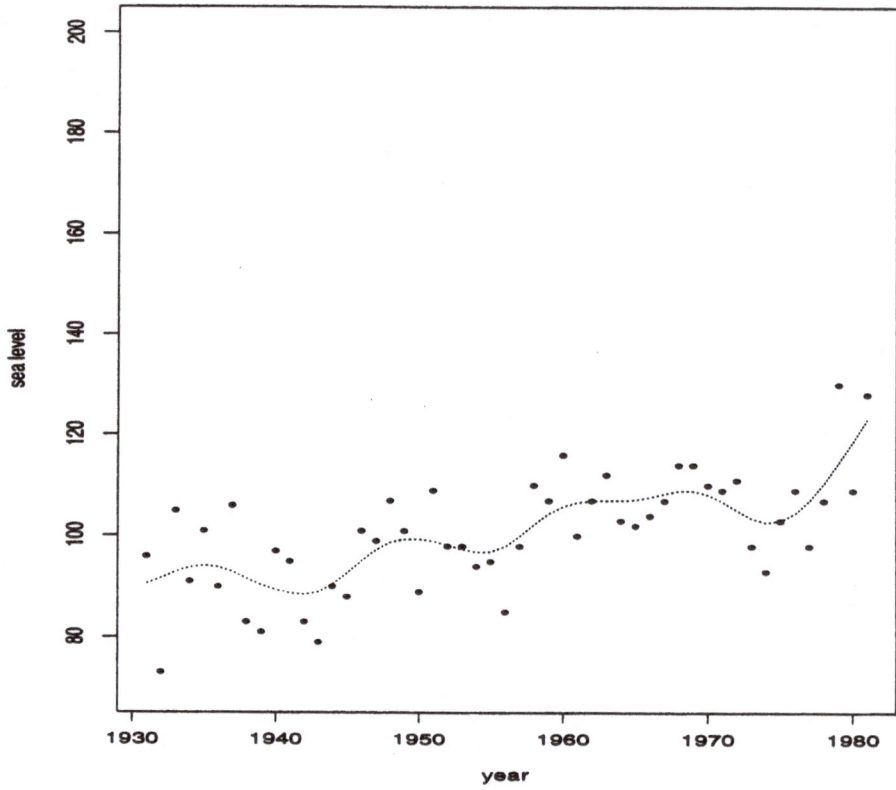

Figure 3: 0.99 empirical percentiles (points) overlaid with a
smooth curve

The points in figure 3 are the 0.99 empirical quantile for each year, that is the fourth largest observation (out of 365), X_{i4}, $i = 1, \ldots, N$. The dashed line is a smooth of these points. This curve is wigglier than the one obtained by the penalized likelihood approach. In this plot, confidence bands are not given. Had the disribution function F been known, it would have been possible to construct pointwise confidence bands using the variance of the sample quantiles.

6   Discussion

In the current study we fit extreme percentile regression for a dependent variable, given a scalar covariate. Our main contribution is the integration of two usually separate statistical areas: extreme value theory and smoothing. This is achieved by the penalized likelihood approach, which is semi parametric in nature. Our emphasis is on extreme percentiles, a subject which is not very common in the literature despite its applicability to a variety of fields. It might seem that our method is restricted to a design in which there are n observations in each of N design points and that the same number r of extremes is used in each design point. In fact, it is possible to base the estimation on the r_i largest observations from a sample of size n_i at each covariate value ($1 \leq i \leq N$). To this end, it is only necessary to substitute r_i rather than r in equation (5). Conditional percentiles will then be estimated at each t_i using equation (4) with corresponding n_i and c_i. In this case, further smoothing of the resulting estimates may be required, unlike the situation where the sample size n is common to all the design points.

There are still several points requiring further reasearch, on which we are currently working:

- Incorporation of several explanatory variables.

- Selection of smoothing parameters by some objective criterion.

- Exploration of the properties of the bootstrap algorithm proposed in section 4.

The computations in this study were performed by S-plus (Becker and Chambers, 1988, Chambers and Hastie, 1993)

Acknowledgement

We would like to thank the referee for helpful comments.

Appendix: calculation of derivatives

The i'th components of $\boldsymbol{u}_\mu$ and $\boldsymbol{u}_\sigma$ are given by:

$$
\begin{aligned}
\boldsymbol{u}_\mu^i &= -\tfrac{1}{\sigma(t_i)}\exp(-(x_{ir}-\mu(t_i))/\sigma(t_i))+\tfrac{r}{\sigma(t_i)}\\
\boldsymbol{u}_\sigma^i &= -\tfrac{r}{\sigma(t_i)}-\tfrac{1}{\sigma^2(t_i)}(x_{ir}-\mu(t_i))\exp(-(x_{ir}-\mu(t_i))/\sigma(t_i))+\\
&\quad \tfrac{1}{\sigma^2(t_i)}\textstyle\sum_{j=1}^{r}(x_{ij}-\mu(t_i))\ .
\end{aligned}
$$

Let $V_j=(X_j-\mu)/\sigma$, where X_j has the limit distribution of the j'th extreme. Then according to Smith (1986):

$$
E[V_j^m\exp(-\alpha V_j)]=(-1)^m\Gamma^{(m)}(j+\alpha)/\Gamma(j)\,,
$$

where $\Gamma^{(m)}$ is the m'th derivative of the gamma function. Using the relations:

$$
\begin{aligned}
x\Gamma(x) &= \Gamma(x+1)\\
\Psi(x) &= \Gamma'(x)/\Gamma(x)\,,
\end{aligned}
$$

where Ψ is the digamma function, the following expressions for the W's are obtained:

$$
W_\mu = \operatorname{diag}\left(\frac{r}{\sigma^2(t_i)}\right)
$$

$$
W_{\mu\sigma} = \operatorname{diag}\left(-\frac{1}{\sigma^2(t_i)}(r\Psi(r)+1)\right)
$$

$$
W_\sigma = \operatorname{diag}\left(\frac{1}{\sigma^2(t_i)}[2(r+1)\Psi(r)+r\Psi'(r)+r\Psi^2(r)+2-r-2\sum_{j=1}^{r}\Psi(j)]\right)
$$

References

[1] Arnold, B.C., Balakrishnan, N. and Nagaraja, H.N. (1992). A first course in order statistics. John Wiley and Sons Inc., New York.

[2] Becker, R.A., Chambers, J.M. and Wilks, A.R. (1988). The new S language. Wadsworth and Brooks, California.

[3] Boos, D.D. (1984). Using extreme value theory to estimate large percentiles. Technometrics, 26, 33–39.

[4] Castillo, E. (1988). Extreme value theory in engineering. Academic Press Inc., New York.

[5] Chambers, J.M. and Hastie, T.J. (1993). Statistical models in S. Chapman and Hall, New York.

[6] Cole, T.J. (1988). Fitting smoothed centile curves to reference data. J. of the Royal Statistical Society Soc. A, 151, 385–418.

[7] Cole, T.J. and Green, P.J. (1992). Smoothing reference centile curves: the LMS method and penalized likelihood. Statistics in Medicine, 11, 1305–1319.

[8] David, H.A. (1981). Order Statistics. John Wiley and Sons Inc., New York.

[9] Efron, B. (1991). Regression percentiles using asymmetric squared error loss. Statistica Sinica ,1, 93–125.

[10] Goldstein, H. and Pan, H. (1992). Percentile smoothing using piecewise polynomials, with covariates. Biometrics, 48, 1057–1068.

[11] Green, P.J. and Silverman, B.W. (1994). Nonparametric regression and generalized linear models: a roughness penalty approach. Chapman and Hall, London.

[12] Härdle, W. (1991). Applied nonparametric regression. Cambridge university press, Cambridge.

[13] Hastie, T.J. and Tibshirani, R.J. (1990). Generalized additive models. Chapman and Hall, London.

[14] Healy, M.J.R., Rasbash, J. and Yang, M. (1988). Distribution-free estimation of age-related centiles. Annals of Human Biology, 15, 17–22.

[15] Hendricks, W. and Koenker, R. (1992). Hierarchical spline models for conditional quantiles and the demand for electricity. J. of the American Statistical Association, 87, 58–68.

[16] Koenker, R. and Basset, G. (1978). Regression quantiles. Econometrica, 46, 33–50.

[17] Koenker, R., Pin, N. and Portnoy, S. (1994). Quantile smoothing splines. Biometrika, 81, 673–680.

[18] McDonald, J.A. (1982). Projection pursuit regression with the ORION I workstation. A 25 minute film, available from Jerome H. Friedman, Computation Research Group, Bin 88 SLAC, P.O. 4349, Stanford, CA 94305.

[19] Smith, R.L. (1986). Extreme value theory based on the r largest annual events. J. of Hydrology, 86, 27–43.

[20] Smith, R.L. (1989). Extreme value analysis of environmental time series: an application to trend detection in ground-level ozone. Statistical Science, 4, 367–393.

[21] Smith, R.L. and Huang, L.S. (1993). Modelling high threshold exceedances of urban ozone. National institute of statistical sciences. Technical report # 6.

[22] Weissman, I. (1978). Estimation of parameters and large quantiles based on the k largest observations. J. of the American Statistical Association, 73, 812–815.

Mean and Dispersion Additive Models

Robert A. Rigby and Mikis D. Stasinopoulos

University of North London, Holloway Road, London, United Kingdom

Summary

This paper presents a flexible model for the mean and variance of a dependent variable. The variance is modelled as a product of a dispersion parameter and a known variance function of the mean. The dependence of each of the mean and dispersion parameter on explanatory variables is modelled using a semi-parametric additive model. We call this model the 'Mean and Dispersion Additive Model' or 'MADAM'. The MADAM is fitted either by maximisation of the penalised extended Quasi-likelihood or by pseudo-maximization of the penalised Normal likelihood. A successive relaxation fitting algorithm is described and is implemented in GLIM4 allowing flexible and interactive modelling of both the mean and dispersion of a dependent variable. Two examples are given to demonstrate the use of the MADAM for modelling overdispersion in each of Poisson regression model and a Binomial logistic regression model.

Keywords: Dispersion parameter, generalised additive models, extended quasi-likelihood, pseudo-likelihood, penalised likelihood, MADAM

1   Introduction

Here we consider models in which both the mean and the dispersion parameter of a generalised model vary smoothly with explanatory variables and are modelled using semi-parametric Additive models (Hastie and Tibshirani, 1990). Specifically, our model assumes that the response variable y_i is independently distributed with mean μ_i and variance $\phi_i V(\mu_i)$, where

$$g_1(\mu_i) = \eta_i = \sum_{j=1}^{p} f_j(x_{ij}), \qquad g_2(\phi_i) = \xi_i = \sum_{k=1}^{q} h_k(z_{ik}) \qquad (1)$$

and where ϕ_i is the dispersion parameter and $V(\mu_i)$ is the known variance function, for $i = 1, 2, \ldots, n$. The function f_j is either a linear or a non-parametric function of explanatory variable x_j for the mean, for $j = 1, 2, \ldots, p$, and the function h_k is either a linear or a non-parametric function of explanatory variable z_k for the dispersion, for $k = 1, 2, \ldots, q$. Further, let $\mu, \phi, \eta, \xi, \mathbf{x}_j, \mathbf{z}_k, f_j(\mathbf{x}_j)$ and $h_k(\mathbf{z}_k)$ be the corresponding vectors of length n where, for example $\mu = (\mu_1, \mu_2, \ldots, \mu_n)$. The $\mathbf{x}$'s and the $\mathbf{z}$'s are assumed to be fixed and g_1 and g_2 are known monotonic link functions, where typically g_2 is a log link. The above set up is subsequently referred to as the Mean and Dispersion Additive Model or **MADAM**.

The MADAM is very flexible and allows a wide range of parametric, semi-parametric or non-parametric models for both the mean and the dispersion. This flexibility can be used diagnostically to detect heterogeneity and to suggest possible parametric models for the mean and dispersion. Typically MADAM model can be used to investigate variance heterogeneity in a Normal error model and overdispersion in a Generalised Linear Model. Parametric submodels of MADAM are well known in the statistical literature and are described in section 2. In this paper we extend these models to include non-parametric functions of the covariates.

Estimation of the non-parametric smooth functions f_j and h_k can be achieved by maximising the Penalised extended Quasi-likelihood (L_{PQ}^+), with penalties for lack of smoothness in the additive functions f_j and h_k in the mean and dispersion models respectively. Alternatively the functions f_j and h_k can estimated by pseudo-maximisation of the Penalised Normal likelihood (L_{PN}) i.e. by pseudo-likelihood estimation. The maximising functions f_j and h_k are cubic splines for both L_{PQ}^+ and L_{PN}. See section 2 for a discussion of extended Quasi-likelihood and pseudo-likelihood estimation and section 3 for their penalised versions.

A general algorithm for fitting the MADAM by maximising L_{PQ}^+ or L_{PN} is provided in section 4. The algorithm is a successive relaxation algorithm combined with local scoring. The successive relaxation algorithm alternates between mean and dispersion model fits. An interactive implementation is available in GLIM4. In section 5, MADAM is used for modelling overdispersion in both a Poisson regression model and a Binomial logistic regression model. Section 6 contains concluding remarks.

2 Parametric submodels of MADAM

2.1 Normal error models with parametric mean and variance

Harvey (1976), Aitkin (1987) and Verbyla (1993) have modelled the mean μ_i and variance $\phi_i = \sigma_i^2$ for $i = 1, 2, \ldots, n$ of a Normal error model using parametric linear models of the form

$$g_1(\mu) = \eta = X\beta, \qquad g_2(\phi) = \xi = Z\gamma \qquad (2)$$

where β and γ are vectors of parameters to be estimated. This model is a special case of MADAM where $V(\mu) = 1$ and where all the f_j's and h_k's in (1) are linear. Estimation of the parameter vectors β and γ may be achieved by maximising the exact Normal likelihood function L or by minimising the Normal deviance D given by

$$D = -2\log L = \sum_{i=1}^{n} \frac{(y_i - \mu_i)^2}{\sigma_i^2} + \sum_{i=1}^{n} \log 2\pi\sigma_i^2. \qquad (3)$$

2.2 Generalised Linear Models with parametric mean and dispersion

Pregibon (1984), Nelder and Pregibon (1987), McCullagh and Nelder (1989), Nelder (1992) and Nelder and Lee (1992) have jointly modelled the mean μ and dispersion parameter ϕ of a Generalised Linear Model using the parametric linear model (2) with a general variance function $V(\mu)$ which may be dependent on parameter(s) α. They estimate β, γ and α by maximising the extended Quasi-likelihood function L_Q^+ (or minimising D_Q^+), given by:

$$
\begin{aligned}
D_Q^+ &= -2\log L_Q^+ \\
&= \sum_{i=1}^{n} \frac{d_i}{\phi_i} + \sum_{i=1}^{n} \log\left[2\pi\phi_i V(y_i)\right]
\end{aligned}
\qquad (4)
$$

where $V(y_i)$ is the variance function evaluated at y_i and d_i is the deviance increment defined, for the particular variance function $V(\mu_i)$ used, by

$$d_i = -2 \int_{y_i}^{\mu_i} \frac{y_i - u_i}{V(u_i)} du_i$$

An alternative estimation approach is by pseudo maximisation of L_N, a Normal likelihood for a general variance function $V(\mu_i)$, i.e. by pseudo-likelihood estimation, Davidian and Carroll (1987,1988) and Carroll and Ruppert (1988, ch3), where

$$
\begin{aligned}
D_N &= -2\log L_N \\
&= \sum_{i=1}^{n} \frac{\chi_i^2}{\phi_i} + \sum_{i=1}^{n} \log\left[2\pi\phi_i V(\mu_i)\right]
\end{aligned}
\qquad (5)
$$

where χ_i^2 is the Pearson χ^2 component for the ith observation given by:

$$\chi_i^2 = \frac{(y_i - \mu_i)^2}{V(\mu_i)}$$

Where there exists a distribution of the exponential family with a given variance function $V(\mu)$, the extended Quasi-likelihood is a saddlepoint approximation to that distribution (McCullagh and Nelder, 1989, ch 9.6). In certain specific cases the extended Quasi-likelihood is an exact likelihood, (e.g. if $V(\mu) = 1$ or μ^3 it is an exact Normal or Inverse Gaussian likelihood respectively), while it is an approximation to Binomial and Poisson likelihoods obtained by replacing all factorials by Stirling approximations, and it differs from a Gamma likelihood by a function of the shape parameter. More generally the extended Quasi-likelihood can be thought of as an approximation to the exact distribution (called a 'Quasi distribution' by Nelder, 1992) that would be obtained by normalising it.

The pseudo-likelihood is an exact Normal likelihood with mean μ and variance $\phi V(\mu)$. However the method of maximising this Normal likelihood is an approximate, i.e. pseudo-maximising, method, by alternating between maximising over μ fixing both ϕ and $V(\mu)$, then updating $V(\mu)$ and maximising over ϕ fixing both μ and $V(\mu)$.

The choice between (4) and (5) is not as yet clear cut. Minimising (5) leads to consistent estimators for β, γ and α while (4) results in inconsistent estimators for γ and α. However in small sample sizes (5) can result in estimators of γ and α with a higher variance than (4) resulting in a higher mean square error despite a smaller bias, (Nelder and Lee, 1992). The estimating equations for γ, obtained from (4) or (5), are equivalent to assuming a Gamma variance function $2\phi_i^2$ for d_i or χ_i^2. However their true variances (especially for χ_i^2) are often higher, leading to non-optimal inflated variances for $\hat\gamma$. Adjustment for the effect of ρ_4, (the kurtosis of the true error distribution of an observation of the dependent variable) on the variance of χ_i^2, which is one reason for higher variances for $\hat\gamma$ from (5), can be made by using prior weights $(1 + \hat\rho_4)^{-1}$ when fitting the variance model using Gamma errors for χ_i^2, (McCullagh and Nelder, 1989, p 362). The effect of leverage in underestimating d_i and χ_i^2 at leverage points can be reduced by adjusting d_i and χ_i^2 for the leverage, Nelder (1992) The different estimators of α from (4) and (5) have been compared by Davidian and Carroll (1988) and Nelder and Lee (1992).

3   MADAM estimation

When all of the functions f_j and h_k in (1) are non-parametric then we have an Additive model for both the mean and the dispersion, and the MADAM is fully non-parametric. The non-parametric functions f_j and h_k can be estimated by maximising the Penalised extended Quasi-likelihood function L_{PQ}^+

(or minimising D_{PQ}^+) with penalties for lack of smoothness in the additive functions f_j and h_k in the mean and dispersion models respectively, where

$$
\begin{aligned}
D_{PQ}^+ &= -2\log L_{PQ}^+ \\
&= D_Q^+ + \left[\sum_{j=1}^p \lambda_{1j} \int_{-\infty}^{\infty} \left[f_j''(t) \right]^2 dt + \sum_{k=1}^q \lambda_{2k} \int_{-\infty}^{\infty} \left[h_k''(t) \right]^2 dt \right] .(6)
\end{aligned}
$$

Alternatively the non-parametric functions f_j and h_k can be estimated by pseudo-maximisation of the penalised Normal likelihood L_{PN} (i.e. by Penalised pseudo-likelihood estimation), where

$$
\begin{aligned}
D_{PN} &= -2\log L_{PN} \\
&= D_N + \left[\sum_{j=1}^p \lambda_{1j} \int_{-\infty}^{\infty} \left[f_j''(t) \right]^2 dt + \sum_{k=1}^q \lambda_{2k} \int_{-\infty}^{\infty} \left[h_k''(t) \right]^2 dt \right] (7)
\end{aligned}
$$

For the semi-parametric MADAM, the functions f_j and h_k which are linear are replaced by linear terms $\beta_j x_j$ and $\gamma_k z_k$ in (1) and the corresponding penalty terms are then omitted from (6) and (7).

The maximising functions f_j and h_k are cubic splines for both L_{PQ}^+ and L_{PN}. The integrated squared second derivative of an additive function f_j or h_k is used as the measure of its curvature (or lack of smoothness) to be penalised in (6) or (7). Higher values of a smoothing parameter λ_{1j} (or λ_{2k}) result in a smoother fitted function $\hat{f}_j$ (or $\hat{h}_k$), so that as λ_{1j} (or λ_{2k})$\to \infty$ the fitted values $\hat{f}_j$ (or $\hat{h}_k$) tend to a linear fit. The penalised likelihood approach has been extensively considered by Green and Silverman (1994) and Hastie and Tibshirani (1990) and provides a more formalised approach to statistical modelling than the alternative kernel smoothing approach. Both methods however suffer from the problem of bias in the estimating functions $\hat{f}_j$ (or $\hat{h}_k$) particularly in regions of the explanatory variables x_j (or z_k) where there is high curvature in the true functions f_j (or h_k), e.g. peaks and troughs.

The special Normal case of the above model, i.e. $V(\mu) = 1$, where both the mean and variance are non-parametric functions of the same single explanatory variable was considered by Silverman (1985) and Muller and Stadmuller (1987). Rigby and Stasinopoulos (1995) formalised this to an arbitrary number of regressor variables by modelling the mean μ_i and variance $\phi_i = \sigma_i^2$ of a Normal model using the semi-parametric Additive models given by equation (1). A method of robustly fitting the above Normal model is given by Rigby and Stasinopoulos (1994).

4   The fitting algorithm

The algorithm proposed here extends that of Rigby and Stasinopoulos (1995) to fit the general MADAM. The implementation of the algorithm requires a Generalised Additive Model (GAM) procedure which is a Local Scoring algorithm, combining a Generalised Linear Model (GLM) procedure and a Backfitting algorithm incorporating a Cubic spline smoother. For details of the GAM procedure see Hastie and Tibshirani (1990, ch 6.3). The algorithm is as follows:

- **INITIALISATION** : set prior weights $\mathbf{w}_1 = 1$, set $f_j(\mathbf{x}_j) = 0$, $h_k(\mathbf{z}_k) = 0$ for $j = 1, 2, \ldots, p$, and $k = 1, 2, \ldots, q$.

- **REPEAT**

    - STEP 1. Fit a GAM model with response variable $\mathbf{y}$ against the $\mathbf{x}$'s using the error distribution which has variance function $V(\boldsymbol{\mu})$, scale parameter 1, link function g_1, and prior weights $\mathbf{w}_1$, to give the current fitted $\hat{f}_j(\mathbf{x}_j)$'s and hence $\hat{\boldsymbol{\mu}}$.

    - STEP 2. Calculate the deviance increment d_i, for the extended Quasi-likelihood, or the Pearson residuals χ_i^2 for the pseudo likelihood, in order to form the 'response' variable for the next step.

    - STEP 3. Fit a GAM model with response variable d_i or χ_i^2 against the $\mathbf{z}$'s using a Gamma error, scale parameter 2, link function g_2, and prior weights $\mathbf{w}_2 = 1$ to give the current fitted $\hat{h}_k(\mathbf{z}_k)$'s and hence $\hat{\boldsymbol{\phi}}$.

    - STEP 4. Use the fitted values $\hat{\boldsymbol{\phi}}$ from step 3 to calculate the new prior weights $\mathbf{w}_1 = 1/\hat{\boldsymbol{\phi}}$ for fitting the mean model.

- **END**

- **UNTIL**: The penalised Quasi-deviance (6) or pseudo-deviance (7) converges.

It can be shown using arguments similar to the ones given by Rigby and Stasinopoulos (1995) that the above algorithm, (subsequently called the MADAM algorithm), maximises L_{PQ}^+ (or L_{PN}) over all functions f_j and h_k with continuous first and second derivatives and integrable second derivatives. The proof involves showing firstly that the penalty integrals in (6) or (7) can be replaced by quadratic forms, secondly that Fisher's scoring leads to a successive relaxation algorithm which alternates between fitting the mean and the dispersion models respectively, and finally that each step of the successive relaxation algorithm is a GAM fit. Note that in the special cases where $V(\boldsymbol{\mu}) = 1$ or $\boldsymbol{\mu}^3$ the above algorithm maximises a proper penalised Normal or Inverse Gaussian likelihood respectively.

The above algorithm has been implemented in GLIM4 using the macro MADAM (available on request). This macro calls the FGAMOD macro of Stasinopoulos and Francis (1993) to perform the GAM procedure (called the Local Scoring algorithm and described by Hastie and Tibshirani, 1990, Ch 6). This in turn calls a Cubic Spline routine (de Boor, 1978, Ch14) for smoothing each non-parametric function during the backfitting part of the GAM procedure. Our implementation uses modified backfitting (Hastie and Tibshirani, 1990, ch 5), where linear parametric terms in the mean and dispersion models are fitted using an extra GLM fit within the backfitting algorithm in the GAM procedures in steps 1 and 3 of the MADAM algorithm respectively..

The amount of smoothing used for each of the $\mathbf{x}$'s or $\mathbf{z}$'s is determined in advance by fixing the effective degrees of freedom to be used for the variable (Hastie and Tibshirani, 1990, Ch 6.8). Different degrees of freedom can easily be tried since the algorithm is relatively fast. In practice four degrees of freedom are often a good starting point for deciding whether the functional form of an f_j or h_k function is non-linear.

The error degrees of freedom df^{err} for any MADAM is approximated by

$$df^{err} = n - df_\mu - df_\phi$$

where

$$df_\mu = d_1 + \sum_{j=1}^{p} df_{1j}, \qquad df_\phi = d_2 + \sum_{k=1}^{q} df_{2k}$$

where d_1 and d_2 are the degrees of freedom used in the parametric part of the models for μ and ϕ respectively, and df_{1j} and df_{2k} are the effective additional degrees of freedom to be used in the smoothed functions f_j and h_k respectively. For a definition and explanation of the effective additional degrees of freedom see Hastie and Tibshirani (1990, ch 5.4).

Having fixed the (effective additional) degrees of freedom for all f_j and h_k functions, the resulting fitted values for μ_i and ϕ_i for $i = 1, 2, \ldots, n$ are substituted into D_Q^+ (or D_N) to give the fitted deviance denoted by $\hat{D}$.

Two nested parametric models H_0 and H_1, with fitted deviances $\hat{D}_0$ and $\hat{D}_1$, can be compared using the test statistic $\Lambda = \hat{D}_0 - \hat{D}_1$, which has an asymptotic Chi-square distribution with $df_0^{err} - df_1^{err}$ degrees of freedom under H_0 for Normal errors. More generally when comparing two fitted 'nested' MADAMs a Chi-square distribution for Λ is used as a rough guide to the selection of fitted model, in the same way that Hastie and Tibshirani (1990, ch 3.9) compare two 'nested' Generalised Additive Model fits. They have found empirical justification for the use of the Chi-square distribution with $df_0^{err} - df_1^{err}$ degrees of freedom for fitted GAM selection. The overall approaches used here for fitted MADAM selection are Forward selection and Akaike's information criterion (AIC), as outlined by Hastie and Tibshirani (1990, ch 9.4).

5 Examples

In this section two examples are used to demonstrate the flexibility of the MADAM. In both examples the MADAM is used to identify overdispersion and to determine whether overdispersion is due to the influence of explanatory variables. The first is a Poisson regression model, while the second is a Binomial logistic regression model.

5.1 The AIDS data

The following 45 quarter yearly frequency counts of reported AIDS cases in the UK from January 1983 to March 1994 were obtained from the Public Health Laboratory Service, Communicable Disease Surveillance Centre, London :

2, 6, 10, 8, 12, 9, 28, 28, 35, 32, 46, 47, 50, 61, 99, 95, 150, 143, 197, 159, 204, 168, 196, 194, 210, 180, 277, 181, 327, 276, 365, 300, 356, 304, 307, 386, 331, 358, 415, 374, 412, 358, 416, 414, 496.

The counts (y) are plotted against time (t) in quarter years in figure 1. The counts y were modelled using an overdispersion Poisson regression model in time t. Specifically y_i is modelled with mean μ_i and variance $\phi_i \mu_i$, for $i = 1, 2, \ldots, n$, where

$$\log \mu_i = f(t_i) + qrt, \qquad \log(\phi_i - 1) = h(t_i)$$

and where qrt is a quarterly seasonal effect. The shifted log link function $g(\phi) = log(\phi - 1)$ was chosen in preference to $g(\phi) = log(\phi)$ in order to ensure that $\phi > 1$, so that the model for the counts is overdispersed (and never underdispersed) Poisson. The derivative of the link function was adjusted to $1/(\phi - 1 + \delta)$ where $\delta = 0.00001$ to avoid overflow when $\hat{\phi}$ is very close to 1.

The model was fitted to the data for a range of smoothing parameters corresponding to a range of degrees of freedom df_μ and df_ϕ, for the mean and dispersion respectively, using the MADAM algorithm (implemented in the GLIM4 macro MADAM), and resulting fitted deviances are given in table 1. Note that the first column in table 1 with $df_\phi = 0$ corresponds to an exact Poisson model having $\phi = 1$ and therefore zero degrees of freedom for the dispersion. The second column with $df_\phi = 1$ corresponds to an overdispersion Poisson model with $\phi = constant$. Subsequent columns, correspond to models for $log(\phi - 1)$ which are non-parametric functions of time $h(t)$ using a total of df_ϕ degrees of freedom **including** the constant. Similarly the rows correspond to models for $\log \mu$ which include non-parametric functions of time $f(t)$ using a total of df_μ degrees of freedom **including** the constant.

Choice between fitted models, corresponding to adjacent entries in the table, used the 5% level of a Chi-squared distribution with $df_0^{err} - df_1^{err}$ degrees of freedom as the critical value for the statistic $\Lambda = \hat{D}_0 - \hat{D}_1$ for rejecting the simpler fitted model (as discussed in section 4). Using a forward selection approach (at a 5% significance level with $\chi^2_{1, 0.05} = 3.8$ and $\chi^2_{2, 0.05} = 6.0$), led to the choice of $df_\mu = 11$ with $df_\phi = 5$, i.e. MADAM (11,5). Using a Akaike criterion with penalty 3 for each additional degree of freedom used in either the mean or dispersion model led to the same choice.

Table 1: Analysis of deviance table for the AIDS data.

df_μ df_ϕ	0	1	3	5	7
9	383.0	370.6	362.2	354.2	350.3
11	372.9	365.0	353.2	346.2	341.1
13	365.8	360.7	348.3	341.8	334.9

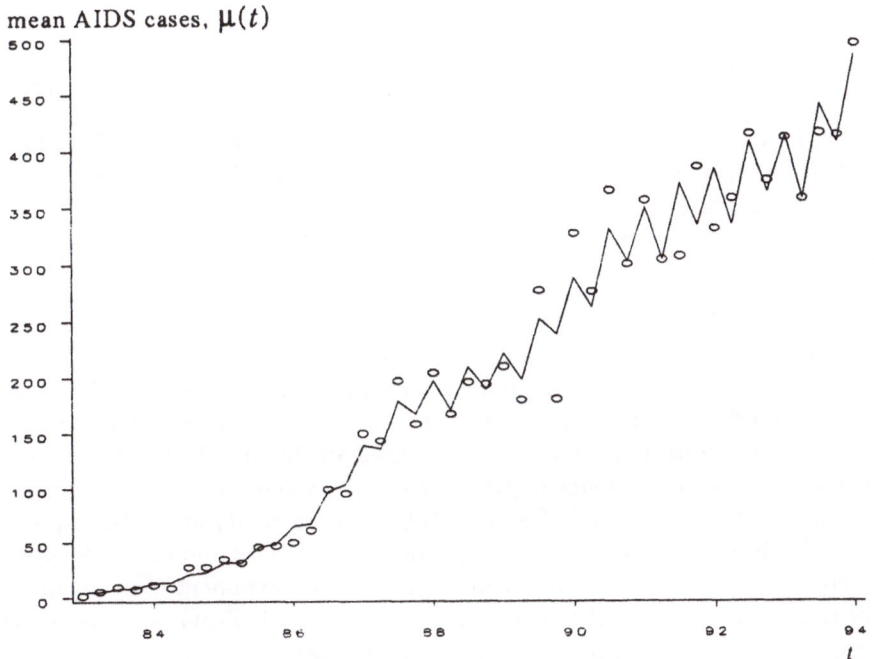

Figure 1: Quarter yearly AIDS frequency counts (x) with the fitted mean from model **MADAM** (11,5).

224

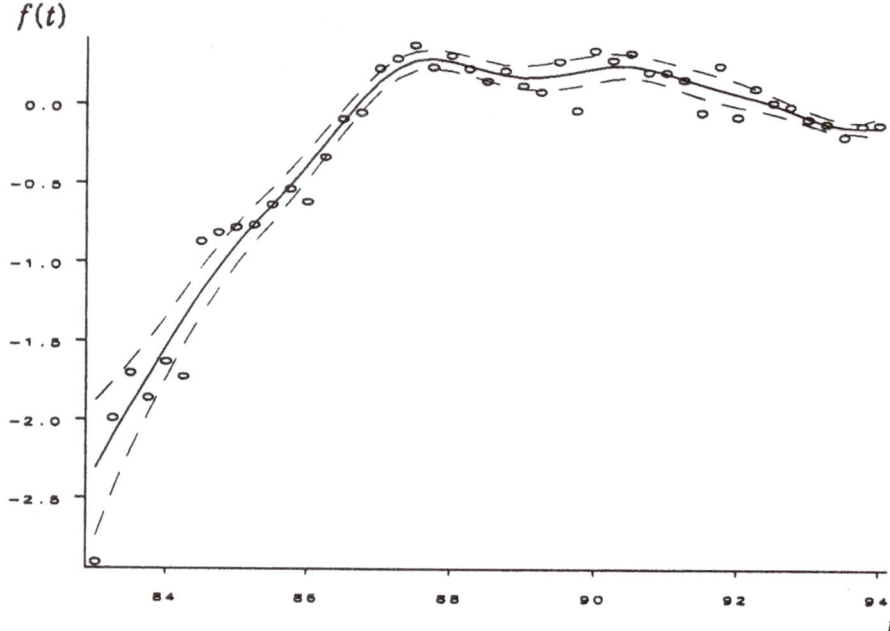

Figure 2: Partial working residuals (x) and fitted function $f(x)$ for model MADAM (11,5) (having removed the linear term in t) together with 95% confidence bands (- - -).

Figure 1 shows a plot of the AIDS counts, together with the fitted mean model (including the fitted quarterly seasonal effect) from model MADAM (11,5), against time. Figure 2 shows a plot of the fitted function $f(t)$ from the same model (having removed the seasonal effect and the linear term in time), together with approximate 95% confidence bands. This shows a clear breakpoint in the dependence of $f(t)$ on time at around July 1987, with linear segments for $f(t)$ before and after July 1987 and continuity at the breakpoint. Since a log link function was used for μ, the corresponding model for the mean or expected AIDS count, would then comprise two exponential segments in time (together with a multiplicative quarterly seasonal effect), a model which was considered by Stasinopoulos and Rigby (1992).

Figure 3 shows a plot of three fitted dispersion models $\phi(t)$ against t for fitted MADAM (df_μ, df_ϕ) with $df_\mu = 11$ and $df_\phi = 3, 4, 5$, together with the fitted deviance residuals for MADAM (11,5). The fitted dispersion models

dispersion models

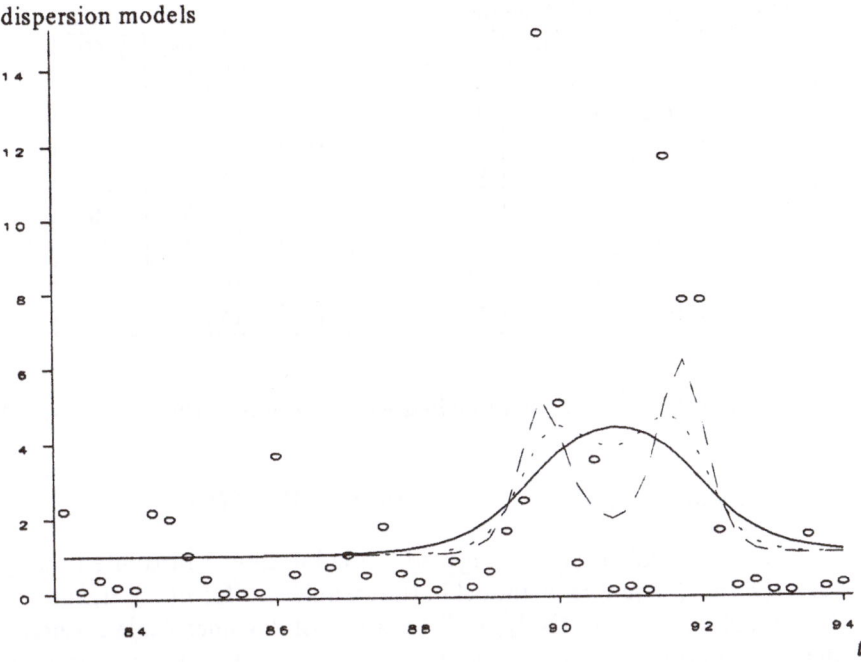

Figure 3: Deviance residuals (x) for MADAM (11,5) with the fitted dispersion models with $df_\phi = 3(—), 4(\cdots), 5(- - -)$.

highlight the increase in dispersion between 1989 and 1992.

5.2 Toxoplasmosis data

These data were used by Efron (1986) to demonstrate the use of his double exponential family. The response variable is the number of subjects who tested positive of the disease toxoplasmosis (y) in 34 cities of El Salvador and it was desired to assess the effect of rainfall on the proportion of subjects testing positive. The sample size (N_i) tested in each city varies from 1 to 82. The data are shown in figure 4 as sample proportions testing positive, where the area of the ellipse is proportional to the sample size tested.

Here y_i is modelled using an overdispersion Binomial regression model, with mean μ_i and variance $\phi_i \mu_i (N_i - \mu_i)/N_i$, where μ_i is modelled as a

Table 2: Analysis of deviance table for the toxoplasmosis data.

models	mean	df_μ	dispersion	df_ϕ	deviance	df^{err}
1	**R**	2	1	0	165.0	32
2	**R**< 3 >	4	1	0	153.5	30
3	**R**(3)	4	1	0	155.5	30
4	**R**	2	**I**	1	151.4	31
5	**R**< 3 >	4	**I**	1	145.6	29
6	**R**(3)	4	**I**	1	146.7	29
7	**R**< 3 >	4	**N**< 2 >	3	142.7	27
8	**R**(3)	4	**N**(2)	3	144.5	27

function of rainfall R_i and ϕ_i is modelled as a function of the sample size N_i respectively by

$$\text{logit}(\mu_i) = f(R_i), \qquad \log(\phi_i - 1) = h(t_i)$$

for $i = 1, 2, \ldots, n$. Efron (1986) suggested a cubic model in rainfall for the function $f(R)$ and a quadratic model in the sample size for $h(N)$, denoted by R< 3 > and N< 2 > respectively in the analysis of deviance table 2 where < d > indicates a polynomial model of order d. Here we also fit non-parametric functions $R(3)$ and $N(2)$, in rainfall R and sample size N respectively, where (d) indicates a non-parametric function with d degrees of freedom **additional** to the constant. Models 1, 2 and 3 are simple Binomial logistic models with dispersion parameter equal to one (**1**). Models 4, 5 and 6 are overdispersed Binomial models where the dispersion parameter is modelled by a constant (**I**), while in models 7 and 8 the log of the dispersion is modelled as a function of the sample size (**N**). Comparison of models 1, 2 and 3 with models 4, 5 and 6 indicates that the data are overdispersed Binomial since the (constant) overdispersion model produces a significant reduction in the deviance over the simple Binomial logistic model with dispersion parameter one.

Comparison of model 1 with either model 2 or 3 indicates the possible need for a cubic or non-parametric term in rainfall. However when the simplest (constant) overdispersion model (**I**) is fitted, the difference in deviance, comparing model 4 with either 5 or 6, is reduced to only 5.71 or 4.64 respectively indicating that the cubic term in rainfall is of borderline significance. Furthermore, comparing model 5 with 7 or model 6 with 8 indicates that modelling the dispersion parameter as a function of N is probably not supported by the data.

Figure 4 shows the observed and fitted proportion testing positive for models 6, 7 and 8. Figure 5 shows the fitted dispersion parameter for models 6, 7 and 8 with the deviance residuals from model 6. The figures and table 2 confirm, firstly, that there is no strong support for modelling the dispersion parameter as a function of the sample size N as Efron had suggested and, secondly, that even the 'cubic' like relationship between the mean and the

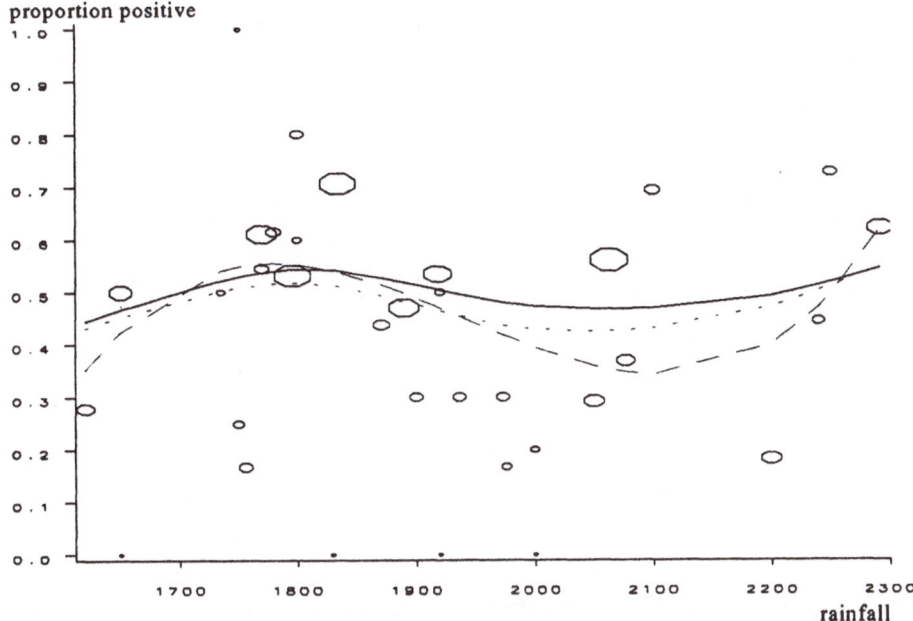

Figure 4: Fitted models for the proportion of subjects testing positive for the disease toxoplasmosis in El- Salvador : model 6 (—), model 7 (- - -), model 8 (···) and the observed sample proportions ◯ with area proportional to sample size, N.

rainfall **R** seems to be dubious.

6   Concluding remarks

The Mean and Dispersion Additive Model (MADAM) provides a very flexible model for the mean and the variance of a dependent variable. The variance is modelled as a product of the dispersion parameter ϕ and the variance function $V(\mu)$. The semi-parametric Additive models used for the mean and the dispersion parameter provides a powerful tool for checking whether the mean μ and the dispersion parameter ϕ are affected by explanatory variables. Careful specification of $V(\mu)$ is also important. If the heterogeneity has random rather than systematic components, a compound or random effects model is appropriate.

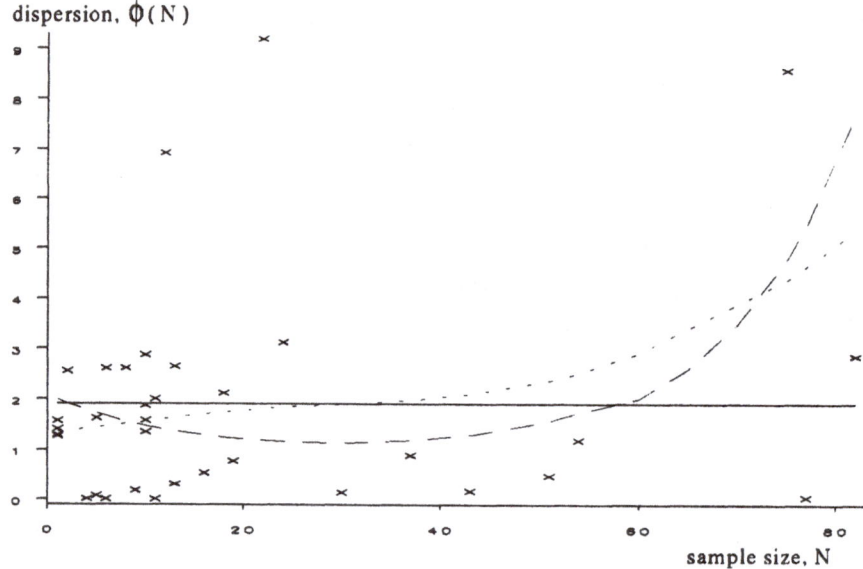

Figure 5: Fitted models for the dispersion for the toxoplasmosis data: model 6 (—), model 7 (- - -), model 8 (· · ·) with the observed deviance residuals (x) from model 6.

Acknowledgement

We would like to thank John Nelder for kindly providing us with his GLIM macros for fitting the parametric models, and to the referees for their helpful comments on the layout of the paper.

References

[1] Aitkin, M. (1987) Modelling variance heterogeneity in normal regression using GLIM. *Appl. Statist.*, **36**, 332-339.

[2] Carroll, R.J. and Ruppert, D. (1988) *Transformations and Weighting in Regression*. London: Chapman and Hall.

[3] Davidian, M. and Carroll, R.J. (1987) Variance function estimation. *J. Am. Statist. Ass.*, **82**, 1079-1091.

[4] Davidian, M. and Carroll, R.J. (1988) A note on extended quasi-likelihood. *J. R. Statist. Soc. B*, **50**, 74-82.

[5] de Boor, C. (1978) *A Practical Guide to Splines*. New York : Springer.

[6] Efron, B. (1986) Double Exponential Families and their use in Generalised Linear Regression. *J. Am. Statist. Ass.*, **81**, 709-721.

[7] Green P.J. and Silverman B.W. (1994) *Nonparametric Regression and Generalised Linear Models*. London: Chapman and Hall.

[8] Harvey, A.C. (1976) Estimating regression models with multiplicative heteroscedasticity. *Econometrica*, **41**, 461-465.

[9] Hastie, T.J. and Tibshirani, R.J. (1990) *Generalized Additive Models*. London: Chapman and Hall.

[10] McCullagh, P. and Nelder, J.A. (1989) *Generalised Linear Models* (2nd ed.), London, Chapman and Hall.

[11] Muller, H. G. and Stadtmuller, U. (1987) Estimation of heteroscedasticity in regression analysis. *Ann. Statist.*, **15**, 610-625.

[12] Nelder, J.A. (1992) Joint modelling of the mean and dispersion. In *Statistical Modelling*, 263-272 (P.G.M. Van der Heijden, W. Jansen, B. Francis and G.U.H. Seeber eds). North Holland: Amsterdam.

[13] Nelder, J.A. and Pregibon, D. (1987) An extended quasi- likelihood function. *Biometrika*, **74**, 211-232.

[14] Nelder, J.A. and Lee, Y. (1992) Likelihood, Quasi-likelihood and Pseudo likelihood: Some Comparisons. *J. R. Statist. Soc. B*, **54**, 273-284.

[15] Pregibon, D. (1984) Review of Generalised Linear Models. *Ann. Statist.*, **12**, 1589-96.

[16] Rigby, R.A. and Stasinopoulos, D.M. (1994) Robust fitting of an Additive model for variance heterogeneity. In *COMPSTAT, Proceedings in Computational Statistics*, eds R. Dutter and W. Grossmann, pp 261-268, Physica-Verlag.

[17] Rigby, R.A. and Stasinopoulos, D.M. (1995) A semi-parametric Additive model for variance heterogeneity. (to appear in *Statistics and Computing*).

[18] Silverman, B.W. (1985) Some aspects of the spline smoothing approach to non-parametric regression curve fitting (with discussion). *J. R. Statist. Soc. B*, **47**, 1-52.

[19] Stasinopoulos, D. M. and Francis, B. (1993) Generalised Additive Models in GLIM4. *GLIM Newsletter*, **22**, 30-36.

[20] Stasinopoulos, D. M. and Rigby, R. A. (1992) Detecting break points in generalised linear models. *Computational Statistics and Data Analysis*, **13**, 461-471.

[21] Verbyla, A.P. (1993) Modelling variance heterogeneity : residual maximum likelihood and diagnostics. *J. R. Statist. Soc. B*, **55**, 493-508.

Interaction in Nonlinear Principal Components Analysis

G. D. Costanzo[1] and J. L. A. van Rijckevorsel[2]

[1] on leave from Dept. of Statistics, Probability and Applied Statistics, University "La Sapienza", Rome, Italy. [2] Department of Statistics, TNO Prevention and Health, Leiden, The Netherlands

Summary

An alternating least squares algorithm based on maximal correlation between variables is proposed to introduce linear and nonlinear interaction terms in PCA. Such algorithm fits a model in which principal components are data driven dimensionality reduction functions. The detection of meaningful interaction is a way to specify such general unknown functions. As an example the highly nonlinear structure of a circle is recovered by the first component of a nonlinear interactive PCA of the circle coordinates.

Keywords: Interaction terms, B-splines, adaptive, PCA, nonlinear.

1 Introduction

Dimensionality reduction of a data set is an important problem in applied statistics. Principal Components Analysis (PCA) is probably the oldest and best known technique for this purpose. It was first introduced by Pearson (1901), and developed independently by Hotelling (1933). The central idea of PCA is to reduce the dimensionality of a data set which consists of a large number of interrelated variables while retaining as much as possible of the variation present in the data set. This is achieved by defining a new set of variables, the Principal Components (PCs), which are uncorrelated, and which are ordered so that the first components retain most of the variation present in all of the original variables. Let $\mathbf{H} = [\mathbf{h}_1, \mathbf{h}_2, \ldots, \mathbf{h}_m]^T$ be the data matrix and $\mathbf{R}$ the associated correlation matrix, the general k-th PC can be defined as

---

[1]The author was partially supported by a grant of the Italian Ministry of University, Scientific and Technological Research (M.U.R.S.T. 40%, n.940326)

$$\mathbf{x}_k = \alpha_k^T \mathbf{H} = \alpha_{k1}\mathbf{h}_1 + \alpha_{k2}\mathbf{h}_2 + \ldots + \alpha_{km}\mathbf{h}_m = \sum_{j=1}^{m} \alpha_{kj}\mathbf{h}_j.$$

While PCs are linear combinations of the original variables; the vector of m components α_k, defining the parameters of such linear combination, is found to be the (unit) k-th eigenvector corresponding to the k-th largest eigenvalue λ_k of $\mathbf{R}$.

The effectiveness of a PCA can be severely limited in the presence of nonlinear relationships among variables as is true for any of the classical techniques derived from normal distribution theory. Due to the use of the linear correlation coefficient as measure of internal dependency to find the smallest dimensional linear manifold giving a heuristic measure of the internal association exhibited by the original variables, PCA is not able to detect and represent nonlinear dependencies among variables. Various nonlinear gener-alizations of PCA have been proposed. For example Van Rijckevorsel and De Leeuw (1988) and Koyak (1987) work with linear manifolds of transformed variables while Hastie and Stuetzle (1989) and LeBlanc and Tibshirani (1994) work with nonlinear manifolds of the original variables. The aim of this arti-cle is to further generalize the PCA by the introduction of interaction terms, which are **linear and nonlinear functions of two (or more) variables, while considering linear manifolds of univariate and bivariate (or multivariate) transformed variables.**

To illustrate the introduction of interaction terms in PCA consider the following example: say we have three chemical substances $\mathbf{h}_i$, $\mathbf{h}_j$ and $\mathbf{h}_k$ with no scientifically relevant interaction. Suppose two of the substances chemically react into a new polluting hazardous substance correlated to the third substance: i.e. $(\mathbf{h}_j \times \mathbf{h}_k)$, the coordinatewise product of two of the variables, is correlated to $\mathbf{h}_i$. If the interaction term is perfectly correlated, the rank of $\mathbf{H}^* = [\mathbf{h}_i, \mathbf{h}_j, \mathbf{h}_k, (\mathbf{h}_j \times \mathbf{h}_k)]^T$ will reveal the existence of the hazardous pollutant because of being correlated with $\mathbf{h}_i$. Our problem in this situation is to identify the correct interaction term out of all possible interaction terms, including nonlinear interaction. In defining PCA in this way we are able to detect more complex interrelations among variables in both, linear and nonlinear PCA.

2 Background

The use of optimally nonlinear transformed variables in PCA has been ex-tensively treated in literature. The first proposal to use transformations on variables in PCA is due to Thurstone (1947); by observing that the met-rics of observed test scores are usually arbitrary, he proposed to "factor" them just by using rank-order considerations and in a such way that the "transformed" correlation matrix is best from some point of view of factor analysis. Guttman recognised that the linear correlation coefficient is not able to reveal nonlinear dependencies like the U-shaped perfect regression of one variable to another (Guttman, 1959, 266-267). When data do not come from a multinormal distribution and thus the linear correlation coefficient does not coincide with the correlation ratio, PCA is stochastically misleading. He suggested to use monotone transformations such that the regressions among the transformed variables were at least linear to obtain a well-defined regression system. Kruskal and Shepard (1974) introduced the explicit use of loss functions to fit what they called the "linear model" to a rectangular data matrix H. Subsequent contributions to nonlinear PCA – nonmetric PCA is a name used as well – are discussed by Gifi (1990) and Jackson (1991). De Leeuw and van Rijckevorsel (1980) introduced the PRINCALS algorithm for NLPCA based on the principle of Alternating Least Squares (ALS); they used the same loss function approach as Kruskal and Shepard. The same argument as Guttman against the linear correlation coefficient was used by Koyak (1987). Koyak's solution to make PCA stochastically significant by maximizing correlation is implicitly present in the homogeneity loss function used by van Rijckevorsel and De Leeuw (1988) for defining a nonlinear variety of PCA based on B-splines called SPLINALS. We shall use in the following the conceptual ideas and the technical methods of De Leeuw (1982) and of van Rijckevorsel et al. (1988). Gnanadesikan and Wilk (1969) attempted to generalize PCA to include both nonlinearly transformed variables and a "new" variable generated by considering (linear) functions of the original variables. Such a new variable is useful in revealing the mathematical law underlying the data, improving the "representativity" of NLPCA. A famous example is trying to recover the equation of a circle by the principal components of its coordinates. The case of the circle data will be extensively treated in section 5.2. Our approach to PCA could allow for the analysis of complicated multivariate functions like a volumetric variable (see Rao, 1958 for an example) or the circle equation without explicitly specifying such variables. So we could be able to introduce multivariate nonlinear relationships automatically.

3   Model

To relax the linearity assumptions underlying PCA and simultaneously introduce interaction terms, we assume the following general model

$$\mathbf{x}_k = F(\gamma_k; \mathbf{h}_1, ..., \mathbf{h}_m).$$

Through this model we interpret the principal components (PCs) as general "data dimensionality reduction functions". The PCs will be general unknown linear combinations in the functional form of the transformations of the variables and in the number of interaction terms and variables involved in these interaction. The dimensionality of the vector γ_k, which defines the parameters of these combinations depends on the specification of F, which is data driven with respect to the univariate transformations and multivariate transformations (interaction). The general specification for F will be

$$\mathbf{x}_k = \sum_{j=1}^{m} \gamma_{kj} f_j(\mathbf{h}_j) + \sum_{r,s \in J} \gamma_{k,rs} f_{rs}(\mathbf{h}_r, \mathbf{h}_s) + \sum_{l,p,q \in J'} \gamma_{k,lpq} f_{lpq}(\mathbf{h}_l, \mathbf{h}_p, \mathbf{h}_q) + \ldots$$

where J is a subset of all pairs of elements $1, \ldots, m$, J' is a subset of all triples of elements $1, \ldots, m$, etc. Such way, the nonlinearity of the analysis, which detects dependencies among variables not noticed in a linear PCA, is introduced by having univariate and bivariate (more in general multivariate) transformations, specified by the unknown function F. For example, for data coming from multinormal distribution optimal transformations should be linear (Koyak, 1987); if there exist no relevant interaction terms in the data matrix, the specification of F is the unit transformation as in linear PCA

$$\mathbf{x}_k = F(\gamma_k; \mathbf{h}_1, ..., \mathbf{h}_m) = \sum_{j=1}^{m} \gamma_{kj} \mathbf{h}_j, \qquad j = 1, \ldots, m.$$

In the case of the pollution example in section 1, F could be

$$\mathbf{x}_k = F(\gamma_k; \mathbf{h}_1, ..., \mathbf{h}_m) = \sum_{j=1}^{3} \gamma_{kj} f_j(\mathbf{h}_j) + \gamma_{k4}(\mathbf{h}_j \times \mathbf{h}_k).$$

The problem is to identify the correct specification of F.
Koyak (1987) shows that to measure dependence in a set of random variables, a multivariate analog of maximal correlation coefficient (i.e. Gebelein, 1941 and Renyi, 1959) consists of transforming each of the variables so that the partial sums of the k largest eigenvalues of the resulting correlation matrix is maximized. Such "maximized" measure of association obtained in this way permits to evaluate the strength of internal dependence exhibited by the random variables. It is also shown that optimizing transformations exist (they satisfy a geometrically interpretable fixed point property) and in particular if the variables are jointly Gaussian then the identity transformation is shown to be the optimal one. An algorithm to construct such optimal transformations for PCA can be found in Koyak (1985). In the same vein optimal transformations have been considered by Van Rijckevorsel and De Leeuw (1988, 1992) and van Rijckevorsel et al. (1993). The latter treat the PCA problem as a "simultaneous regression problem".

Let $\mathbf{f}_j = f_j(\mathbf{h}_j)$, the simultaneous regressions of transformed variables on fixed principal components $\mathbf{X}$, being $(n \times p)$ the dimension of $\mathbf{X}$ and a fixed $p \leq m$, simultaneously define the nonlinear PCA problem

$$\mathbf{X} = \mathbf{f}_j \mathbf{a}_j^T + \varepsilon, \qquad with \qquad \mathbf{a}_j^T = (\mathbf{f}_j^T \mathbf{f}_j)^{-1} \mathbf{f}_j^T \mathbf{X}, \qquad j = 1, \ldots, m$$

Each transformed variable's regression on principal components is optimal if the quadratic error term $(\mathbf{X} - \mathbf{f}_j \mathbf{a}_j^T)^T (\mathbf{X} - \mathbf{f}_j \mathbf{a}_j^T)$ is minimal for $\mathbf{f}_j$ and $\mathbf{a}_j$. A principal component is defined as the latent variable on which all regressions are simultaneously optimal and the first eigenvalue consists of the averaged goodness of fit $tr \sum_j \mathbf{a}_j \mathbf{a}_j^T, j = 1, \ldots, m$ of all these regressions on the first component. Variables are transformed by B-spline transformation functions while $\mathbf{G}_j, j = 1, \ldots, m$ is a $(n \times r_j)$ pseudoindicator matrix of B(asis)-splines with $(r_j \times 1)$-vector of coefficients $\mathbf{z}_j$. This determines the general PCA problem with error term

$$\sigma_1 = \sigma_1(\mathbf{X}, \mathbf{G}_1, \ldots, \mathbf{G}_m, \mathbf{Z}, A) = m^{-1} \sum_{j=1}^{m} SSQ(\mathbf{X} - \mathbf{G}_j \mathbf{z}_j \mathbf{a}_j^T).$$

The dimension r_j of the spline basis $\mathbf{G}_j$'s is determined by the fixed degree of the B-spline transformations and by the number of knots (for a complete review on B-spline functions see i.e. Schumaker, 1981 or De Boor, 1978). For a fixed dimension p these solutions are such that

$$m^{-1}(\sum_{j=1}^{m} \lambda_j^2) \leq m^{-1}(\sum_{j=1}^{m} \psi_j^2).$$

Here λ^2's are the eigenvalues of the original correlation matrix and ψ^2's are the eigenvalues of the univariate transformed correlation matrix. The kind of PCA obtained minimizing this loss is named nonmetric and adaptive PCA. Actually the use of this loss can be seen as a more general way to introduce various multivariate (nonlinear) analysis. Its minimization comprises Multiple Correspondence Analysis when $\mathbf{z}_j \mathbf{a}_j^T = \mathbf{y}_j$ is a quantification vector for categorical variables and comprises the ordinary PCA when $\mathbf{G}_j \mathbf{z}_j = \mathbf{h}_j$. In nonmetric PCA the vector of B-spline coefficients $\mathbf{z}_j$ has to be estimated. We can handle the problem of introducing interaction terms in PCA in this general framework. In our case we will refer to the following modified version of the previous loss

$$\begin{aligned} \sigma_2 &= \sigma_2(\mathbf{X}, \mathbf{G}_1, \ldots, \mathbf{G}_m, \mathbf{G}_1^*, \ldots, \mathbf{G}_{m^*}^*, \bar{\mathbf{Z}}, \bar{\mathbf{A}}) \\ &= m^{-1} \sum_{j=1}^{m} SSQ(\mathbf{X} - \mathbf{G}_j \mathbf{z}_j \mathbf{a}_j^T) + m^{*-1} \sum_{k=1}^{m^*} SSQ(\mathbf{X} - \mathbf{G}_k^* \mathbf{z}_k^* \mathbf{a}_k^{*T}) \\ &= \sigma_1 + \sigma_1^*. \end{aligned}$$

Here, $\mathbf{G}_k^* = \mathbf{G}_i \otimes \mathbf{G}_j; i = 1, \ldots, m; j = i+1, \ldots, m; k = 1, \ldots, m_*, \bar{\mathbf{Z}} = [\mathbf{Z}|\mathbf{Z}^*]$ and $\bar{\mathbf{A}} = [\mathbf{A}|\mathbf{A}^*]$. The number of interaction terms m^* has to be determined for every specific data context. Interaction (namely bivariate interaction) is represented by tensorial products of the univariate B-basis functions (Schumaker, 1981). In fact the introduction of interaction terms in PCA implies looking for functions of two (or more) variables $f(\mathbf{h}_i, \mathbf{h}_j)$ showing substantial correlations with the original variables in the data matrix. The problem we face is to find a way to specify these unknown functions (if they exist) with respect to the functional form. We have decided to adopt a two-dimensional spline representation, such that the functions $f(.)$ will be formalized as linear combinations of basis functions of tensorial products. By estimating the parameters $\{z^*\}$ of such linear combination inside the global minimization of the loss σ_2, we will obtain the kind of functions we are interested in, namely functions maximally correlated with the original variables in the data matrix. The optimal bivariate transformations that minimize the previous loss will be such that

$$
m^{-1}(\sum_{j=1}^{m} \lambda_j^2) \leq m^{-1}(\sum_{j=1}^{m} \psi_j^2) < (m + m^*)^{-1}(\sum_{j=1}^{m+m^*} \phi_j^2).
$$

Here ϕ^2's are the eigenvalues of matrix of the optimal univariate transformations and the interaction terms. Since the loss σ_2 can be written as the sum of the two losses σ_1 and σ_1^* we can use the "alternating least squares" (ALS) approach and actually split up the problem of finding optimal univariate transformations and optimal bivariate transformations (interaction) in two following steps. We will have also an evaluation criterion in this way. In fact we will test the interaction term most correlated with the manifold of the first p principal components each time. The last will be obtained solving the loss σ_1 without any interaction term. The selection of the most correlated interaction will be based on the loss σ_1^* instead. The introduction of an interaction term we evaluate by comparing the value of σ_2 against the value of σ_1, according to the criterion $\| \sigma_2 - \sigma_1 \| > \epsilon$, for a fixed value of ϵ.

4 Algorithm

In the loss σ_2 (and in the loss σ_1, as well) corresponding to the choice of univariate and bivariate B-spline transformations the knot location is a free parameter that has to be adaptively optimized. van Rijckevorsel et al. (1993) optimize knot location by an adaption optimizing algorithm, namely Friedman's algorithm for recursive partitioning. Actually we could use in the bivariate B-spline case the optimal knots found in the univariate case, minimizing the loss σ_1. However such knot location can lead to the presence of empty areas in the bivariate scatter of the interaction variable; that is adaptive bivariate B-splines are troubled by the curse of dimensionality. We have decided to start with a fixed knot location for both univariate and bivariate B-spline transformations instead. To introduce interaction terms in PCA, we use two algorithms. One of them is to generate and select the interaction terms (Forward or Generating Algorithm), the other one is to test their usefulness (Backward Algorithm). A sketch of the Forward Algorithm in the simpler case of fixed knots is

```
do t = 1 to T;
    do j = 1 to m;
        do k = j + 1 to m;
            run LOF(X₀, A₀, Gⱼₖ, lofⱼₖ);
            if lof*ⱼₖ = min{lofⱼₖ; j = 1, ..., m; k = j + 1, ..., m; } then
                G_{m+1} = G*ⱼₖ;
        end;
    end;
    run PRINCALS(G₁, G₂, ..., Gₘ, G_{m+1}; X, A, Z);
    X⁺ = X;
    A⁺ = A;
end.
```

Here $\mathbf{G}_{jk} = \mathbf{G}_j \otimes \mathbf{G}_k$, $j = 1, \ldots, m$ and $k = j + 1, \ldots, m$.

The algorithm combines the "alternating least squares" (ALS) approach of the PRINCALS algorithm with a selection procedure defined in a LOF subroutine. Given a knot location – either fixed or adaptively found – let $\mathbf{D}_j = \mathbf{G}_j^T \mathbf{G}_j$, the PRINCALS algorithm minimizes the loss σ_1 over $\mathbf{X}$, $\mathbf{z}_j$ and $\mathbf{a}_j$ simultaneously with normalizing conditions $\mathbf{X}^T \mathbf{X} = I$, $\mathbf{z}^T \mathbf{D} \mathbf{z} = 1$ and $\mathbf{e}^T \mathbf{X} = 0$, and a general objective function equal to $tr \sum_j \mathbf{a}_j \mathbf{a}_j^T$ $j = 1, \ldots, m$, $\mathbf{e}$ being a n-vector with unit elements. If we pose $\mathbf{Y}_j = \mathbf{z}_j \mathbf{a}_j^T$, given an estimate $\hat{\mathbf{Y}}_j$ for fixed $\mathbf{X}$, $\mathbf{z}_j$ and $\mathbf{a}_j$ can be iteratively estimated from $\hat{\mathbf{Y}}_j$ (Bekker and De Leeuw, 1988). The estimates are respectively: $\hat{\mathbf{a}}_j = \hat{\mathbf{Y}}_j \mathbf{D}_j \mathbf{z}_j$ and $\hat{\mathbf{z}}_j = (\mathbf{a}_j^T \mathbf{a}_j)^{-1} \hat{\mathbf{Y}}_j \mathbf{a}_j$ and the estimate of $\mathbf{X}$ is equal to $\mathbf{U}(\mathbf{U}^T \mathbf{U})^{-1/2}$ while $\mathbf{U} = m^{-1} \sum_j \mathbf{G}_j \mathbf{z}_j \mathbf{a}_j^T$. The LOF routine previously outlined tries to find every time the best interactive term by regressing each one of the candidate terms on the fixed first p principal components space (represented at the starting step by $\mathbf{X}^0$ the coordinate orthogonal system obtained by the ordinary PCA). Let $\mathbf{X} = \mathbf{f}_{jk} \mathbf{a}_{jk}^{*T} + \varepsilon$ be the equation defining the regression problem corresponding to the generic interaction term, where

$$\mathbf{f}_{jk} = f_{jk}(\mathbf{h}_j, \mathbf{h}_k) = (\mathbf{G}_j \otimes \mathbf{G}_k) \mathbf{z}_{jk} = \mathbf{G}_j^* \mathbf{z}_j^* \qquad j = 1, \ldots, m^*.$$

The vector $\mathbf{a}_{jk}^{*T} = (\mathbf{f}_{jk}^T \mathbf{f}_{jk})^{-1} \mathbf{f}_{jk}^T \mathbf{X}$ will be the projection coordinates p-vector of the generic interaction term on $\mathbf{X}$. The LOF routine, for each one of the candidate interaction and for fixed $\mathbf{G}_{jk}$ and $\mathbf{X}$, is used to produce estimates of $\mathbf{z}_{jk}^*$ - the coefficient vectors of bivariate basis - by which estimates of $\mathbf{a}_{jk}^*$ are obtained. The latter are then used to compute the lack of fit (lof) "measures" $lof_{jk} = [1 - tr(\mathbf{a}_{jk}^* \mathbf{a}_{jk}^{*T})]$, $j = 1, \ldots, m$ and $k = j + 1, \ldots, m$. The best interaction is every time the one with the minimum lof. Next a nonlinear PCA on the existing data with the selected interaction term is executed. A new coordinate system $\mathbf{X}$, is computed and used to update the LOF routine: the selection is now repeated using this new coordinate system, $\mathbf{X}^+$. The iterative procedure has to be continued till the selection is stable from a previous step to the next. Then the projection (or regression) coordinates vector ($\mathbf{a}^*$) on which the choice of the particular term is based, has to be either the same or in certain proportional relation with the component loadings vector ($\mathbf{a}$) of the analysis performed with this "new variable". The decision whether to include a certain interaction, is then based on the percentage of increment in the explained variability; the last is evaluated by comparing the first time (that is for the first interaction) the value of σ_2 against the value of σ_1, accordingly to the criterion $\|\ \sigma_2 - \sigma_1\ \| > \epsilon$ for a value of say $\epsilon = 5\%$. If the decision is to include a certain interaction, the last term will become a (new) variable in the data matrix and in the successive interaction the "inclusion" criterion $\|\ \sigma_2 - \sigma_1\ \| > \epsilon$ will be based on an updated version of $\sigma_1 : \sigma_1^+$.

Further, with respect to the successive interaction, we leave out the variables involved in the interaction previously chosen and allow only for selection of second order interaction (involving just two variables). We could decide to leave in such variables as well, and in such way allow for selection of "higher order" interaction (involving more than two variables). The problem of empty areas previously mentioned is linked to the choice of the B-spline basis functions. This could suggest for the use of the truncated power basis instead. In fact, the use of these basis functions combined with the optimal knot location obtained solving the loss, has worked. However the resulting spline functions do not give a good approximation of the true bivariate functions. A correct approach could be to use the knot location obtained for the truncated power basis, to construct the first degree B-spline basis for example.

5 Examples

5.1 Multinormal data

Three samples of three variables each have been drawn from a standardized multinormal distribution with 25, 100, 1000 observations respectively. The variables $\mathbf{h}_1$, $\mathbf{h}_2$ and $\mathbf{h}_3$ are drawn in such a way that they are not correlated but that the product ($\mathbf{h}_2 \times \mathbf{h}_3$) is substantially correlated with $\mathbf{h}_1$: firstly we draw from a multinormal distribution the three variables $\mathbf{h}_1$, $\mathbf{h}_2$ and $\mathbf{h}^*$ in such a way that just $\mathbf{h}_1$ and $\mathbf{h}^*$ are (substantially) correlated; then we leave out $\mathbf{h}^*$ and consider the variables $\mathbf{h}_1$, $\mathbf{h}_2$ and $\mathbf{h}_3 = \mathbf{h}^*/\mathbf{h}_2$. The aim is to use these data to show if and to which extent the selection procedure previously outlined is able to detect the proper interaction. As a first step a linear PCA with interaction (linear) terms along the previously outlined strategy correctly identifies the proper term with a quite uniform increment of approximately 5% independent of sample size (see Table 1).

In all the three samples, the algorithm selects the proper product ($\mathbf{h}_2 \times \mathbf{h}_3$) after the first iteration cycle: after including the interaction term the coordinate system $\mathbf{X}$ is updated and all possible candidates are reconsidered. The

	no interaction term	with interaction term
n = 25	70.57%	75.43%
n = 100	72.63%	77.65%
n = 1000	71.70%	78.28%

Table 1: Explained variability by the first two PCs, with and without the inclusion of the proper interaction term for three sample sizes.

iterative process stops because with this new update $\mathbf{X}$, the same interaction term is selected. Or in other words: the component loadings weighed for the eigenvalues stabilize after one cycle. In the linear case the ALS approach combined with our selection procedure works quite well, while at the same time it has to be noticed that its ability to discriminate among the different interaction terms and then the probability that the right term is chosen, generally increases with the number of observations; this can be easily verified in the next table (see Table 2) comparing the different values of the lof in the three different sample sizes.

	lof_{12}	lof_{13}	lof_{23}
n = 25	.80	.56	.04
n = 100	.99	.60	.01
n = 1000	.99	.99	.007

Table 2: Lack of fit (lof) of three interaction terms in linear PCA for three sample sizes.

On the same set of data the nonlinear PCA with interaction – which includes linear PCA with linear interaction as particular case – has been performed. For $n = 1000$ the nonlinear PCA with nonlinear interaction terms using first degree B-splines and fixed knots, selects the proper interaction term. It however fails to do so for smaller samples (see Table 3).

	lof_{12}	lof_{13}	lof_{23}
n = 25	.0	.3	.29
n = 100	.0	.35	.05
n = 1000	.4	.31	.0

Table 3: Lack of fit (lof) of three interaction terms in nonlinear PCA with first degree B-splines for three sample sizes.

The sensibility to small samples is remarkably larger in the nonlinear case. The tendency to overfit noise, if there are not enough data and/or sufficiently structured data, to which nonlinear data driven transformation functions are prone, is observed.

This result tells us that in large samples ($n \geq 1000$) from multinormal distributions the knot location is not extremely influent. This relates to similar results obtained in the univariate case by van Rijckevorsel and De Leeuw (1992). At the same time we can say that the mixing of approaches used in constructing the algorithm- that is the maximization of the partial sum of eigenvalues as maximized measure of association, the ALS approach with partial regressions, the selection criterion to discover significant interaction and the use of the tensorial product of univariate splines as mathematical tool to represent them - is able to discover the correct interaction and consequently the right functional form (linear in such case) of interaction. However with real data the knot location is an important problem because the ability of the algorithm to choose a certain interaction rather than another one also depends on it.

5.2 Circle data

The circle data have been often used in literature to illustrate the limitations of linear PCA. The principal problem of these data is that in a traditional PCA, the analysis would conclude that the two variables defining the circle are uncorrelated (see Jackson, 1991, for a complete analysis). To overcome this problem Gnanadesikan and Wilk (1969) proposed the following generalisation: let the two variables $\mathbf{h}_1$ and $\mathbf{h}_2$ defining the circle and add three more variables: $\mathbf{h}_1^2, \mathbf{h}_2^2$ and $\mathbf{h}_1 \times \mathbf{h}_2$. The number of eigenvalues of the expanded covariance matrix equals four and this means that the covariance matrix is only of rank four. Nevertheless they showed that the eigenvector corresponding to the zero eigenvalue is important because using this vector the fifth principal component would be the equation of the unit circle: $\mathbf{x}_5 = \mathbf{h}_{12} + \mathbf{h}_{22} = 1$. In fact zero roots represent singularities in the data and the vectors associated with them may be used to characterize what these singularities may represent. In this case the vector defines the nature of the nonlinearity, namely "the circle structure". Hastie and Stuetzle (1989) and LeBlanc and Tibshirani (1994) use the circle data to show that their "principal curves" are able to recover the circle structure in a satisfactory way. These authors use the non linear manifold of a principal curve to recover the mathematical law underlying the data and the generalized analysis proposed by Gnanadesikan uses the fifth principal component. LeBlanc and Tibshirani (1994, p. 57) use the circle data in conjunction with a noise term. We use the same kind of data as the one by LeBlanc et al. and have generated respectively 100 and 1000 observations from the circle model

$$\mathbf{h}_1 = 5sin(\gamma) + \epsilon_1$$
$$\mathbf{h}_2 = 5sin(\gamma) + \epsilon_2$$
$$\mathbf{h}_3 = \epsilon_3,$$

where $\gamma \sim U[0, 2\pi]$, $\epsilon \sim N[0, .25]$ and h_3 being the low variance "noise variable" added to the data. The same sensitivity to small sample size as in the previous example is observed by inspecting Table 4.

	linear PCA	nonlin. PCA	nonlin. PCA with interaction
n = 100	71%	97%	97.5%
n = 1000	68%	68%	97.5%

Table 4: Explained variability by the first two PCs, with and without the inclusion of the proper interaction term for two sample sizes of the circle data.

The nonlinear PCA without interaction term for n = 100 clearly overfits. The explained variability should approximately be the same as in the linear case and should in any case never approximate 100 % that closely. For n = 1000 nonlinear PCA with interaction term selects the proper interaction term between the circle-coordinates h_1 and h_2; the very term that Gnanadesikan and Wilk had to add in order to reproduce the original data adequately. The eigenvalues corresponding to the first two dimensions are about 1 and 1, showing that the two variables h_1 and h_2 are almost perfectly correlated. In adding the new term the first two eigenvalues are equal to 3 and 1 respectively. The automatic inclusion of interaction leads to an increment of 29.5 % in the explained variability (from 68 % to 97.5 %). The component loadings corresponding to the two circle variables on the first dimension are in all cases approximately equal to 1 and -1, respectively, and the component loading of the interaction term is equal to 1 as well.

Gnanadesikan and Wilks were able, after some simple calculations to recover the circle equation by using the fifth principal component. In a nonlinear PCA with automatic interaction detection the first PC cannot reveal such an equation in an exact way, due to the data driven character of the transformation functions. The eigenvectors are equal 0.57 for the two circle coordinates and the interaction term and approximately zero for the noise variable. This means that we still can write the first PC as

$$x_1 = 0.57 f_1(h_1) + 0.57 f_2(h_2) + 0.57 f_{12}(h_1, h_2)$$

but we do not know the mathematical specification of f. In Figure 1 we can inspect the shape of f_1, f_2 and f_{12}. All scatterplots with the third variable are left out because they form random scatters. Transformations of the circle coordinates are modulus functions. The transformation of the interaction term, when plotted versus the original product of the coordinates is a circle.

Note in Figure 2 that the scatters between transformed circle coordinates and the transformed interaction are monotone. Regressions are linearized to a considerable extent by transformations and the scatter looks much more linear than the highly nonlinear circle.

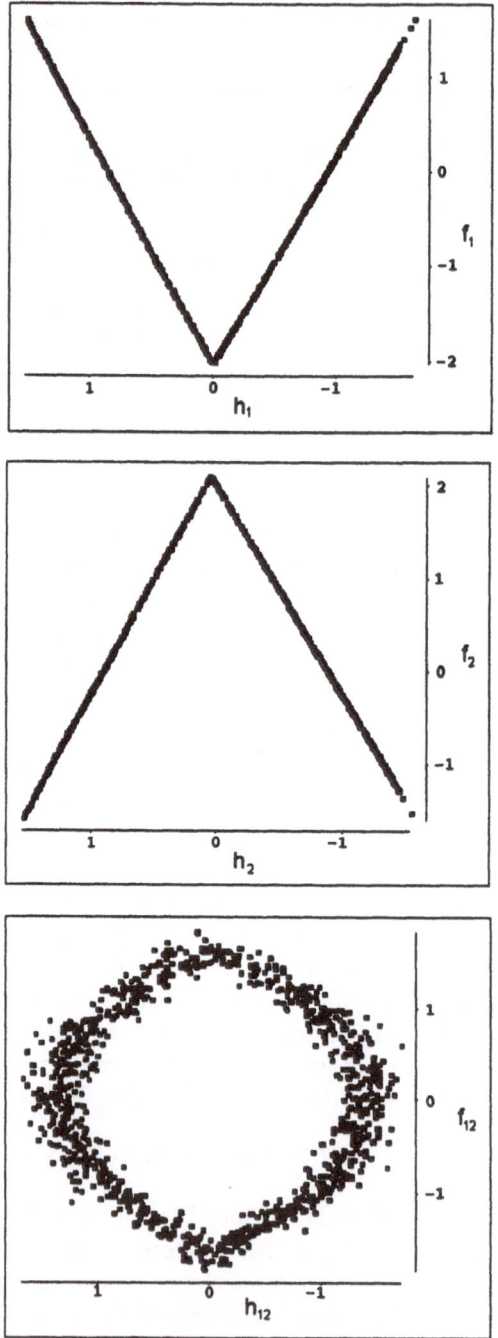

Figure 1: Circle data: the original coordinates and their product on the Y-axis plotted against their transformations on the X-axis.

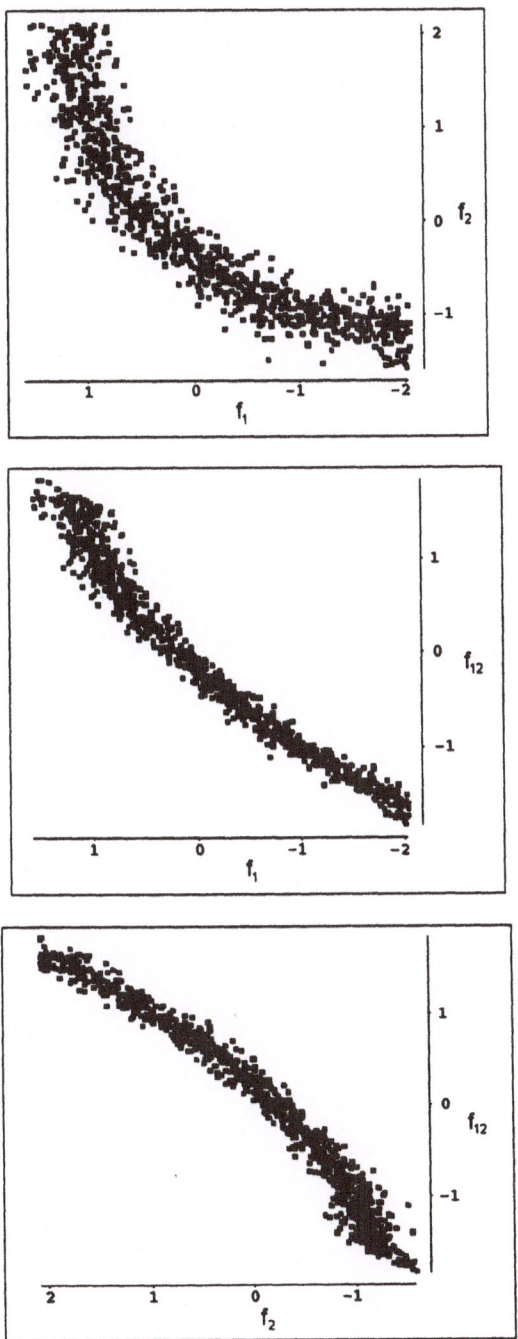

Figure 2: Circle data: Scatter diagrams of the transformed variables.

In our case allowing for a PCA with nonlinear transformations and (automatic) selection of interaction we are able to visualize the mathematical relation which the data represent by means of just the first principal component instead of five like Gnanadesikan and Wilk and staying within the linear manifold, which is the hallmark of Principal Component Analysis.

References

[1] Bekker, P. and De Leeuw, J. (1988). Relations between variants of non-linear Principal Component Analysis. In: J. van Rijckevorsel and J. De Leeuw (Eds), Component and Correspondence Analysis, 1-31. New York: Wiley

[2] De Boor , C. (1978). A practical Guide to Splines. New York: Springer Verlag

[3] De Leeuw, J. (1982). Nonlinear principal component analysis. In H. Caussinus et al.(eds), COMPSTAT 82, 77-86. Proceedings in Computational Statistics, part I. Wien: Physika-Verlag.

[4] De Leeuw, J. and van Rijckevorsel, J.L.A. (1980). HOMALS and PRINCALS. In Diday E. et al. (eds) Data Analysis and Informatics, 231-241. Amsterdam: North Holland.

[5] Gebelein,H. (1941). Das statistische Problem der Korrelation als Varations-und Eigenwertproblem und sein Zusammenhang mit der Ausgleichungsrechnung. Z. Angew. Math. Mech. 21, 364-379.

[6] Gifi, A. (1990). Nonlinear Multivariate Analysis. New York: Wiley

[7] Gnanadesikan, R. and Wilk, M.B. (1969). Data analytic methods in multivariate statistical analysis. Multivariate Analysis, II (P.R. Krishnaiah, editor), Academic Press, New York.

[8] Guttman, L. (1959). Metricizing rank-ordered or unordered data for a linear factor analysis. Sankhya, A, 21, 257-68.

[9] Hastie, T. and Stuetzle, W. (1989). Principal Curves. J. of the American Statistical Association, 84, 502-516.

[10] Hotelling, H. (1933). Analysis of a complex of statistical variables into principal components. J. Educ. Psychol., 24, 417-441, 498-520.

[11] Jackson, J.E. (1991). A User's Guide to Principal Components. New York: Wiley.

[12] Koyak, R. (1985). Optimal Transformations for Multivariate Linear Reduction Analysis. Unpublished PhD thesis. University of California, Berkeley, California: Dept. of Statistics.

[13] Koyak, R. (1987). On measuring internal dependence in a set of random variables. The Annals of Statistics, 15, 1215-1228.

[14] Kruskal, J.B. and Shepard, R.N. (1974). A nonmetric variety of linear factor analysis. Psychometrika, 39, 123-57.

[15] LeBlanc, M. and Tibshirani, R. (1994). Adaptive Principal Surfaces. J. of the American Statistical Association, 89, 53-64.

[16] Pearson, K. (1901). On lines and planes of closest fit to systems of points in space. Phil. Mag. (6), 2, 559-572.

[17] Rao, C.R. (1964). The use and interpretation of principal component analysis in applied research. Sankhya, A, 26, 329-358.

[18] Renyi, A. (1959). On measures of dependence. Acta. Math. Acad. Sci. Hungar., 10, 441-451.

[19] van Rijckevorsel, J.L.A. and De Leeuw, J. (eds)(1988). Component and Correspondence Analysis. New York: Wiley.

[20] van Rijckevorsel, J.L.A. and De Leeuw, J. (1992). Some results about the importance of knot selection in non linear multivariate analysis. The Italian Journal of Applied Statistics, vol. 4 (4), 429-451.

[21] van Rijckevorsel, J.L.A. and Tessitore, G. (1993). An algorithm for multivariate adaptive component and correspondence analysis. Bulletin of the International Statistical Institute, 49th session. Contributed papers, 2, 513-14, Firenze, 1993.

[22] Schumaker, L. (1981). Spline Functions: Basic Theory. New York: Wiley.

[23] Thurstone, L.L. (1947). Multiple factor analysis. Chicago:University of Chicago press.

Nonparametric Estimation of Additive Separable Regression Models

R. Chen[1], W. Härdle[2], O.B. Linton[3] and E. Severance-Lossin[4]

[1]Department of Statistics, Texas A& M University, College Station, TX 77843, U.S.A.
[2]Humboldt-Universität zu Berlin, Wirtschaftswissenschaftliche Fakultät, Institut für Statistik und Ökonometrie, Spandauer Str. 1, D-10178 Berlin, Germany
[3]Cowles Foundation for Research in Economics, Yale University, New Haven, CT 06520, U.S.A.
[4]Humboldt-Universität zu Berlin, Wirtschaftswissenschaftliche Fakultät, Institut für Statistik und Ökonometrie, Spandauer Str. 1, D-10178 Berlin, Germany

Summary

Additive regression models have been shown to be useful in many situations. Numerical estimation of these models is usually done using the iterative back-fitting technique. This paper proposes an estimator for additive models with an explicit 'hat matrix' which does not use iteration. The asymptotic normality of the estimator is proved. We also investigate a variable selection procedure using the proposed estimator and prove that asymptotically the procedure finds the correct variable set with probability 1. A simulation study is presented investigating the practical performance of the procedure.

1 Introduction

An additive nonparametric regression model has the form

$$m(x) = E(Y \mid X = x) = c + \sum_{\alpha=1}^{d} f_\alpha(x_\alpha), \tag{1}$$

where Y is a scalar dependent variable, $X = (X_1, \cdots, X_d)$ is a vector of explanatory variables, c is a constant and $\{f_\alpha(\cdot)\}_{\alpha=1}^{d}$ is a set of unknown functions satisfying $E[f(X_\alpha)] = 0$, and $x = (x_1, \cdots, x_d)$. Additive models of this form have been shown to be useful in practice: they naturally generalize the linear regression models and allow interpretation of marginal changes i.e.

the effect of one variable on the mean function m holding all else constant. They are also interesting from a theoretical point of view since they combine flexible nonparametric modeling of many variables with statistical precision that is typical for just one explanatory variable. This paper is concerned with variable selection and direct estimation of the functions $f_\alpha(\cdot)$ and $m(\cdot)$ in an additive regression model (1).

To our knowledge model (1) has been first considered in the context of input-output analysis by Leontief (1947) who called it *additive separable*. In the statistical literature the additive regression model has been introduced in the early eighties, and promoted largely by the work of Buja, Hastie and Tibshirani (1989) and Hastie and Tibshirani (1990). It has lead to the development of a variety of theoretical results and to many applications implemented using modern software. Stone (1985, 1986) proved that model (1) can be estimated with a one-dimensional rate of convergence typical for estimating a single function f_α of one regressor only.

Buja, Hastie and Tibshirani (1989, eq (18)) consider the problem of finding the projection of m onto the space of additive functions representing the right hand side of (1). Replacing population by sample, this leads to a system of normal equations with $nd \times nd$ dimensions. To solve this in practice, the backfitting or Gauss-Seidel algorithm, is usually used, see Venables and Ripley (1994). This technique is iterative and depends on the starting values and convergence criterion. It converges very fast but has, in comparison with the direct solution of the large linear system, the slight disadvantage of a more complicated 'hat matrix', see Härdle and Hall (1993). Unfortunately, not many statistical measures of this procedure like bias and variance have been fully derived in closed form.

We assume that model (1) holds exactly, i.e. the regression function is additive. For this case, Linton and Nielsen (1995) proposed a method of estimating the additive components f_α. Their method is to estimate a functional of m by marginal integration; under the additive structure this functional is f_α up to a constant. Their analysis is restricted to the case of dimension $d = 2$. Tjøstheim and Auestad (1994) proposed a similar estimator, mistakenly called 'projector', for time series but did not fully derive its asymptotic properties, specifically its bias.

The same model has been examined by Härdle and Tsybakov (1995) for general d under the assumption that the covariates are mutually independent. They introduced a principal component-like procedure for selecting important variables based on the variance of the estimated components, see also Maljutov and Wynn (1994).

The present paper improves upon these earlier results in various ways. First, a direct estimator based on marginal integration is proposed thereby avoiding iteration. Second, the explanatory variables are allowed to be correlated with a joint density p that does not factorize. This improves upon the paper by Härdle and Tsybakov (1995). Third, the dimension of X is not restricted to dimension $d = 2$ as in Linton and Nielsen (1995). Fourth, the 'hat matrix' of the proposed estimator is of less complicated form than in backfitting. Fifth, we give the exact asymptotic bias of our estimator thereby improving Tjøstheim and Auestad (1994). In addition to extending results on the estimator a procedure is given for selecting significant regressors.

The 'integration idea' is based on the following observation. If $m(x) = E(Y \mid X = x)$ is of the additive form (1), and the joint density of $X_{i\underline{\alpha}} = X_{i1}, \cdots, X_{i(\alpha-1)}, X_{i(\alpha+1)}, \cdots, X_{id}$ is denoted as $p_{\underline{\alpha}}$, then for a fixed $x_\alpha \in \mathbf{R}$,

$$f_\alpha(x_\alpha) + c = \int m(x_1, \cdots, x_\alpha, \cdots, x_d) p_{\underline{\alpha}}(x_{\underline{\alpha}}) \prod_{\beta \neq \alpha} dx_\beta, \qquad (2)$$

where $x_{\underline{\alpha}} = (x_1, \cdots, x_{\alpha-1}, x_{\alpha+1} \cdots, x_d)$, provided $E[f_\beta(X_\beta)] = 0$, $\beta = 1, \cdots, d$. The idea is to estimate the function $m(\cdot)$ with a multidimensional kernel estimator and then to integrate out the variables other than X_α. We shall establish the asymptotic normal distribution of the estimator for f_α and derive explicitly its bias and variance. In obtaining this result we shall see that the rate of convergence for estimating the mean function m is $n^{2/5}$, typical for regression smoothing with just one explanatory variable.

The variable selection problem is important for practical use of additive regression modeling. It has been addressed by many authors. Often there are many predictor variables and we wish to select those components that contribute much explanation. We analyze here a procedure based on the size of $S_\alpha = E[f_\alpha^2(X_\alpha)]$. A component function f_α will be called significant if $S_\alpha \geq s_0$, where s_0 is a defined threshold level. We give an estimator for the set of significant functions and derive an upper bound (inverse to the sample size n) of the probability of misspecifying this set of significant component functions.

The rest of the paper is organized as follows. In section 2, we introduce the technique of estimating the functions in the additive model. In section 3, we propose a variable selection procedure which uses the estimator proposed in section 2 and prove that asymptotically it finds the correct variable set with probability 1. Section 4 provides a simulation study and a real example. The detailed conditions and proofs of the theorems are given in the appendix.

2 The Estimator

Let $(X_{i1}, \cdots, X_{id}, Y_i)$, $i = 1, \cdots, n$ be a random sample from the following additive model

$$Y_i = c + \sum_{\alpha=1}^{d} f_\alpha(X_{i\alpha}) + \epsilon_i, \qquad (3)$$

where the ϵ_i have mean 0, finite variance $\sigma^2(X_i)$, and are mutually independent conditional on the $X's$. The functions $f_\alpha(\cdot)$ are assumed to have zero mean $\int f_\alpha(w)p_\alpha(w)dw = 0$, where $p_\alpha(\cdot)$ is the marginal density of X_α. Let $m(x_1,\cdots,x_d) = c + \sum_{\alpha=1}^d f_\alpha(x_\alpha)$ be the mean function. Then for a fixed x, the functional

$$\int m(x)p_{\underline{\alpha}}(x_{\underline{\alpha}}) \prod_{\beta \neq \alpha} dx_\beta$$

is $f_\alpha(x_\alpha) + c$. Let $K(\cdot)$ and $L(\cdot)$ be kernel functions with finite support. Let $K_h(\cdot) = h^{-1}K(\cdot/h)$ and define $L_g(\cdot)$ similarily. Using a multidimensional Nadaraya-Watson estimator (Nadaraya 1964, Watson 1964) to estimate the mean function $m(\cdot)$, we average over the observations to obtain the following estimator. For $1 \leq \alpha \leq d$ and any x in the domain of $f_\alpha(\cdot)$, define, for $h > 0$, $g > 0$,

$$
\begin{aligned}
\hat{f}_\alpha(x_\alpha) &= \tfrac{1}{n}\sum_{i=1}^n \tilde{m}(X_{i1},\cdots,X_{i(\alpha-1)},x_\alpha,X_{i(\alpha+1)},\cdots,X_{id}) \\[2mm]
&= \tfrac{1}{n}\sum_{i=1}^n \left[\frac{\sum_{l=1}^n L_g(X_{i\underline{\alpha}}-X_{l\underline{\alpha}})K_h(X_{i\alpha}-x_\alpha)Y_l}{\sum_{t=1}^n L_g(X_{i\underline{\alpha}}-X_{t\underline{\alpha}})K_h(X_{t\alpha}-x_\alpha)} \right] \\[2mm]
&= \sum_{l=1}^n \left\{ \tfrac{1}{n}\sum_{i=1}^n \left[\frac{L_g(X_{i\underline{\alpha}}-X_{l\underline{\alpha}})K_h(X_{i\alpha}-x_\alpha)}{\sum_{t=1}^n L_g(X_{i\underline{\alpha}}-X_{t\underline{\alpha}})K_h(X_{t\alpha}-x_\alpha)} \right] \right\} Y_l.
\end{aligned}
$$

(4)

If the X's were independent, we might use $\tfrac{1}{n}\sum_{l=1}^n Y_l \frac{K_h(X_{l\alpha}-x_\alpha)}{\tfrac{1}{n}\sum_{t=1}^n K_h(X_{t\alpha}-x_\alpha)}$ to estimate $f_\alpha(x_\alpha)$. This is a one-dimensional Nadaraya-Watson estimator. However, this estimator has larger variance in comparison to our estimator even in this restricted situation, see Härdle and Tsybakov (1995).

To illustrate our method, we simulated a data set $(X_{i1},\cdots,X_{id},Y_i)$, $i = 1,\cdots,200$, according to model (3) with ϵ_i distributed as $N(0,0.5^2)$, $f_1(x_1) = x_1^2 - 1$, $f_2(x_2) = x_2/2$ and $X_1, X_2 \sim N(0,1)$ with $cov(X_1,X_2) = 0.2$ and $c = 0$. The data points are shown on top of the needles in Figure 1. The parabolic shape of f_1 is quite visible but the linear form of f_2 is less evident from the projection onto the (x_2,y) plane. This becomes more clear from Figure 2 where we apply the estimator (4) to estimate the additive component functions. In the Figure we show the estimated curves dashed lines and the true curves as solid lines. Both $\hat{f}_1$ and $\hat{f}_2$ show some smoothing bias at the boundary of the support but capture the general form of the component functions quite well. In both cases we used the bandwidth $h = 0.5$ and $g = 1.5$ and a Normal kernel. We also applied the backfitting procedure to this data set and obtained almost identical curves, see Härdle, Klinke and Turlach (1995, chapter 1).

The bias and variance of the integration estimator are given in the following theorem. Denote by $p(x_1,\cdots,x_d)$ the joint density of $X_{11},\cdots,X_{1d}$.

THEOREM 1. *Suppose that conditions (A1) - (A6) given in section 5 hold. Let $h = \gamma_0 n^{-1/5}$. Assume that the bandwidths g and h satisfy $nhg^{d-1}/\log n \to \infty$, and that the order of L is $q > \frac{d-1}{2}$. Then*

$$n^{2/5}\{\hat{f}_\alpha(x_\alpha) - f_\alpha(x_\alpha) - c\} \xrightarrow{L} N\{b_\alpha(x_\alpha), v_\alpha^2(x_\alpha)\}$$

where

$$b_\alpha(x_\alpha) = \gamma_0^2 \mu_2(K) \left\{ \tfrac{1}{2} f_\alpha''(x_\alpha) + f_\alpha'(x_\alpha) \int \frac{p_\alpha(x_\alpha)}{p(x)} \frac{\partial p}{\partial x_\alpha}(x) dx_{\underline{\alpha}} \right\}$$

$$v_\alpha^2(x_\alpha) = \gamma_0^{-1} \|K\|_2^2 \int \frac{\sigma^2(x) p_\alpha^2(x_\alpha)}{p(x)} dx_{\underline{\alpha}}.$$

From this theorem, we see that the rate of convergence to the asymptotic normal limit distribution does not suffer from the 'curse of dimensionality.' To achieve this rate of convergence, though, we must impose some restrictions on the bandwidth sequences. This condition is needed for bias reduction of components $\beta \neq \alpha$. Note that the above bandwidth condition does not exclude the 'optimal one dimensional smoothing bandwidth' $g = h = n^{-1/5}$ for $d \leq 4$. More importantly one can take $g = o(n^{-1/5})$, leaving only the first two terms in the bias expression. For higher dimensions, $d \geq 5$, though we can no longer use g at the rate $n^{-1/5}$ and the terms involving g dominate. To avoid this problem and obtain the one-dimensional rate of convergence we must reduce bias in the directions not of interest. This can be done by taking L to be a higher order kernel.

Define $\hat{m}(x_1, \cdots, x_d) = \sum_{\alpha=1}^d \hat{f}_\alpha(x_\alpha) - (d-1)\hat{c}$, where $\hat{c} = n^{-1} \sum_{i=1}^n Y_i$. The following theorem gives the asymptotic distribution of $\hat{m}$ and shows that asymptotically the covariance between $\hat{f}_\alpha(x_\alpha)$ and $\hat{f}_\beta(x_\beta)$ is of smaller order than the variances of each component function.

THEOREM 2. *Under the assumptions of Theorem 1,*

$$n^{2/5}\{\hat{m}(x_1, \cdots, x_d) - m(x_1, \cdots, x_d)\} \xrightarrow{L} N(b(x), v^2(x)),$$

where $b(x) = \sum_{\alpha=1}^d b_\alpha(x_\alpha)$ and $v^2(x) = \sum_{\alpha=1}^d v_\alpha^2(x_\alpha)$.

3   Variable Selection Procedure

To establish a variable selection procedure we first show that the integral $S_\alpha = \int f_\alpha^2(w) p_\alpha(w) dw$ can be estimated $n^{1/2}$ consistently. The following theorem establishes this result.

THEOREM 3. *Suppose that conditions (A1) - (A6) given in section 5 hold. Let $h = \gamma_0 n^{-1/4}$. Assume that the bandwidths g and h satisfy $nhg^{d-1}/\log n \to \infty$, and that the order of L is $q > \frac{d-1}{2}$. Then,*

$$\hat{S}_\alpha = \frac{1}{n} \sum_{i=1}^{n} \{\hat{f}_\alpha(X_{i\alpha})\}^2 = S_\alpha + O_p(n^{-1/2}).$$

As in the linear regression, it is important to select a suitable subset among all the available predictors for building an additive model. We propose the following variable selection procedure. Let A be a subset of $\{1, \cdots, d\}$ such that $A = \{\alpha : S_\alpha > 0\}$ so that for $\alpha \notin A$, $S_\alpha = 0$. Note that since A is finite $\{S_\alpha \mid \alpha \in A\}$ is bounded away from zero. Since $\hat{S}_\alpha$ estimates the functional S_α a large $\hat{S}_\alpha$ implies that the variable X_α should be included in the model. Our variable selection procedure selects the indices α such that $\hat{S}_\alpha > b_n$ where b_n is some prescribed level. Denote by $\hat{A} = \{\alpha : \hat{S}_\alpha \geq b_n\}$, the set of estimated coefficients. The following theorem states the asymptotic correctness of this selection procedure.

THEOREM 4. *Under the assumptions of Theorem 3 and $b_n = O(n^{-1/4})$,*

$$P(\hat{A} \neq A) \leq \frac{C}{n},$$

for some constant C.

Since $\hat{S}_\alpha$ estimates S_α we can view $\hat{S}_\alpha / \sum_{\beta=1}^{d} \hat{S}_\beta$ as an estimate of the portion of variation in Y explained by X_α. This allows for a meaningful finite sample interpretation of the test statistic.

4 Simulation Study and Application

In this section we investigated some small sample properties of our estimator through a simulation study. We concentrated on the following questions. First, how variable is the estimator and how much bias do we have to expect? Second, how does the precision depend on the bandwidth choice? Third, how much more variable is the estimator in higher dimensions. We also applied the additive estimator in an econometric context investigating livestock production of Wisconsin farms.

First we continue with our introductory example with parabolic - linear functions. We simulated 250 data sets of size $n = 200$ with $f_1(x_1) = x_1^2 - 1$ and $f_2(x_2) = x_2/2$. The covariance structure of X and the error distribution were as in the introductory example. The simulated 90 % confidence intervals for f_1 and f_2 are shown as thin lines in figure whereas the mean values are shown as thick line. For all simulations we used the bandwidths $h = 0.5$ and $g = 1.5$ and a Normal kernel. In both graphs the true curve has been subtracted for better bias judgment. The smoothing bias becomes visible in figure at the boundaries due to the parabolic shape. It is less a problem for the linear f_2 as Theorem 1 suggests. In both cases the true curve lies well within the computed confidence limits and the shape of the true curve is well reflected by the confidence intervals. The intervals become wider at the boundaries since we have less observations there. We also investigated the effect of bandwidth choice on the above findings and found that, for example, at $x = 0$, the center of the confidence interval increases as the bandwidth increases and the band becomes smaller. This is in full accordance with Theorem 1 showing the dependence of the asymptotic bias on the curvature and the variance as a function of bandwidth. We observe the same phenomenon for dimension $d = 5$, through simulation.

Next we consider a real example. We consider the estimation of a production function for livestock in Wisconsin. A typical economic model in this context is a Cobb-Douglas production function,

$$log(Y) = \alpha_1 log(X_1) + \ldots + \alpha_d log(X_d).$$

The model is additive with parametric linear components. We replace the linear components in the model with arbitrary, up to smoothness conditions, nonlinear functions. This gives a very flexible model of a strongly separable production function.

We use a subset (250 observations) of an original data set of over 1000 Wisconsin farms collected by the Farm Credit Service of St. Paul, Minnesota in 1987. The data were cleaned, removing outliers and incomplete records and selecting only farms that only produce animal outputs. The data consists of farm level inputs and outputs measured in dollars. The output (Y) used in this analysis is livestock, and the input variables used are family labor, hired labor, miscellaneous inputs (repairs, rent, custom hiring, supplies, insurance, gas, oil, and utilities), animal inputs (purchased feed, breeding, and veterinary services), and intermediate run assets (assets with a useful life of one to 10 years) resulting in a five dimensional X variable.

We applied the additive kernel estimator to the farm data set. The results are displayed in Figure 4. The curves are shown together with their marginal scatterplots. They are all quite linear except for the component X_2 (hired labor). Note that the smooth curves do not reflect the form of the scatterplots since our estimator does not use marginal smoothing.

A Appendix

We assume the following conditions hold:

(A1) *The kernel function $K(\cdot)$ is bounded, nonnegative, compactly supported, Lipschitz continuous, with $\int K(u)du = 1$ and $\int uK(u) = 0$. Let $\|K\|_2^2 = \int K^2(u)du < \infty$ and $\mu_2(K) = \int u^2 K(u)du < \infty$.*

(A2) *The kernel function $L(\cdot)$ is bounded, nonnegative, compactly supported, Lipschitz continuous and $\int L(u)du = 1$. Let $\mu_i(L) = \int u^i L(u)du$, then $\mu_i(L) = 0$, $i = 1, 2, \cdots, q-1$, while $\mu_q(L) < \infty$ and $\|L\|_2^2 = \int L^2(u)du < \infty$.*

(A3) *The densities $p_\alpha(\cdot)$, $p_{\underline{\alpha}}(\cdot)$ and $p(\cdot)$ are bounded, Lipschitz continuous and bounded away from zero by a constant p_0.*

(A4) *$E(\epsilon_i^4) < \infty$.*

(A5) *The variance function, $\sigma^2(\cdot)$, is Lipschitz continuous.*

(A6) *The functions $f_\alpha(\cdot)$ have q Lipschitz continuous derivatives.*

Proof of Theorem 1: We use the following notation. Define $E_i[W] = E[W \mid X_i]$ and $E_*[W] = E[W \mid X_1, \cdots, X_n]$. Let

$$p_i = p(X_{i1}, \cdots, X_{i(\alpha-1)}, x_\alpha, X_{i(\alpha+1)}, \cdots, X_{id}),$$

and

$$\widehat{p}_i = \frac{1}{n} \sum_{l=1}^{n} L_g(X_{i\underline{\alpha}} - X_{l\underline{\alpha}}) K_h(X_{l\alpha} - x_\alpha).$$

To simplify the notation we always write the α'th component first.

Note that by (2)

$$\frac{1}{n} \sum_{i=1}^{n} m(x_\alpha, X_{i\underline{\alpha}}) = f_\alpha(x_\alpha) + c + O_p(n^{-1/2}), \tag{5}$$

since $m(x_\alpha, X_{i\underline{\alpha}})$ are i.i.d. random variables with finite second moments. Then we can write

$$\widehat{f}_\alpha(x) - f_\alpha(x) - c = \frac{1}{n} \sum_{i=1}^{n} \frac{\widehat{r}_i - \widehat{p}_i m(x_\alpha, X_{i\underline{\alpha}})}{\widehat{p}_i} + O_p(n^{-1/2})$$

$$\equiv \frac{1}{n} \sum_{i=1}^{n} \frac{\widehat{a}_i}{p_i} + O_p(n^{-1/2}), \tag{6}$$

where $\hat{r}_i = \frac{1}{n}\sum_{l=1}^{n} L_g(X_{i\underline{\alpha}} - X_{l\underline{\alpha}})K_h(X_{l\alpha} - x_\alpha)Y_l$. It suffices to work with the first term on the right hand side, ignoring the $O_p(n^{-1/2})$ remainder. We separate this into a systematic "bias" and a stochastic "variance".

$$\frac{1}{n}\sum_{i=1}^{n}\frac{E_i(\hat{a}_i)}{\hat{p}_i} + \frac{1}{n}\sum_{i=1}^{n}\frac{\hat{a}_i - E_i(\hat{a}_i)}{\hat{p}_i} \equiv T_{1n} + T_{2n}.$$

Then,

$$T_{1n} = \frac{1}{n}\sum_{i=1}^{n}\frac{E_i(\hat{a}_i)}{p_i}\{1 + o_p(1)\}$$

$$T_{2n} = \frac{1}{n}\sum_{i=1}^{n}\frac{\hat{a}_i - E_i(\hat{a}_i)}{p_i}\{1 + o_p(1)\},$$

since, by Silverman (1986), $\sup\left|\frac{\hat{p}_i - p_i}{p_i p_i}\right| = o_p(1)$. It remains to work with the first order approximations.

Let

$$\tilde{T}_{1n} = \frac{1}{n}\sum_{i=1}^{n}\frac{E_i(\hat{a}_i)}{p_i} \quad ; \quad \tilde{T}_{2n} = \frac{1}{n}\sum_{i=1}^{n}\frac{\hat{a}_i - E_i(\hat{a}_i)}{p_i}.$$

We prove the theorem by showing:

I. $\quad \tilde{T}_{1n} = b_\alpha(x_\alpha) + o_p(n^{-2/5})$

II. $\quad \tilde{T}_{2n} = \sum_{j=1}^{n} w_{j\alpha}\epsilon_j + o_p(n^{-2/5}),$

where $w_{j\alpha} = \frac{1}{n}K_h(x_\alpha - X_{j\alpha})\frac{p_\alpha(X_{j\alpha})}{p(x_\alpha, X_{j\underline{\alpha}})}$, and $n^{2/5}\sum_{j=1}^{n} w_{j\alpha}\epsilon_j$ obeys a Central Limit Theorem with asymptotic variance as stated in Theorem 1. To see this note that

$$E\left[\left\{n^{2/5}\sum_{j=1}^{n} w_{j\alpha}\epsilon_j\right\}^2\right] = n^{4/5}\sum_{j=1}^{n} E\left[w_{j\alpha}^2\epsilon_j^2\right] = n^{9/5}E\left[w_{1\alpha}^2\epsilon_1^2\right],$$

since $w_{j\alpha}\epsilon_j$ are mean zero and i.i.d., and

$$E\left[w_{1\alpha}^2\epsilon_1^2\right] = \frac{1}{n^2}\int \sigma^2(z, w)K_h^2(x_\alpha - z)\frac{p_\alpha^2(w)}{p^2(x_\alpha, w)}p(z, w)\,dz\,dw.$$

Changing variables to $u = \frac{x_\alpha - z}{h}$ gives

$$
E\left[w_{1\alpha}^2 \epsilon_1^2\right] = \frac{1}{n^2 h} \int \sigma^2(x_\alpha + hu, w) K^2(u) \frac{p_\alpha^2(w)}{p^2(x_\alpha, w)} p(x_\alpha + hu, w)\, du\, dw
$$

$$
= n^{-9/5} \mu_2(K) \int \sigma^2(x_\alpha, w) \frac{p_\alpha^2(w)}{p(x_\alpha, w)}\, dw + o(n^{-9/5}),
$$

by assumption (A5) and the bandwidth conditions. The Lindeberg condition, required for the CLT, follows from the existence of the fourth moments and the compact support of the kernels.

We now establish the approximations in **I** and **II**.

I. Consider $p_i^{-1} E_i(\hat{a}_i)$, which is, in fact, an approximation of the conditional bias of the Nadaraya-Watson estimator at $(x_\alpha, X_{i\underline{\alpha}})$. This is,

$$
p_i^{-1} E_i(\hat{a}_i) = E_i\left[\frac{1}{p_i} n^{-1} \sum_{l=1}^{n} L_g(X_{l\underline{\alpha}} - X_{i\underline{\alpha}}) K_h(X_{l\alpha} - x_\alpha)\right.
$$
$$
\left. \times \{Y_l - m(x_\alpha, X_{i\underline{\alpha}})\}\right]
$$

$$
= \frac{1}{p_i} \int L_g(w - X_{i\underline{\alpha}}) K_h(z - x_\alpha)\left\{ f_\alpha(z) - f_\alpha(x_\alpha) + \sum_{\beta \neq \alpha} f_\beta(w) - f_\beta(X_{i\beta})\right\} p_{\underline{\alpha}}(w) p_\alpha(z)\, dw\, dz,
$$

since $E_*[\epsilon_i] = 0$. We now change variables to $u = \frac{z - x_\alpha}{h}$ and $v = \frac{w - X_{i\underline{\alpha}}}{g}$, where v is a $d - 1$-dimensional vector with typical component v_β, and find

$$
p_i^{-1} E_i(\hat{a}_i) = \frac{1}{p_i} \int L(v) K(u)\left\{ f_\alpha(x_\alpha + hu) - f_\alpha(x_\alpha) + \sum_{\beta \neq \alpha} \{f_\beta(x_\beta + g v_\beta)\right.
$$
$$
\left. - f_\beta(X_{i\beta})\}\right\} p_\alpha(x_\alpha + hu) p_{\underline{\alpha}}(X_{i\underline{\alpha}} + gv)\, du\, dv
$$

$$
= h^2 \mu_2(K)\left\{\frac{1}{2} f_\alpha''(x_\alpha) + \frac{f_\alpha'(x_\alpha)}{p_i} \frac{\partial p}{\partial x_\alpha}(x_\alpha, X_{i\underline{\alpha}})\right\} + O_p(g^q),
$$

by assumptions (A1), (A2), (A3), and (A6). Since the $p_i^{-1}E_i(\widehat{a}_i)$ are independent and bounded we have

$$
\begin{aligned}
\widetilde{T}_{1n} &= h^2 \mu_2(K) \left\{ \tfrac{1}{2} f''_\alpha(x_\alpha) + f'_\alpha(x_\alpha) \int \tfrac{p_\alpha(z)}{p(x_\alpha,z)} \tfrac{\partial p}{\partial x_\alpha}(x_\alpha, z) dz \right\} \\
&\quad + o_p(h^2) + O_p(g^q) + O_p(n^{-1/2}) \\
&= b_\alpha(x_\alpha) + o_p\left(n^{-2/5}\right).
\end{aligned}
$$

II. We now turn to the stochastic term,

$$
\widetilde{T}_{2n} = \frac{1}{n} \sum_{i=1}^{n} \frac{\widehat{a}_i - E_i(\widehat{a}_i)}{p_i}.
$$

We further write

$$
\widehat{a}_i - E_i(\widehat{a}_i) = \widehat{a}_i - E_*(\widehat{a}_i) + E_*(\widehat{a}_i) - E_i(\widehat{a}_i).
$$

II.1. We show that $\frac{1}{n}\sum_{i=1}^{n} \frac{\widehat{a}_i - E_*(\widehat{a}_i)}{p_i} = \sum_{j=1}^{n} w_{j\alpha}\epsilon_j + o_p\left(n^{-2/5}\right)$, where

$$
\widehat{a}_i - E_*(\widehat{a}_i) = n^{-1} \sum_{j=1}^{n} K_h(x_\alpha - X_{j\alpha}) L_g(X_{i\underline{\alpha}} - X_{j\underline{\alpha}}) \epsilon_j.
$$

Therefore,

$$
\begin{aligned}
\frac{1}{n}\sum_{i=1}^{n} \frac{\widehat{a}_i - E_*(\widehat{a}_i)}{p_i} &= \frac{1}{n}\sum_{i=1}^{n} \frac{1}{p_i} n^{-1} \sum_{j=1}^{n} K_h(x_\alpha - X_{j\alpha}) L_g(X_{i\underline{\alpha}} - X_{j\underline{\alpha}}) \epsilon_j \\
&= n^{-1} \sum_{j=1}^{n} K_h(x_\alpha - X_{j\alpha}) \epsilon_j \\
&\quad \times \left\{ \frac{1}{n}\sum_{i=1}^{n} \frac{1}{p_i} L_g(X_{i\alpha} - X_{j\alpha}) \right\} \\
&= n^{-1} \sum_{j=1}^{n} w_{j\alpha}\epsilon_j \left\{ 1 + o_p(1) \right\}.
\end{aligned}
$$

$$(7)$$

The last equality is demonstrated as follows. Let

$$
\eta_j = \frac{1}{n} \sum_{i=1}^{n} \frac{1}{p_i} L_g(X_{i\underline{\alpha}} - X_{j\underline{\alpha}}),
$$

and break η_j into $E_j[\eta_j] + \{\eta_j - E_j[\eta_j]\}$. Then,

$$E_j\left[\eta_j\right] = \int \frac{1}{p(x_\alpha, z)} L_g(z - X_{j\underline{\alpha}}) p_{\underline{\alpha}}(z)\, dz$$

$$= \int \frac{1}{p(x_\alpha, X_{j\underline{\alpha}} + gu)} L(u) p_{\underline{\alpha}}(X_{j\underline{\alpha}} + gu)\, du$$

$$= \frac{p_{\underline{\alpha}}(X_{j\underline{\alpha}})}{p(x_\alpha, X_{j\underline{\alpha}})} + O_p(g^q).$$

Also,

$$E_j\left[\{\eta_j - E_j[\eta_j]\}^2\right] \leq \frac{1}{n} \int \left\{ \frac{1}{p(x_\alpha, z)} L_g(z - X_{j\underline{\alpha}}) - \frac{p_{\underline{\alpha}}(X_{j\underline{\alpha}})}{p(x_\alpha, X_{j\underline{\alpha}})} \right\}^2 p_{\underline{\alpha}}(z)\, dz$$
$$+ O_p\left(n^{-1} g^{2q}\right)$$

$$= \frac{1}{n} \int \left\{ \frac{1}{p(x_\alpha, z)} L_g(z - X_{j\underline{\alpha}}) \right\}^2 p_{\underline{\alpha}}(z)\, dz + O_p\left(n^{-1}\right).$$

By a change of variables we get

$$E_j\left[\{\eta_j - E_j[\eta_j]\}^2\right] = \frac{1}{n g^{d-1}} \int \left\{ \frac{1}{(p x_\alpha, X_{j\underline{\alpha}} + gv)} L(v) \right\}^2 p_{\underline{\alpha}}(X_{j\underline{\alpha}} + gv)\, dv$$
$$+ O_p\left(n^{-1}\right)$$

$$= \frac{1}{n g^{d-1}} \frac{p_{\underline{\alpha}}(X_{j\underline{\alpha}})}{p^2(x_\alpha, X_{j\underline{\alpha}})} \|L\|_2^2 + O_p\left(n^{-1}\right) = o_p(1),$$

by the assumptions (A1), (A2), (A3) and the conditions on the bandwidths. Thus the last line in (7) is shown.

II.2. Next we show $\frac{1}{n} \sum_{i=1}^n \frac{E_*(\widehat{a}_i) - E_i(\widehat{a}_i)}{p_i} = o_p(n^{-2/5})$. Let

$$U_{0n} = \frac{1}{n} \sum_{i=1}^n \frac{E_*(\widehat{a}_i) - E_i(\widehat{a}_i)}{p_i} = \sum_{i=1}^n \sum_{j=1}^n \widetilde{\zeta}_{ij},$$

where $\widetilde{\zeta}_{ij} = \zeta_{ij} - \overline{\zeta}_i$ and $\overline{\zeta}_i = E_i(\zeta_{ij})$, with

$$\zeta_{ij} = \frac{1}{n^2 p_i} K_h(x_\alpha - X_{j\alpha}) L_g(X_{i\underline{\alpha}} - X_{j\underline{\alpha}}) \left\{ m(X_{j\alpha}, X_{j\underline{\alpha}}) - m(x_\alpha, X_{i\underline{\alpha}}) \right\}.$$

The double sum U_{0n} is mean zero. When $i = j$, we have

$$\zeta_{ii} = \frac{1}{n^2 g^{d-1}} L(0) \frac{1}{p_i} K_h(x_\alpha - X_{i\alpha}) \left\{ f(X_{i\alpha}) - f(x_\alpha) \right\}, \tag{8}$$

and $\sum_{i=1}^{n} \zeta_{ii} = O_p((nh)^{-1/2}(ng^{d-1})^{-1/2})$. We now calculate the variance of $\sum \sum_{i \neq j} \tilde{\zeta}_{ij}$; this involves the following calculations

$$\sum \sum_{i \neq j} E(\tilde{\zeta}_{ij}^2), \quad \sum \sum_{i \neq j} E(\tilde{\zeta}_{ij} \tilde{\zeta}_{ji}), \quad \sum \sum \sum_{i \neq j, i \neq k, j \neq k} E(\tilde{\zeta}_{ij} \tilde{\zeta}_{ik}),$$

since all other terms are mean zero by a conditioning argument. Now $E(\tilde{\zeta}_{ij} \tilde{\zeta}_{ik})$ $= E[E_i^2(\tilde{\zeta}_{ij})]$, for $i \neq j, i \neq k, j \neq k$ using conditional independence. But

$$
\begin{aligned}
E_i(\zeta_{ij}) &= \tfrac{1}{n^2} E_i \left[\tfrac{1}{p_i} K_h(x_\alpha - X_{j\alpha}) L_g(X_{i\underline{\alpha}} - X_{j\underline{\alpha}}) \{ m(X_{j\alpha}, X_{j\underline{\alpha}}) \right. \\
&\qquad \left. - m(x_\alpha, X_{i\underline{\alpha}}) \} \right] \\
&= \tfrac{1}{n^2} O(h^2 + g^q),
\end{aligned}
$$

and so

$$\sum \sum \sum_{i \neq j, i \neq k, j \neq k} E(\tilde{\zeta}_{ij} \tilde{\zeta}_{ik}) = O(n^{-1}) O(h^4 + g^{2q}). \tag{9}$$

Also,

$$
\begin{aligned}
E(\tilde{\zeta}_{ij} \tilde{\zeta}_{ji}) &= \tfrac{1}{n^4} E\left(\tfrac{1}{p_i p_j} L_g^2(X_{i\underline{\alpha}} - X_{j\underline{\alpha}}) K_h(x_\alpha - X_{j\alpha}) K_h(x_\alpha - X_{i\alpha}) \right. \\
&\quad \times \left. \{ m(X_{i\alpha}, X_{i\underline{\alpha}}) - m(x_\alpha, X_{j\alpha}) \} \{ m(X_{j\alpha}, X_{j\underline{\alpha}}) - m(x_\alpha, X_{i\alpha}) \} \right) \\
&= O(\tfrac{1}{n^4 g^{d-1}}).
\end{aligned}
\tag{10}
$$

Then the total contribution to the variance of U_{0n} from these terms is $O(\tfrac{1}{n^2 g^{d-1}})$. By similar arguments we get

$$\sum \sum_{i \neq j} E(\tilde{\zeta}_{ij}^2) = O(\tfrac{1}{n^2 g^{d-1} h}). \tag{11}$$

Then by (8), (9), (10) and (11) and the assumptions on the bandwidths,

$$E_j \left[\{ \eta_j - E_j [\eta_j] \}^2 \right] = o_p \left(n^{-2/5} \right).$$

This completes the proof of **II**. ∎

Proof of Theorem 2: To simplify the notation we always write the αth component of the density first and the βth component second. In order to prove the theorem we show that the asymptotic covariance between $n^{2/5}\widehat{f}_\alpha(x_\alpha)$ and $n^{2/5}\widehat{f}_\beta(x_\beta)$ is $o(1)$. By **II** in the proof of Theorem 1 we need to show that

$$E\left[\left\{\sum_{j=1}^{n} w_{j\alpha}\epsilon_j\right\}\left\{\sum_{j=1}^{n} w_{j\beta}\epsilon_j\right\}\right] = nE\left[w_{1\alpha}w_{1\beta}\epsilon_1^2\right] = o(n^{-4/5}),$$

since $E[\epsilon_i\epsilon_j] = 0$ for $i \neq j$ and $w_{j\alpha}w_{j\beta}\epsilon_j^2$ are i.i.d.

$$
\begin{aligned}
E\left[w_{1\alpha}w_{1\beta}\epsilon_1^2\right] &= \tfrac{1}{n^2}\int \sigma^2(z_\alpha, z_\beta, w)K_h(x_\alpha - z_\alpha)K_h(x_\beta - z_\beta)\\
&\quad \times \frac{p_\alpha(z_\beta,w)p_\beta(z_\alpha,w)}{p(x_\alpha,z_\beta,w)p(z_\alpha,x_\beta,w)}p(z_\alpha, z_\beta, w)dz_\alpha dz_\beta dw\\[2mm]
&= \tfrac{1}{n^2}\int \sigma^2(x_\alpha + hu, x_\beta + hv, w)K_h(u)K_h(v)\\
&\quad \times \frac{p_\alpha(x_\beta+hv,w)p_\beta(x_\alpha+hu,w)}{p(x_\alpha,x_\beta+hv,w)p(x_\alpha+hu,x_\beta,w)}p(x_\alpha + hu, x_\beta + hv, w)dudvdw\\[2mm]
&= O\left(n^{-2}\right),
\end{aligned}
$$

by a change of variables argument and assumptions (A1), (A2), (A3), and (A5). This establishes the negligible asymptotic covariance of $n^{2/5}\widehat{f}_\alpha(x_\alpha)$ and $n^{2/5}\widehat{f}_\beta(x_\beta)$, thus proving the theorem.

Proof of Theorem 3 : We break $\widehat{S}_\alpha$ into the following terms,

$$
\begin{aligned}
\widehat{S}_\alpha &= \tfrac{1}{n}\sum_{i=1}^{n}\left\{\widehat{f}_\alpha(X_{i\alpha})\right\}^2\\[2mm]
&= \tfrac{1}{n}\sum_{i=1}^{n}\left\{f_\alpha(X_{i\alpha}) + c + \widehat{f}_\alpha(X_{i\alpha}) - f_\alpha(X_{i\alpha}) - c\right\}^2\\[2mm]
&= \tfrac{1}{n}\sum_{i=1}^{n}\left\{[f_\alpha(X_{i\alpha}) + c]^2 + 2[\widehat{f}_\alpha(X_{i\alpha}) - f_\alpha(X_{i\alpha}) - c]f_\alpha(X_{i\alpha})\right.\\
&\quad\left. + [\widehat{f}_\alpha(X_{i\alpha}) - f_\alpha(X_{i\alpha}) - c]^2\right\}\\[2mm]
&\equiv U_{1n} + U_{2n} + U_{3n}.
\end{aligned}
$$

Since the X_i's are i.i.d.,

$$U_{1n} = S_\alpha + O_p(n^{-1/2}).$$

By Theorem 1, U_{2n} can be approximated by

$$U_{2n} \approx \frac{2}{n} \sum_{i=1}^n \left\{ b_\alpha (X_{i\alpha}) + \sum_{j=1}^n w_{j\alpha} (X_{i\alpha}) \epsilon_j \right\} f_\alpha(X_{i\alpha})$$

$$= \frac{2}{n} \sum_{i=1}^n b_\alpha (X_{i\alpha}) f_\alpha(X_{i\alpha}) + \frac{2}{n} \sum_{i=1}^n \sum_{j=1}^n w_{j\alpha} (X_{i\alpha}) \epsilon_j f_\alpha(X_{i\alpha}) \quad (12)$$

$$= O_p (h^2) + O_p (n^{-1/2}) = O_p (n^{-1/2}),$$

by the bandwidth conditions. From the proof of Theorem 1 and an application of the Cauchy-Schwarz inequality, $U_{3n} = O_p(n^{-3/4})$. Hence,

$$\widehat{S}_\alpha = S_\alpha + O_p(n^{-1/2}).$$

$\blacksquare$

Proof of Theorem 4: If $S_\alpha = 0$, then note that $E\left[\widehat{S}_\alpha\right]^2 = O\left(\frac{1}{n^2 h^2}\right) = O\left(n^{-6/4}\right)$ and $E\left[\widehat{S}_\alpha^2\right] = O\left(n^{-6/4}\right)$. Therefore, there exists n such that $b > E\left[\widehat{S}_\alpha\right]$. Then,

$$\Pr\left[\widehat{S}_\alpha > b\right] = \Pr\left[\widehat{S}_\alpha - E\left[\widehat{S}_\alpha\right] > b - E\left[\widehat{S}_\alpha\right]\right]$$

$$\leq \Pr\left[\left|\widehat{S}_\alpha - E\left[\widehat{S}_\alpha\right]\right| > b - E\left[\widehat{S}_\alpha\right]\right]$$

$$\leq \frac{E\left[\widehat{S}_\alpha^2\right] - E\left[\widehat{S}_\alpha\right]^2}{\left(b - E\left[\widehat{S}_\alpha\right]\right)^2} \quad (13)$$

$$= O\left(n^{-6/4}\right) \frac{1}{\left(b - E\left[\widehat{S}_\alpha\right]\right)^2} = O\left(n^{-1}\right).$$

If $S_\alpha > 0$ then there exists an n such that $b < E\left[\widehat{S}_\alpha\right]$. Then,

$$\Pr\left[\widehat{S}_\alpha < b\right] = \Pr\left[\widehat{S}_\alpha - E\left[\widehat{S}_\alpha\right] < b - E\left[\widehat{S}_\alpha\right]\right]$$

$$\leq \Pr\left[\left|\widehat{S}_\alpha - E\left[\widehat{S}_\alpha\right]\right| > E\left[\widehat{S}_\alpha\right] - b\right]$$

$$\leq \frac{E\left[\widehat{S}_\alpha^2\right] - E\left[\widehat{S}_\alpha\right]^2}{\left(E\left[\widehat{S}_\alpha\right] - b\right)^2} \tag{14}$$

$$= O\left(n^{-1}\right)\frac{1}{\left(E\left[\widehat{S}_\alpha\right] - b\right)^2} = O\left(n^{-1}\right).$$

The theorem follows from (13) and

$$\Pr\left[\widehat{A} \neq A\right] \leq d\left\{\Pr\left[\widehat{S}_\alpha < b \mid \alpha \in A\right] + \Pr\left[\widehat{S}_\alpha > b \mid \alpha \notin A\right]\right\}.$$

■

Acknowledgement: We would like to thank Enno Mammen for pointing out an error in an earlier version of this paper, and Ray Carroll for fruitful discussions. Special thanks goes to Stephen Sperlich for computation help.

References

[1] BUJA, A, HASTIE, T.J. AND TIBSHIRANI, R.J. (1989), Linear smoothers and additive models, *Ann. Statist.*, 17, 453-510.

[2] HÄRDLE, W. (1990). *Applied Nonparametric Regression*. Econometric Society Monograph 19, Cambridge University Press.

[3] HÄRDLE, W. AND HALL, P. (1993) On the backfitting algorithm for additive regression models. *Statistica Neerlandica*, 47, 43-57.

[4] HÄRDLE, W. KLINKE, S. AND TURLACH, B.A. (1995) XploRe - an interactive statistical computing environment. Springer Verlag, New York

[5] HÄRDLE, W AND MARRON, J.S. (1985), Optimal bandwidth selection in nonparametric regression function estimation. *Ann. Statist*, 13, 1465-1481

[6] HÄRDLE, W. AND TSYBAKOV, A.B. (1995) Additive nonparametric regression on principal components, *J. Nonparametric Statistics*, in press

[7] HASTIE, T. J. AND R. J. TIBSHIRANI (1990), Generalized additive models, Chapman and Hall: London

[8] LINTON, O. AND J. P. NIELSEN (1995) A kernel method of estimating structured nonparametric regression based on marginal integration, *Biometrika* **82**, 93-101.

[9] LEONTIEF, W. (1947) Introduction to a theory of an internal structure of functional relationships. *Econometrika*, 15, 361-373.

[10] MALJUTOV, M.B. AND WYNN, H.P. (1994) Sequential screening of significant variables of an additive model. in : Markov processes and applicazions. Festschrift for Dynkin, Birkhäuser Progress.

[11] MASRY, E. AND TJØSTHEIM, D. (1993), Nonparametric estimation and identification of ARCH and ARX nonlinear time series. Convergence properties and rates. *working paper, University of Bergen, Dept. of Math.*

[12] NADARAYA, E.A. (1964), On estimating regression. *Theor. Probab. Appl.*, 9, 141-142.

[13] PARZEN, E. (1962), On estimation of a probability density and mode. *Ann. Math. Stat.*, 35,1065-76.

[14] SEVERANCE-LOSSIN, E. (1994), Nonparametric Testing of Production assertions in Data with Random Shocks. , PhD Thesis, University of Wisconsin-Madison .

[15] SILVERMAN, B.W. (1986), Density Estimation for Statistics and Data Analysis, Chapman and Hall, London.

[16] STONE, C.J. (1985), Additive regression and other nonparametric models, *Ann. Statist.* 13, 689 - 705

[17] STONE, C.J. (1986), The dimensionality reduction principle for generalized additive models, *Ann. Statist.* 14, 590- 606

[18] TJØSTHEIM, D. AND AUESTAD, B.H. (1994), Nonparametric identification of nonlinear time series: projections, *J. American Statistical Association*, 89, 1398-1409

[19] VENABLES, W.N. AND RIPLEY, B. (1994) Modern applied statistics with S-Plus. Springer Verlag, New York.

[20] WATSON, G.S. (1964), Smooth regression analysis. *Sankhyā A*, 26, 359-372

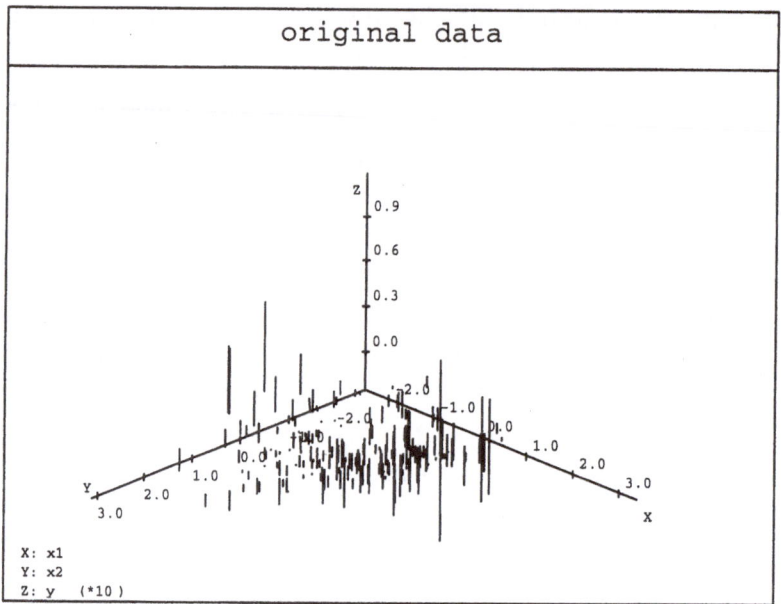

Figure 1: Raw data from an additive model

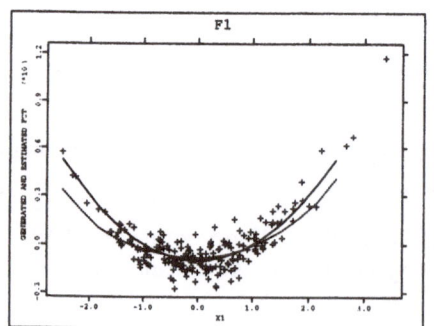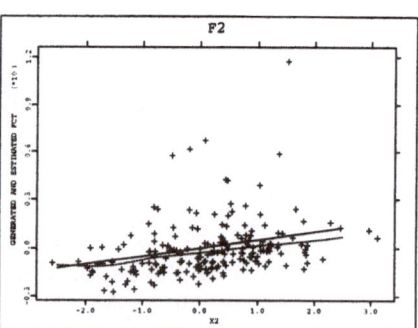

Figure 2: Estimated and true functions

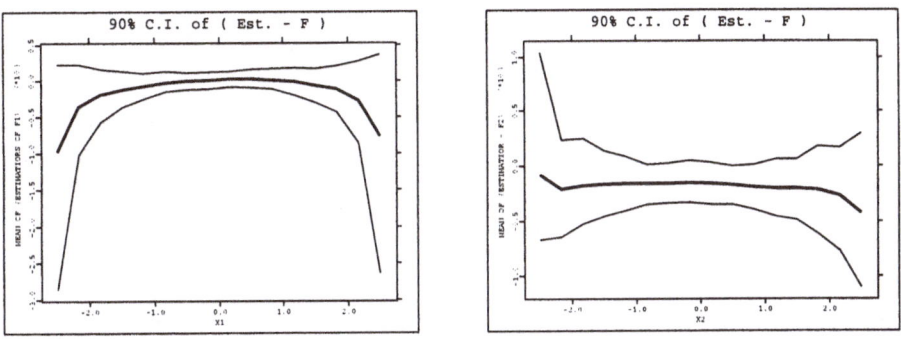

Figure 3: Confidence Intervals for the estimated functions

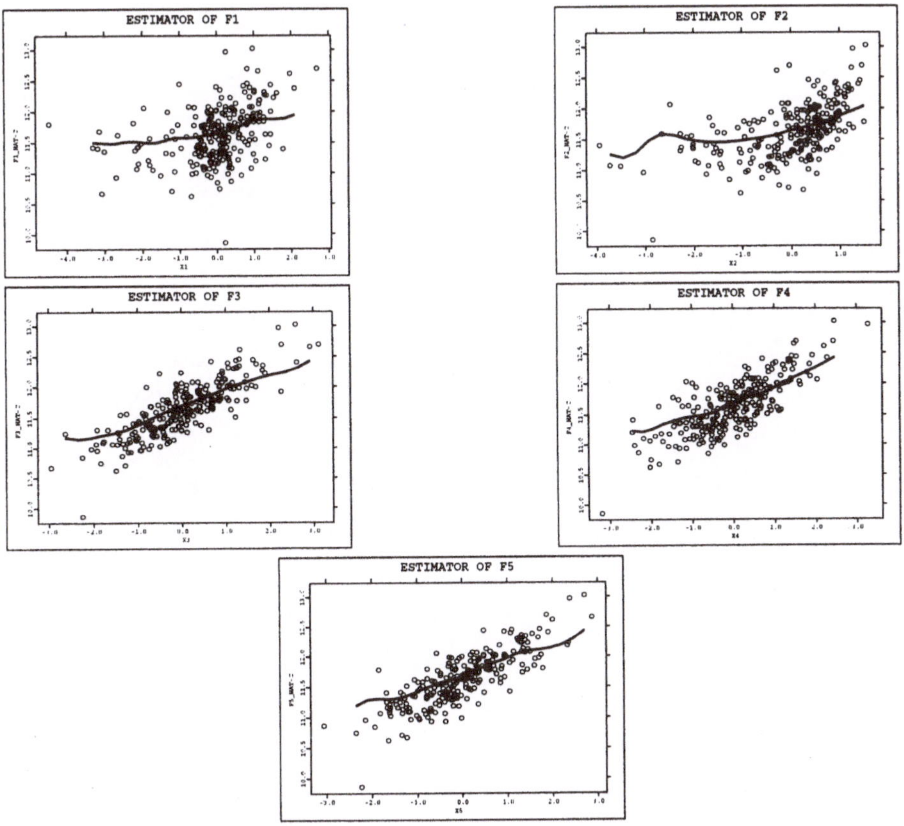

Figure 4: Estimated functions for Wisconsin farm data